# Herbicide Resistance
## In Plants

# Herbicide Resistance
# In Plants

*Edited By*

## HOMER M. LeBARON

CIBA-GEIGY Corporation
Greensboro, North Carolina

## JONATHAN GRESSEL

The Weizmann Institute of Science
Rehovot, Israel

A WILEY–INTERSCIENCE PUBLICATION
JOHN WILEY & SONS
New York • Chichester • Brisbane • Toronto • Singapore

*Library of Congress Cataloging in Publication Data:*

Main entry under title:
Herbicide resistance in plants.

    "A Wiley-Interscience publication."
    Includes bibliographical references and index.
    1. Herbicide resistance—Congresses. 2. Weeds—
Congresses. 3. Plants, Effect of herbicides on—Con-
gresses. I. LeBaron, Homer M. II. Gressel, Jonathan.
III. Weed Science Society of America.

| | | |
|---|---|---|
| SB951.4.H44 | 632′.954 | 81-16381 |
| ISBN 0-471-08701-7 | | AACR2 |

Printed in the United States of America

10  9  8  7  6  5  4  3  2  1

Rc A

# Contributors

**H. U. Ammon**

*Swiss Federal Research Station for Agronomy*
*Zurich-Reckenholz*
*Switzerland*

**C. J. Arntzen**

*MSU/DOE Plant Research Laboratory*
*Michigan State University*
*East Lansing, Michigan*

**J. D. Bandeen (deceased)**

**A. D. Bradshaw**

*Department of Botany*
*University of Liverpool*
*Liverpool, United Kingdom*

**P. S. Carlson**

*Department of Crop and Soil Sciences*
*Michigan State University*
*East Lansing, Michigan*

**H. A. Collins**

*Biological Research Department*
*Agricultural Division*
*CIBA-GEIGY Corporation*
*Greensboro, North Carolina*

**E. R. Cowett**

*Technical Sales Service*
*Agricultural Division*
*CIBA-GEIGY Corporation*
*Greensboro, North Carolina*

**J. S. Faulkner**

*Northern Ireland Plant Breeding Station*
*Loughgall, Armagh, United Kingdom*


**H. Fogelfors**

*Department of Ecology and Environmental Research*
*Swedish University of Agricultural Sciences*
*Uppsala, Sweden*


**J. Gasquez**

*Laboratoire de Malherbologie*
*I.N.R.A.*
*Dijon, France*


**J. Gressel**

*Department of Plant Genetics*
*The Weizmann Institute of Science*
*Rehovot, Israel*


**H. Haas**

*Department of Crop Husbandry and Plant Breeding*
*The Royal Veterinary and Agricultural University*
*Copenhagen, Denmark*


**D. B. Harper**

*Department of Agriculture*
*Belfast, Northern Ireland*


**B. M. R. Harvey**

*Faculty of Agriculture and Food Science*
*The Queen's University of Belfast*
*Newforge Lane*
*Belfast, Northern Ireland*


**J. R. Hensley**

*Biological Research Department*
*Agricultural Division*
*CIBA-GEIGY Corporation*
*Greensboro, North Carolina*

**R. J. Hill**

*Department of Biology*
*York College of Pennsylvania*
*York, Pennsylvania*

**R. J. Holliday**

*Planning Department*
*Merseyside County Council*
*Metropolitan House*
*Liverpool, United Kingdom*

**J. S. Holt**

*Department of Botany*
*University of California*
*Davis, California*

**K. I. N. Jensen**

*Agriculture Canada*
*Kentville, Nova Scotia, Canada*

**Q. O. N. Kay**

*Department of Botany and Microbiology*
*University College*
*Swansea, United Kingdom*

**H. Kees**

*Department of Plant Protection*
*Bavarian State Institute for*
  *Soil and Plant Cultivation*
*Munich, West Germany*

**H. M. LeBaron**

*Biochemistry Department*
*Agricultural Division*
*CIBA-GEIGY Corporation*
*Greensboro, North Carolina*

**C. P. Meredith**

*Department of Viticulture and Enology*
*University of California*
*Davis, California*

**J. V. Parochetti**

*United States Department of Agriculture*
*Science and Education Administration-Extension*
*Office of the Administrator, Extension*
*Washington, District of Columbia*

**K. Pfister**

*Basic Botany Laboratories*
*CIBA-GEIGY Ltd.*
*Ch-4003*
*Basle, Switzerland*

**P. D. Putwain**

*Department of Botany*
*University of Liverpool*
*Liverpool, United Kingdom*

**S. R. Radosevich**

*Department of Botany*
*University of California*
*Davis, California*

**G. F. Ryan**

*Western Research and Extension Center*
*Washington State University*
*Puyallup, Washington*

**M. G. Schnappinger**

*Biological Research Department*
*Agricultural Division*
CIBA-GEIGY *Corporation*
*Greensboro, North Carolina*

**K. R. Scott**

*Department of Botany*
*University of Liverpool*
*Liverpool, United Kingdom*

**L. A. Segel**

*Department of Applied Mathematics*
*The Weizmann Institute of Science*
*Rehovot, Israel*


**V. Souza Machado**

*Department of Horticultural Science*
*University of Guelph*
*Guelph, Ontario, Canada*


**K. E. Steinback**

MSU/DOE *Plant Research Laboratory*
*Michigan State University*
*East Lansing, Michigan*


**G. R. Stephenson**

*Department of Environmental Biology*
*University of Guelph*
*Guelph, Ontario, Canada*


**J. C. Streibig**

*Department of Crop Husbandry and Plant Breeding*
*The Royal Veterinary and Agricultural University*
*Copenhagen, Denmark*


**B. Truelove**

*Agricultural Experiment Station*
*Department of Botany, Plant Pathology and Microbiology*
*Auburn University*
*Auburn, Alabama*

# Preface

The recent advent of herbicide-resistant weeds in increasing and significant numbers and broad distributions has compelled us to remember that such genetic–ecological adaptability is almost universally present in all classes of living organisms. Even though the appearance of resistance to pesticides in previously susceptible insects and diseases has become quite common, some had assumed that weeds would be different. Plants have a much longer life cycle, are not so mobile, and are often less prolific than insects and plant pathogens.

The exact number of species that have become resistant to various pesticides is difficult to determine for several reasons, but the following data are relatively accurate and current.

*Insects and related anthropods:* According to a recent report, 428 species have become resistant to one or more insecticides that were once effective against them.* About 60% of these are agricultural pests and many of the others affect human health.

*Plant pathogens:* Data are less complete because resistance is relatively recent (mostly within the past 12 years) but there are now 81 known cases of pathogen resistance to the benzimidazole fungicides and eight cases of resistance to the bactericide streptomycin.* A few cases of resistance have been reported for some of the other new systemic fungicides. The older multiple sites of action fungicides have shown little or no loss of effectiveness.

*Plants:* To date, 30 common annual weed species in 18 genera, including 23 dicots and 7 monocots, previously susceptible to the triazine herbicides have been found to be resistant. There have been local and rather isolated occurrences of resistance or increased tolerance in various weed species to several other types of herbicides, including phenoxys (e.g., 2,4-D), trifluralin, paraquat, and ureas (e.g., diuron).

The loss in effectiveness applies to all pest control measures; it is not unique to chemicals. When the countermeasure used is a crop variety bred to resist a pest, new strains of the attacking organisms usually evolve to renew the injury. Over the centuries, the plowing of land to control weeds has led to the evolutionary selection of plants that not only survive such action but prosper only when so abused. Pests develop tolerance to introduced biological agents such

---

*G. P. Georghiou, Department of Entomology, University of California, Riverside, California, personal communication, 1981.

as diseases, parasites, and predators, and to control measures based on physical factors and mechanical action. There is no doubt that the ancient practice of cooking food has contributed to the appearance of food-destroying bacteria able to withstand scalding temperatures. The swish of the cow's tail and quick stroke of the flyswatter have selected for a more nimble housefly. The evolution of resistance is universal. It is an expected outcome of natural or artificial selection and is the nature of all living things.

In reality, the appearance of herbicide-resistant weeds had long been predicted and should have been expected. As with all living organisms exposed to other adverse environmental influences, all the genetic and biochemical flexibility inherent in plants have worked to escape or counteract the phytotoxic effects of applied herbicides. The plants are not reacting to herbicides any differently than they do when they gradually evolve to a more prostrate growth habit to escape the cutter bar of a mower. Those species having the capacity to escape the specific adverse effect evolve and survive; those that do not are eliminated. Our lives, as well as those of plants and all other organisms, have always depended on such characteristics.

Our battle against pests is not inevitably one we are going to lose; it is dynamic and must be fought, as a complex war, with all available weapons. It is a constant interplay of measure and countermeasure by all organisms concerned, including man. Common sense and the laws of nature tell us that it is a game we can never entirely win. Yet there is no reason to believe that we cannot maintain a satisfactory level of crop protection indefinitely. We have simply learned again that we must never become complacent or assume that our present herbicides are the ultimate, the final solution. We must keep available all the tools we have ever had, including the hoe, while we continue searching for new and better answers.

Interspecific resistance, or selectivity, of plant species to herbicides has always existed, and is the very reason for the phenomenal success and value of modern weed control chemicals in crop production. Screening methods for new herbicides are designed for selecting chemicals that will kill some plants (weeds) but to which other plants (crops) are resistant. Extensive research has been conducted to understand and better utilize the mechanisms involved in plant selectivity and modes of action of herbicides.

Differential intraspecific responses of weeds to herbicides and the discovery of the existence of resistant biotypes within weed species have also been observed and studied. These were not distinguishable as different biotypes until the herbicide was applied. In this respect, herbicides have uncovered some characteristics about plants and natural selection that were not previously known. One must wonder what other unknown features for adaptability and survival against adverse factors are present in plants.

The recognition of the scientific and practical importance of herbicide resistance has stimulated a multidisciplinary interest in its many aspects. The implications of herbicide resistance extend well beyond the scope of weed science and control into the realms of pesticide chemistry and design; plant physiology and biochemistry; classical genetics and breeding, together with genetic engineering; theoretical population biology, ecology, and taxonomy; as well as agronomy, horticulture, and agricultural economics. More than half the pesticides used are herbicides; it is thus imperative to wonder about, to assess, and to evaluate the problems of herbicide resistance.

The plan for preparing this book for publication as an overview of herbicide resistance developed over the past few years. The seeds were sown when several scientists met informally at the Weed Science Society of America meetings on February 8, 1978, in Dallas, Texas, to discuss and coordinate their research on triazine-resistant weeds. This was received with such enthusiasm that everyone agreed we must plan a formal symposium for the following WSSA meeting. The senior editor organized a one-day symposium on herbicide resistance in plants, with 14 scientific papers, on February 8, 1979, in San Francisco. Eight of the papers were presented by U.S. scientists, four were from Canada, and one each from Germany and Israel. It was interesting to all who attended and exciting to those involved to see how much had become known in so short a time and the abundance of good research begun on this subject. We agreed that most of the papers presented should be combined into a publication that would reach the various scientific disciplines. It soon became apparent that, in order to do the job properly and, at the same time, help stimulate further research within several related areas of science, the chapters should be expanded and published as a coherently edited book, not a collection of papers. In late 1979, Dr. Jonathan Gressel was co-opted to assist in expanding the chapters. The authors expanded their chapters considerably to bring them up to date in the intervening period and to adapt them to the broad audience who wants and needs to know.

This book deals mainly with these herbicide-resistant biotypes, their origin, discovery, their taxonomical and physiological nature, the mechanisms of selectivity, how to avoid or delay their occurrence, the practical consequences, and their control once they have developed. The major message of this book is not gloom and doom, but, rather, optimism and excitement. Our herbicides for the most part are continuing to do a good job. The occurrence of resistant weeds is relatively limited, and there are several alternate means of preventing most of them from becoming serious problems. Even when weeds that are resistant to an important herbicide develop, that herbicide seldom needs to be wholly abandoned. Effective strategies to avoid the development and spread of resistant weeds are discussed. Mathematical models are presented to help predict the rate of infestation of resistant biotypes in an area under certain chemical and alternate control programs, and the consequences of that infestation.

One of the principal themes in this book is that the limited development of herbicide resistance thus far could be more of a blessing than a curse. It has had only very limited and localized adverse economic consequences to date. However, the research conducted over recent years on the mechanisms of resistance or selectivity, especially with triazine-herbicide-resistant biotypes, has led to a much greater understanding of photosynthesis and its inhibition, and of herbicide binding sites in the thylakoid membranes of chloroplasts. This new knowledge is continuing to be developed, and herbicide-resistant plants have great potential as tools for future studies on herbicide mechanisms of action, plant biochemical processes, screening for new herbicides, and other research in theoretical genetics and applied breeding.

Another potential benefit from herbicide-resistant plants is in the possible transfer of resistance from one plant (e.g., weed) to another plant (e.g., crop). Considering the cost and time required to develop new herbicides, the limitations of effective herbicides for most minor and some major crops, and the advantage of flexibility in being able to use the same, a more effective, or more economical herbicide in several crops, such efforts could yield great benefits.

This book presents methods and progress in the use of single cell and tissue culture research to screen for or develop herbicide-resistant crops. It also reviews the use of conventional genetics and breeding techniques to transfer resistance to closely related crop plants (e.g., from *Brassica campestris,* wild bird's rape, to *B. napus,* cultivated rape). Modern technology in genetic engineering, such as recombinant DNA, opens up much greater potential for the transfer of specific genetic characteristics or markers (e.g., herbicide resistance) to crop plants. Some progress has already been reported and is presented here.

It is the sincere hope of the editors and authors that this book will achieve its purpose to be a valuable source of basic scientific information and ideas on the many facets of herbicide resistance in plants; to motivate further good research toward understanding weeds—both as individual plants and, as the weeds interact and evolve with each other, in plant societies; and to aid in discovering the short and long-term impact of herbicides on plants, how to use and manipulate "weeds" and their rich source of genetic material and ecological adaptability, and other potentially valuable knowledge for the future benefit of mankind.

Those involved with this book and with research on the resistance of plants to herbicides were saddened by the untimely death on August 14, 1979 of Dr. John D. Bandeen, of the University of Guelph. John had written and presented the paper on the distribution of triazine resistance for the WSSA meetings earlier that year and had agreed to prepare Chapter 2 of this book. He had conducted some fine research on weed resistance to herbicides and other topics, and his passing was a personal and scientific loss to us all. We appreciate the work of his colleague, Dr. Gerald R. Stephenson, in taking over this assignment.

*January 1982*
*Greensboro, North Carolina*                                    HOMER M. LEBARON

*Rehovot, Israel*                                    JONATHAN GRESSEL

# Editor's Notes

## RESISTANCE VERSUS TOLERANCE

Among scientists, as with the population in general, there are not uniform and consistent definitions of the terms "resistance" and "tolerance." They are sometimes misused or used interchangeably. Tolerance, as used throughout this book, refers to the natural and normal variability to pesticides and other agents which exists within a species and can easily and quickly evolve. Even though tolerance refers to the variability within a species, it may also be used to make comparisons between species. One can say that one species has become more tolerant to a given herbicide than another. Tolerance usually refers to relatively minor or gradual differences in intraspecific variability.

As near as possible, we have tried to use the term "resistance" as defined by the FAO: "Resistance is defined as a decreased response of a population of animal or plant species to a pesticide or control agent as a result of their application. It has been found convenient to distinguish between physiological and behavioristic resistance. The former involves resistance in the presence of the control agent on or in the pest organism, whereas the latter involves resistance because of some behavioristic factor which decreases the probability of contact between pest and control agent. Resistance should not be confused with natural tolerance or (single) low susceptibility due to a normal physiological or behavioristic property of an unselected population."*

In our specific case, a working definition of a resistant weed is one that survives and grows normally at the usually effective dose of a herbicide. Resistance is the maximum tolerance that can be achieved. Resistant individuals are usually found in much lower frequencies than tolerant ones in natural, untreated populations.

## OTHER DEFINITIONS AND NOMENCLATURE

Throughout this book, we have tried to be consistent in the use of terms, definitions, abbreviations, and names of plants and chemicals. Whenever possible, we have followed the WSSA (Weed Science) nomenclature.

---

*Anonymous, Report of the First Session of the FAO Working Party of Experts on Resistance of Pests to Pesticides, *Rome,* October 4-9, 2, (1965).

The authors have followed the Standarized Names of Weeds listed in *Weed Science,* **19**, 435–476 (1971), using the scientific names of weeds throughout. It was decided, however, to use the common names for crops. This presents a few minor problems, such as:

**1** Some plant species can be both a weed and a crop [e.g., perennial ryegrass (*Lolium perenne*) and annual bluegrass (*Poa annua*) in Chapter 12]. In such cases, we have tried to be consistent in the use of common names when we refer to crops and Latin names when we refer to weeds.

**2** Some crops are referred to by different common names in various countries. In such cases, we have somewhat arbitrarily followed the nomenclature most commonly used in the United States (e.g., corn and rutabaga, rather than maize and sweeds, respectively).

**3** Even different scientific names or spellings are sometimes used for the same plant species depending on taxonomic background or the bias of the scientist. For example, some would insist that *Amaranthus powellii* should be written as *Amaranthus Powellii.* The WSSA nomenclature agrees with the rules of the International Code of Botanical Nomenclature, which states:

> All specific epithets . . . should be written with a small initial letter, although authors desiring to use capital initial letters may do so when the epithets are directly derived from the names of persons (whether actual or mythical), or are vernacular (or non-Latin) names, or are former generic names.*

The Appendix includes some tables that help clarify or properly identify all plant species mentioned in the book. Table A2 lists all species alphabetically by family, identifying each by Latin name and common name, and indicating the herbicide(s) to which they have developed resistance or tolerance.

Table A1 lists all reported cases of intraspecific differences in herbicide resistance and tolerance by common names of herbicides. Throughout the book, we have generally omitted the abbreviations for authors following the scientific names (e.g., L. for Linnaeus). These have been included with the names in Appendix A1.

Common and chemical names of herbicides referred to in this book have been taken from the latest issues of *Weed Science,* except for a very few not listed there. Table B in the Appendix provides a complete alphabetical listing of the herbicides by common names, corresponding trade names, and major manufacturers. Throughout the book, the terms "triazines" and "*s*-triazines" are normally used interchangeably in referring to this class of herbicides. If reference is made to an asymmetrical triazine (e.g., metribuzin), it is so specified.

Wherever possible, measurements are given in metric units, using standard

---

*The International Code of Botanical Nomenclature,* A. Oosthoek's Uitgeversmaatschappij N.V., Recommendation 73F, Utrecht, Netherlands, 1972, p. 66.

abbreviations. All rates of applied herbicides are in terms of active ingredient per unit area (e.g., kg/ha).

Rather than compile a separate glossary, the editors preferred to have terms that may be unfamiliar properly explained or clarified in the text where they are used. They have tried to provide a comprehensive subject index with sufficient cross-referencing to provide easy access to all important topics.

Although we have published no acknowledgment section or author index, the editors and authors are aware of and express appreciation for the very significant contributions made by other scientists and colleagues not directly involved with this book. With the separate and rather comprehensive references at the end of each chapter, it is hoped that all appropriate citations and most published contributions can be easily found.

# Contents

**3  DISCOVERY AND DISTRIBUTION OF HERBICIDE-**          **31**
   **RESISTANT WEEDS OUTSIDE NORTH AMERICA**
   *J. Gressel, H. U. Ammon, H. Fogelfors, J. Gasquez,*
   *Q. O. N. Kay, and H. Kees*

# CHAPTER ONE

# Introduction

**H. M. LeBARON**

Biochemistry Department
CIBA-GEIGY Corporation
Greensboro, North Carolina

Biological flexibility and ecological adaptability have been recognized as laws of nature for a long time. The ability of living organisms to compensate for or adapt to adverse or changing environmental conditions is remarkable. Regardless of how and when the various living species began, the "survival of the fittest" has been and still is going on.

Some organisms are much more flexible and adaptable than others. Many species of animals and plants have disappeared as the environmental conditions or the competitive nature of life turned against them. Ecologists have been effective at identifying the many endangered species that are on their way out of existence or are being threatened today.

If life was not such a serious game and so many of the most adaptable species such mortal enemies of human beings, our often feeble efforts to eradicate or control these pests would be rather exciting and fun. In view of such flexibility among our most serious pests, only the most extreme optimist or pessimist would dare predict with confidence the eventual outcome.

The advent of modern organic chemicals as tools to help human beings tip the balance of nature in their favor was heralded with great success and even greater expectations. Actually, the extensive arsenal of pesticides that has been developed and made available to humankind has brought a tremendous revolution in agricultural technology and crop production. In spite of the limitations and problems associated with their use, no one who understands the adverse consequences to humanity in its present numbers can seriously recommend that chemicals be immediately replaced by alternative means of pest control. The net effect of pesticides as with drugs has been the same; an increasing life expectancy with medicines and an increasing food production with pesticides. Chemical pesticides must certainly be a major and increasing part of the agricultural technology in the decades ahead, which will be needed to provide the constantly greater supply of food, fiber, and shelter, with greater cost effectiveness.

Soon after chemicals became part of the environments to which crop pests were exposed, pests began to demonstrate ecological and biochemical adaptability. Insects were probably the first to survive pesticidal chemicals. Those insects within a given species which were somewhat better adapted, survived and multiplied, often tending to produce even more tolerant offspring. This pattern has become very common, leading to hundreds of cases of species resistant to the originally effective organic insecticides. The first recorded example was a strain of *Aspidibtus perniciosus* (San Jose scale) which became resistant to lime sulfur in 1908.[1] By 1957, researchers had identified 76 insect species resistant to the newer organic pesticides, and this number increased to 228 species by 1967. In 1976, there were 364 insects and acarines in which resistant strains had evolved to more than 60 different insecticides. Of these, 225 are of agricultural importance and 139 of medical or veterinary importance.[2] A recent survey showed that through 1980, 428 species of arthropods have become resistant to one or more insecticides that were once effective against them.[2a]

The evolution of plant pathogens to show resistance to fungicides was first reported in 1940. However, up to 1967, there were so few cases of tolerance to fungicides that plant pathologists considered them of little consequence.[3] As more effective fungicides were developed with more specific sites of action, and these were used more exclusively and intensively, the frequency and seriousness of disease resistance has increased markedly. Most plant pathologists, entomologists, and industrial scientists would agree that resistance is a great problem in the continuous use of most currently used chemicals. Thus the further commercial development of new insecticides, acaracides, and fungicides is a continuing requirement.

In the past, some weed scientists and plant physiologists have felt "guilt by association" with entomologists because of the bad press. Their critics have claimed that the continuous and exclusive use of chemicals to control pests were creating resistant superbugs that were worse than those we started with. To some extent, nature has now caught up with all of us. Although there are important differences in the nature and consequences of resistance among insects, diseases, and weeds, there are similarities and opportunities for us to learn from each other.

Some scientists predicted or speculated long ago that resistance would develop in plants to herbicides.[4,5] Harper,[6] in 1956, was the first to seriously consider weeds, which were then susceptible to control by herbicides, would become resistant, presenting serious problems. His assumptions were based on the current theories and preliminary data from other biological systems, but had no firm foundation in plant–herbicide studies. Although some of his expectations have been realized, there have been many exceptions or modifications over the years. By coincidence, his paper was presented very close to the time that atrazine was first tested as a herbicide, but still some years before its commercial introduction.

Whereas entomologists and plant pathologists have had to contend with pesticide resistance for about 30 years, weed scientists have been relatively free of herbicide resistance as a major problem. With herbicides, we have learned to adapt to plant selectivity, interspecific tolerances, and even varietal differences. Depending on soil and climate factors, cultural practices, the crop, and in a few cases, even taxa of the same weed species, new or heavier infestations of more tolerant weeds would evolve following repeated applications of the same herbicide. Only in recent years have the sharp distinctions between susceptible and resistant biotypes been observed.

Our message in this book is not to alarm the public or our fellow scientists that herbicides are failing and that we must now look for new means of weed control. To the contrary, we are indeed fortunate to possess the available herbicides, and we will welcome promising new herbicides that will help the farmer and others keep ahead of the evolving weed problems of the future. One of our major inducements in preparing this book was to tell the exciting story of all that has been learned from the excellent research with herbicide-resistant plants, mostly in the last five years, and how we can profit from its application. We probably already know more about the mechanism of resistance to triazine herbicides in plants than we do about any other case of pesticide resistance, and it will almost certainly result in a greater eventual benefit to mankind.

In Chapters 2 and 3 the authors review the discovery, identification, and distribution of herbicide-resistant weeds throughout the world. The countries affected include mostly those using high levels of herbicides, often the same or similar herbicides frequently, and monocultures or limited crop rotations. The major areas affected are the United States, Canada, and the southern parts of western Europe. However, even in some areas of western Europe and other highly developed countries where herbicides have been utilized extensively for many years, no herbicide resistance has been reported. Although communications with some of the other areas of the world are not as reliable, we know of no case of resistance developing to any herbicide in other countries except for Hungary, Egypt and Japan.

Although 2,4-D and related phenoxy-type herbicides have been used intensively and often exclusively in the same areas for many years, relatively few cases of intraspecific resistance have developed to this group. Although the lack of such resistance cannot be fully explained, the possible multiple mechanisms of action are assumed to be part of the reason.

However, soon after the commercial introduction of the phenoxy herbicides, intraspecific variations in sensitivity of the target weeds were observed. As early as 1950, two recognizable strains of *Commelina diffusa* were claimed to be important perennial weed problems in Hawaiian sugarcane; one was easily controlled with 1 kg/ha of 2,4-D, whereas more than 5 kg/ha was required to control the other.[7] Over the past 30 years, a large number of species have been added to the list of weeds having biotypes with varying degrees of resistance or tolerance to 2,4-D and other postemergence herbicides (e.g., 2,4,5-T, dicamba, dalapon, and TCA).

In some species, a high degree of tolerance was found for more than one herbicide (e.g., *Kochia scoparia* to 2,4-D and dicamba). However, the differential tolerances to these two herbicides within this species were found to be independent.[8] Many of these cases of intraspecific variation in sensitivity to herbicides involved biotypes that were visually distinguishable by morphology prior to the use of the herbicide. Some of the observed differential tolerances, therefore, could be due to differences in physical or phenotypic characteristics (e.g., cuticle thickness or pubescence, erect vs. prostrate growth habit, time of germination, etc.). However, at least one unique case of 2,4-D-resistant *Daucus carota* (wild carrot) has been reported in Ontario, Canada, where the biotypes are morphologically indistinguishable from each other.[9]

In Chapter 4 the authors present some interesting insight into the ecological, genetic, and agronomic factors that have affected the weed flora both before and since herbicides came onto the scene. It is important to recognize that all these factors interact on the types and density of weed species, which have changed in distribution over the years. Indeed, the authors found no way to separate the effects of herbicides on changing weed patterns from those of the other agronomic factors which were introduced simultaneously. Recognizing trends and being able to predict the pest problems of the future are essential aspects of research and crop production. In this case, the authors had a massive data base, a situation quite rare.

One of the academic consequences of the dramatic shifts from mechanical to chemical means of weed control in crop production has been the general tendency for universities to require more courses and experience in chemistry, with a concurrent reduction in certain other required training, such as botany and plant taxonomy. Partly for this reason, and the need to identify weeds in early stages, many weed scientists may have become a bit haphazard in their identification of weed species. This has especially become a problem in the positive identification of closely related species, with similar phenotypic characteristics, which have developed herbicide resistance. Even professional taxonomists have disagreed on the proper identification of some of the resistant weed biotypes. When they agree on the distinguishing characteristics, they have not always agreed on the taxonomy of the biotype. Chapter 5 presents some examples of these problems, the best means of identifying certain of the resistant biotypes, and emphasizes the importance of proper identification, especially when conducting basic studies on differences between the susceptible and resistant biotypes which are assumed to be within the same species. Although the author has made extensive efforts to find some phylogenetic connections or evolutionary relationships between those weed genera and families known to develop herbicide (i.e., triazine) resistance, no such relationships seem to exist. At this time it is impossible to draw any conclusions except that resistance to triazine herbicides has developed independently, almost simultaneously, in a variety of unrelated plants. It appears that the inherent ability or similar genetic basis for this resistance is widespread within the plant kingdom and may be present in many species, and will "come out of the closet" if ecological and environmental conditions and sufficient herbicide selection pressure prevail. Some of the common characteristics that can be recognized based on the triazine-resistant or triazine-tolerant species include (a) herbaceous annuals, (b) at least partially self-fertile, (c) rapid development to maturity, and (d) normal biotype very sensitive to triazine herbicides.

Chapter 6 reviews some interesting case histories on population dynamics and the lack, so far, of phenological resistance to herbicides. The genetic variability of weeds is substantial, allowing immense adaptability. Thus the importance of studying this evolution in action is emphasized.

In a comprehensive book on the subject of herbicide resistance, which needs to consider not only intraspecific resistance to herbicides thus far observed but also possible resistance to other herbicides in the future, it is necessary to provide methods to facilitate distinguishing resistant from susceptible biotypes. This is the subject covered in Chapter 7. The unique characteristics of some of the methods that have been developed to differentiate between biotypes susceptible and resistant to the triazine herbicides are that they are rapid, can be used on vegetative parts (even isolated chloroplasts), and do not result in the destruction of useful plants to determine their selectivity. These procedures should be easily adaptable to some other photosystem II inhibitors.

Before the high degree of resistance developed to the triazine herbicides, there had been much evidence and numerous examples of differential intraspecific responses to various herbicides. Many of these were observed as biotype or taxa variations in tolerances due to differential root uptake, foliar absorption, translocation, or metabolism. Several of these weed biotypes have been of practical importance in weed control programs with triazine herbicides. Most of these examples are reviewed and discussed in Chapter 8.

The first recorded evidence of intraspecific resistance to the s-triazine herbicides was in 1970, when Ryan[10] reported that *Senecio vulgaris* could no longer be controlled in a western Washington nursery after several years of satisfactory results with simazine. Alternate herbicides and other means were found to give good results and the resistant biotype did not appear to spread outside of western Washington. It was not then considered of practical importance.

Within the following few years, cases of triazine resistance in other previously susceptible weed species (e.g., *Amaranthus* spp. and *Chenopodium album*) were reported, also in northwestern Washington. Although these weed species were of major economic importance, the resistant biotypes were found only in limited and relatively isolated areas, far from the corn belt, where atrazine had become the major herbicide. Alternative means of their control were available and quickly put to use. The significance of these developments was considered, therefore, to be primarily of academic interest.

The frequent discoveries over the past five years of completely independent cases of triazine resistance in several species of weeds in Ontario, Quebec, and other areas of Canada, followed by outbreaks on the northeastern seaboard, and finally two important corn belt states, Wisconsin and Michigan, have rapidly brought the matter to a much greater level of importance. Over this same period, a number of similar cases of triazine resistance have been documented and studied in Europe.

In 1975, Radosevich,[11] studying the resistant *Senecio vulgaris* biotype, made a discovery that had a great impact and was more unexpected than the original occurrence of triazine resistance itself. As differential tolerance to many herbicides, including triazines, based on plant uptake, translocation, and metabolism had already been recognized, Radosevich had studied these factors as the possible cause of atrazine resistance. After noting that there were no differences within the biotypes based on these factors, he found that the herbicide caused no inhibition of photosynthesis in isolated chloroplasts from the resistant biotype. Isolated chloroplasts from the susceptible biotype were totally inhibited.[12] Previous studies had led to the conclusion that isolated chloroplasts from both triazine-susceptible and triazine-resistant crop plants were equally susceptible.[13] Some of these studies with crops and weeds are reviewed in Chapters 8 and 9, respectively.

Within the past three years, the practical potential and scientific significance of this discovery of resistance at the plastid level has attracted much attention and research by plant physiologists and geneticists. These extensive and excellent research efforts have helped to uncover important information on the mechanism of resistance and selectivity to *s*-triazine herbicides. Triazine-resistant plants can also serve as valuable tools in the study of the processes within photosynthesis. Considerable progress in our understanding of triazine binding and mechanisms of action has already occurred. Much of the recent research on the mechanism of biotype selectivity, the nature of the triazine binding site, and other valuable information on the process of inhibition of photosynthesis are reviewed and discussed in Chapter 10.

The bipyridylium herbicides (e.g., diquat and paraquat) were found to produce some interesting and important interspecific and intraspecific differences in plant tolerance. The factors influencing these differences and the mode of action of these herbicides have been intensively studied and are reviewed in Chapter 11.

Based on the past history and the presently increasing problems and consequences from insect and disease resistance to pesticides, it is natural to be apprehensive and concerned about the development of weed resistance to herbicides. Certainly, we cannot ignore the undesirable consequences. It has now become obvious that the alleles for triazine resistance are more generally distributed (albeit at very low frequencies) than had been anticipated, both in terms of species of weeds as well as geographical distribution. Considerable research has been conducted to counteract or prevent herbicide-resistant weeds from becoming serious problems, and farmers and other users of herbicides have successfully adapted available technology for this purpose. Although Chapter 16 does not attempt to review all that has been done on this subject, it gives pertinent examples to explain why weeds resistant to a particular herbicide have not become and are not likely to be serious problems because of the availability of alternative means of control.

The possibility of discovering new types of herbicides that are much superior to and that would replace our present herbicides seems to be limited. All the research, development, registration, and production costs for new pesticides make it more economically desirable to find new uses for those already available. In addition, most of our major crops are rotated; corn or sorghum treated with atrazine is usually followed by soybeans, small grains, or other crops which are susceptible even to small amounts of atrazine residues remaining in the soils the next season. It would be of great benefit to growers if they could actually apply triazines to control weeds in both corn and soybeans, or if they could plant these crops without fear of carryover injury. Of course, using the same herbicide on each of the crops in rotation would greatly speed up the evolution for resistance, but it would be necessary to use herbicide combinations, to rotate herbicides, or to use other means to avoid such problems. Furthermore, in spite of the wide usage of triazines, many of the world's fields have never seen this group, which has been used heavily only for corn, sorghum, and perennial woody crops. There is still much potential for use of this family of herbicides in other crops.

Chapter 15 reviews the research and experience with the evolution of plant resistance to heavy metals and the relationship of this to herbicide resistance. High concentrations of heavy metals and their contamination of soil have occurred for thousands of years. The processes by which plants have evolved, survived, and reproduced in such adverse and toxic environments can be informative and analogous to what we can anticipate from herbicide exposure.

In Chapter 17 the authors have taken a new look at the factors responsible for herbicide resistance and their relative importance and interactions, and have developed mathematical models to predict what the future holds. This chapter is especially important in that it provides models that can be used to consider how to avoid, or at least delay, the appearance of herbicide resistance. The authors also provide useful suggestions on the type and biological characteristics of those herbicides that are most desirable, not from a short-term point of view, but in the long-term perspective.

A most exciting and optimistic aspect of weed resistance to herbicides, which should actually make it more a blessing and benefit to humankind than a curse, is the possibility of transferring the resistance characteristic from weeds to crops. Several approaches to develop herbicide-tolerant or herbicide-resistant crop plants are described in Chapters 12, 13, and 14.

Chapter 12 reviews some of the principles of breeding for herbicide-tolerant or herbicide-resistant crops and a few examples of their application. Recent breeding and inheritance studies using triazine-resistant weeds, especially *Brassica campestris,* and the significant progress made toward the commercial development of several triazine-resistant crops, are summarized in Chapter 13.

Numerous attempts have also been made to find resistant or more tolerant crop varieties by screening or by treating breeding stock with various herbicides. Even before resistant weeds provided the present sources of germ plasm, scientists have tried to develop resistance in crops through the use of single cells and plant tissues in nutrient cultures containing the appropriate herbicide. Although no herbicide-resistant commercial variety has yet come from tissue culture technology, significant progress has been achieved, and even greater milestones can be expected in the future. Chapter 14 reviews some of the techniques, problems, and minor successes thus far developed in the use of plant cell and tissue cultures to generate herbicide-resistant crops.

There has been an increasing interest in recent years, with the rapid advances in biotechnology and with the convenient genetic marker of triazine resistance, in transferring this potentially very desirable characteristic to crop plants by protoplast fusion, recombinant DNA, and other genetic engineering. Although weed species generally have a broader and more versatile genetic base than do crop species, several of them are closely related. Among the 29 species of weeds known to have developed resistance to triazine herbicides, *Brassica campestris, Solanum nigrum, Bromus tectorum, Poa annua,* and others are in the same families of important crop species. Preliminary results from some of the current research in this area are referred to in Chapter 18.

## REFERENCES

1   A. L. Melander, Can insects become resistant to sprays? *J. Econ. Entomol.,* **7**, 167 (1914).

2   G. P. Georghiou and C. A. Taylor, Pesticide resistance as an evolutionary phenomenon, *Proc. 15th Int. Congr. Entomol.,* 759 (1976).

2a  G. P. Georghiou, Dept. of Entomology, University of California, Riverside, Calif., personal communication, 1981.

3   C. J. Delp, Resistance to plant disease control agents—how to cope with it, *Proc. Symp., 9th Int. Congr. Plant Protection,* T. Kommendal (Ed), Burgess, Minneapolis, 1981 Vol. 1, p. 253.

4   G. E. Blackman, Selective toxicity and the development of selective weed killers, *J. R. Soc. Arts,* **98**, 500 (1950).

5   R. Pfeiffer and A. Zeller, Practical aspects of weed control by plant growth—regulating substances in cereals, *Proc. Br. Weed Control Conf.,* 70 (1953).

6   J. L. Harper, The evolution of weeds in relation to resistance to herbicides, *Proc. Br. Weed Control Conf.,* **3**, 179 (1956).

7   B. McCall, Are our weeds becoming more resistant to herbicides? *Hawaiian Sugar Technol. Rep.,* **146** (1954).

8   A. R. Bell, J. D. Nalewaja, and A. B. Schooler, Response of *Kochia* selections to 2,4-D, dicamba and picloram, *Weed Sci.,* **20**, 458 (1972).

9   C. W. Whithead and C. M. Switzer, The differential responses of strains of wild carrot to 2,4-D and related herbicides, *Can. J. Plant Sci.,* **43**, 255 (1963).

10  G. F. Ryan, Resistance of common groundsel to simazine and atrazine, *Weed Sci.,* **18**, 614 (1970).

11  S. R. Radosevich, Department of Botany, University of California, Davis, Calif., personal communication, 1975.

12  S. R. Radosevich and O. T. DeVilliers, Studies on the mechanism of *s*-triazine resistance in common groundsel, *Weed Sci.,* **24**, 229 (1976).

13  D. E. Moreland and K. L. Hill, Interference of herbicides with the Hill reaction of isolated chloroplasts, *Weeds,* **10**, 229 (1962).

# Discovery and Distribution of Herbicide-Resistant Weeds in North America

**J. D. BANDEEN (deceased)**

Department of Crop Science
University of Guelph
Guelph, Ontario, Canada

**G. R. STEPHENSON**

Department of Environmental Biology
University of Guelph
Guelph, Ontario, Canada

**E. R. COWETT**

CIBA-GEIGY Corporation
Greensboro, North Carolina

## 1 INTRODUCTION

Botanists have long recognized the existence of morphologically different strains or biotypes of the same weed species. Not long after many of the major classes of herbicides were introduced, weed scientists established that morphologically different weed biotypes often varied in sensitivity to herbicides. Many such examples will be discussed in the third part of this chapter.

The seemingly spontaneous appearance of herbicide-resistant weed biotypes are possibly more dramatic than the small differences in tolerance. They are often visually undistinguishable from the more common biotypes that are highly sensitive to particular herbicides. These biotypes were not known to be present or different in any way prior to the introduction of the herbicide(s). More important, they may never have developed into noticeable weed problems if it were not for the new selective pressure provided by repeated use of a particular herbicide or a particular class of herbicides, as discussed in detail in Chapter 17.

For practical as well as theoretical reasons, it is important to document the distribution of herbicide-resistant weed biotypes on both a local and a continental basis. It is important to know when intraspecific weed resistance or differential tolerance is first apparent for particular herbicides and weed species.

It is especially important to develop accurate case histories of previous cropping, tillage, and herbicide use patterns so that factors responsible for the appearance and spread of herbicide-resistant biotypes can be defined. This may be of considerable assistance in designing strategies for prevention of future occurrences. Another important fact to determine is whether or not there is cross resistance to other herbicides of the same and other chemical groups.

Variations in the control of a particular weed with the same herbicide could be due to differences in application of the herbicide, soil type, rate of herbicide disappearance from the biosphere, depth and time of seed germination, climate, and many factors other than intraspecific variations in the tolerance of the weed to the herbicide. If resistance is suspected, it is absolutely essential to compare the toxicity of the herbicide to both the biotype suspected to be resistant and the more common susceptible biotype under the same field, greenhouse, or laboratory conditions.

It is the intent of this chapter to describe representative examples of weeds that exhibit intraspecific variations in their tolerance to herbicides and to provide an up-to-date documentation of the discovery and distribution of triazine-resistant weed biotypes within North America. In many instances there is a noticeable lack of information, and one function of this volume is to illustrate the type of information that is useful so that future discoveries will be better documented and confirmed. The documentation of herbicide-resistant plants could provide information on the genetic, ecological, phenological, and evolutionary characteristics of plants.

## 2  DISCOVERY, DEVELOPMENT, AND DISTRIBUTION OF TRIAZINE-RESISTANT WEED BIOTYPES

### 2.1  *Senecio vulgaris*

The first North American report of triazine resistance in an initially triazine-sensitive weed species was in 1970 by Ryan[1] for *Senecio vulgaris* in Washington State (Table 2.1). It was actually discovered in the mid-1960s when the owner of an ornamental nursery near Olympia complained to extension personnel that frequent and massive doses of simazine to a conifer nursery had failed to control *S. vulgaris*. By 1968, his plant beds were heavily infested with pure stands of this weed, and other herbicides had to be suggested. Seeds were collected and resistance to simazine, atrazine, and other triazines was documented in greenhouse studies by Ryan[1] and by Radosevich and Appleby.[2]

During the early 1970s, the triazine-resistant biotype of *S. vulgaris* was found in conifer plantations in all of western Washington where atrazine or simazine had been the traditional herbicide used. By 1974, it had become a problem in field corn in the three northwestern counties of Washington and now is prevalent in all cornfields in western Washington, with a total of approximately 250,000 ha being infested (Fig. 2.1). In these areas, atrazine had been the standard herbicide used for weed control in the corn. The resistant biotype is also prevalent in nurseries in eastern Washington.[3] Whether dispersal has occurred by transportation with nursery stock or by concurrent evolution due to selection pressure of the same herbicides is not known. It cannot have been by pollen transport, as the trait is inherited maternally (Chapters 3, 13, and 17).

Distribution of *S. vulgaris* in the northwest is no longer confined to Washington state. Freeman[4] reported that it has been a serious problem on about 2000 ha of field and sweet corn land in the Fraser Valley near Agassiz, British

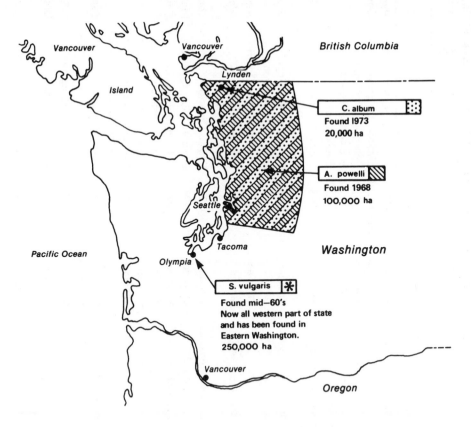

**Figure 2.1** Initial infestations and spread (as of 1980) of *Senecio vulgaris, Chenopodium album,* and *Amaranthus powellii* in Washington State.

Table 2.1 North American Distribution of Triazine-Resistant Weed Biotypes

| Weed Species | Herbicides | Earliest Reports | | | Confirmed by Comparative Studies |
|---|---|---|---|---|---|
| | | Year | Location(s) | Reference | |
| *Amaranthus hybridus* L. | Atrazine | 1972 | Westminster, Maryland | Parochetti[9] | Parochetti[9] |
| | Atrazine, Prometon, Simazine | 1976–1977 | Virginia, Delaware, New York, Pennsylvania, Massachusetts, and Connecticut | Cowett and Hensley[5] | Cowett and Hensley[5] |
| *Amaranthus arenicola* S. Wats. | Atrazine | 1978 | Yuma, Colorado | LeBaron and Hensley[13] | LeBaron and Hensley[13] |
| *Amaranthus powellii* S. Wats. | Atrazine | 1968 | Stanwood, Washington, and western Oregon | Peabody[8] | Peabody[8] |
| *Amaranthus retroflexus* L. | Atrazine | 1977 | West Montrose, Ontario | Alex and Weaver[10] | Weaver and Warwick[11] |
| | Atrazine | 1979 | Ayr, Ontario | Alex and Weaver[10] | Souza Machado and Alex[12] |
| *Ambrosia artemisiifolia* L. | Atrazine | 1977 | Fingal, Ontario | Souza Machado et al.[20] | Souza Machado et al.[20] |
| *Brassica campestris* L. | Atrazine | 1978 | Bromptonville, Quebec | Maltais and Bouchard[21] | Souza Machado et al.[22] |
| *Bromus tectorum* L. | Atrazine, Simazine, Propazine, Ametryn | 1977 | Isolated locations in Kansas, northern Oregon, Washington, northeastern Montana, Nebraska (railroads and fallow land) | Cowett and Hensley[5] | Cowett and Hensley[5] |
| *Chenopodium album* L. | Atrazine and most triazines | 1973 | Ripley, Ontario | Bandeen and McLaren[14] | Bandeen and McLaren[14] |
| | Atrazine | 1973 | Lynden, Washington | Peabody[8] | Peabody[8] |
| | Atrazine | 1975 | Marathon County, Wisconsin | Doll[16] | Doll[16] |
| | Atrazine | 1977 | Wayne County, New York | Duke[17] | Duke[17] |

| Species | Herbicide | Year | Location | Reference | Reference |
|---|---|---|---|---|---|
| | Atrazine | 1978 | Kentville, Nova Scotia | Jensen[15] | Jensen[15] |
| | Atrazine | 1979 | Wythe County, Virginia | Kates[18] | No |
| *Chenopodium missouriense* | Atrazine | 1980 | Huron County, Michigan | LeBaron and Hensley[13] | LeBaron and Hensley[13] |
| | Atrazine | 1978 | Northumberland County, Pennsylvania | Hill (Chap. 5) | Cowett and Hensley[5] |
| *Chenopodium strictum* Roth. | Atrazine | 1978 | Paris, Ontario | Warwick et al.[19] | Warwick et al.[19] |
| *Echinochloa crus-galli* (L.) Beauv. | Atrazine, simazine | 1978 | Northern Maryland | Cowett and Hensley[5] | Cowett and Hensley[5] |
| *Kochia scoparia* L. | Atrazine, simazine | 1976 | Burlington Northern Railroad, Nebraska to Washington | Burnside[23] | Burnside[23] |
| | Atrazine, simazine, prometon | 1976–1977 | Railroads in Oregon, Oklahoma, Kansas, Utah, Colorado | Cowett and Hensley[5] | Cowett and Hensley[5] |
| *Panicum capillare* L. | Atrazine, simazine, ametryn | 1976 | Michigan railroads | Cowett and Hensley[5] | Cowett and Hensley[5] |
| *Poa annua* L. | Simazine | 1977 | San Joaquin Valley, California | LeBaron and Hensley[13] | LeBaron and Hensley[13] |
| *Senecio vulgaris* L. | Simazine | 1976 1978 | Olympia, Washington Fraser Valley, Agassiz, British Columbia | Ryan[1] Freeman[4] | Radosevich and Appleby[2] No |
| | | early 1970s | California, Oregon | Cowett and Hensley[5] | Cowett and Hensley[5] |

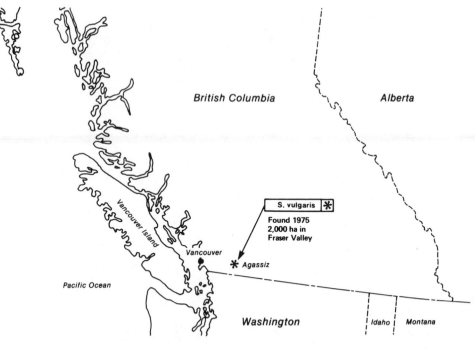

**British Columbia**

**Alberta**

S. vulgaris ✱

Found 1975
2,000 ha in
Fraser Valley

*Vancouver Island*

Vancouver

✱ *Agassiz*

Pacific Ocean

**Washington**     Idaho    Montana

**Figure 2.2**   Initial infestation and spread (as of 1980) of *Senecio vulgaris* near Agassiz, British Columbia, Canada.

Columbia, since 1978 (Fig. 2.2). Cowett and Hensley[5] have found that the triazine-resistant biotype is distributed along the west coast of Oregon and into California, particularly in areas treated repeatedly with simazine.

There is now one unconfirmed report of triazine-resistant *S. vulgaris* on Long Island in the state of New York. According to Bing,[6] this weed is becoming a serious problem in container-grown nursery stock treated with simazine even though simazine originally gave effective control.

## 2.2  *Amaranthus* spp.

The first report of resistance to triazine herbicides in the genus *Amaranthus* was from a farmer near Stanwood, Washington, in Snohomish County (Table 2.1). High rates of atrazine in a sweet corn field failed to control a strain of *Amaranthus,* and in further comparative tests its resistance to several triazine herbicides was confirmed.[7,8] The biotype was initially thought to be *Amaranthus retroflexus,* but it has since been identified by Hill as *Amaranthus powellii* (see Chapter 5). It is present in most corn fields of northwestern Washington, including the counties of King, Snohomish, Skagit, and Whatcom, and over 100,000 ha are now infested (Fig. 2.1).

The failure of triazine herbicides to control normally sensitive *Amaranthus* sp. was also reported over a 4-year period beginning in 1972 in Maryland. The resistance was confirmed by Parochetti[9] in 1976. The first incidence of this *Amaranthus* was near Westminster, Maryland. At first it was thought to be *A. retroflexus,* but it has now been identified as a triazine-resistant biotype of *Amaranthus hybridus* (see Chapter 5). In 1980 it infested about 50,000 ha in Maryland (Fig. 2.3). Cowett and Hensley[5] have now confirmed infestations of triazine-resistant *A. hybridus* in Virginia, New York, Delaware, Pennsylvania, and Massachusetts. Most of these locations were first reported in 1976. Triazine-resistant *A. hybridus* in these eastern states, including Maryland, has usually been found in corn-growing areas, where repeated use of atrazine and "no-till" production methods are common.

Resistance of *Amaranthus* to triazine herbicides in Ontario was first documented in 1977 near West Montrose. The problem had been increasing for

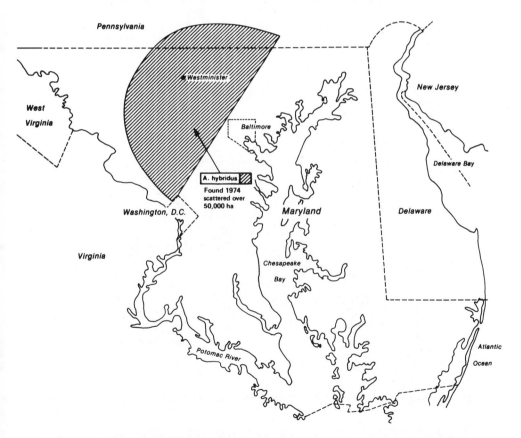

**Figure 2.3**  Initial infestation and spread (as of 1980) of triazine-resistant *Amaranthus hybridus* in Maryland and Pennsylvania.

approximately 4 years, scattered over about 100 ha of monoculture cornland. This resistant biotype has been identified by Alex and Weaver[10] as *A. powellii,* and during the summer of 1979, Weaver and Warwick[11] documented its presence at seven other Ontario locations. One other collection of a triazine-resistant *Amaranthus* was made from a cornfield near Ayr, Ontario, in 1979. This has now been identified as *A. retroflexus* by Alex and Weaver,[10] and its resistance has been confirmed in laboratory as well as field studies by Souza Machado and Alex.[12] Without exception, triazine-resistant biotypes of *Amaranthus* in Ontario have been discovered in fields with a history of continuous corn production and repeated atrazine use. Ahrens[12a] recently confirmed that an *Amaranthus* sp. from an atrazine-treated corn field in Connecticut was resistant, and it was tentatively identified as a hybrid between *A. retroflexus* and *A. hybridus.*

Thus far, there has been only one report of a triazine-resistant biotype of *Amaranthus arenicola.* According to LeBaron and Hensley,[13] it too was discovered on an atrazine-treated cornfield near Yuma, Colorado, in 1978.

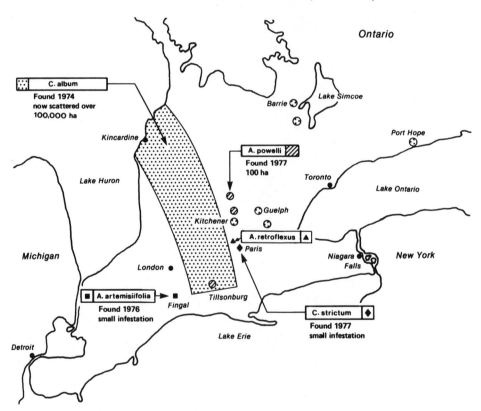

**Figure 2.4** Initial infestations and spread (as of 1980) of *Chenopodium album, Chenopodium strictum, Amaranthus powellii, Amaranthus retroflexus,* and *Ambrosia artemisiifolia* in southern Ontario, Canada.

## 2.3 *Chenopodium* spp.

The selection of a triazine-resistant biotype of *Chenopodium album* was also first reported in the northwestern part of Washington state. It was found near Lynden, Washington, in 1973.[8] It is now present in approximately 20,000 ha of field corn land in the same Washington counties as mentioned for *A. powellii* (Fig. 2.1).

The next documentation of a triazine-resistant *C. album* biotype was by Bandeen and McLaren[14] in Ontario. Seeds were first collected in 1974 from a farm near Ripley in Bruce County. Later, its resistance to atrazine was confirmed in growth chamber studies. This biotype now infests at least 100,000 ha of field corn in Ontario (Fig. 2.4). Atrazine was lethal to the more common sensitive biotype but only stunted the growth of the resistant biotype collected from Ripley, Ontario, when both were treated with 3.4 kg/ha. This illustrates one type of comparative study that should be done to confirm the existence of resistant biotypes.

Triazine-resistant *C. album* has also been reported on several hundred hectares in Nova Scotia (Fig. 2.5). According to Jensen,[15] it was first found

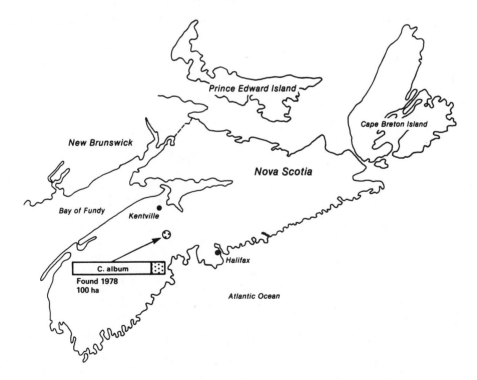

**Figure 2.5** Initial infestations and spread (as of 1980) of triazine-resistant *Chenopodium album* in Nova Scotia, Canada.

near Kentville, Nova Scotia, in 1978. In one field that was planted to potatoes in 1979, it was also found to be resistant to 1 kg/ha of metribuzin applied postemergence. Other recent discoveries of triazine-resistant *C. album* have been documented by Doll[16] on 1000 ha in Marathon county, Wisconsin; by Duke[17] in Wayne county, New York; and most recently in 1979 by Kates[18] in Wythe county, Virginia. As far as can be determined with studies conducted to date, the triazine-resistant *Chenopodium* sp. at all of the locations mentioned above is the same species, *C. album.* Furthermore, these many infestations appear to have evolved separately, as there is little evidence of long-distance transport from one of the areas to another.

Hill (see Chapter 5) has confirmed the existence of a further resistant *Chenopodium* sp. (i.e., *C. missouriense*) in Northumberland county, Pennsylvania. It is of interest that farmers in this relatively isolated area, consisting of two valleys of the Appalachian Mountains, insist that atrazine was never effective for control of this species, which had been originally identified as *C. album.* They learned early to follow atrazine with 2,4-D if this weed was a problem.[13]

At Paris, Ontario, a different type of *Chenopodium* has been reported to have a biotype that is triazine-resistant. It has been identified as *Chenopodium strictum* Roth var. *glaucophyllum* (Aellen) Wahl. by Warwick et al.[19] Infestations of this weed are currently confined to a few hectares (Fig. 2.4).

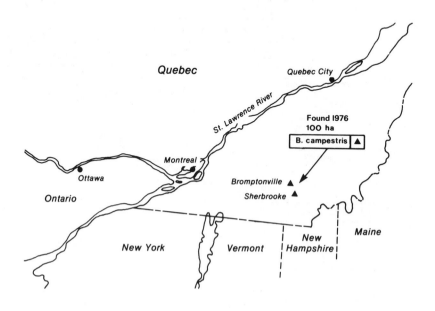

**Figure 2.6** Initial infestation and spread (as of 1980) of triazine-resistant *Brassica campestris* in Quebec, Canada.

## 2.4  *Ambrosia artemisiifolia*

Failure of triazine herbicides to control *A. artemisiifolia* in field corn on a farm near Fingal, Ontario, was first reported to extension personnel in 1976. Growth chamber studies confirmed that this *A. artemisiifolia* was resistant to atrazine, and that collections from other areas were highly sensitive.[20] Thus far, the spread of this resistant biotype has been contained and the infested area has been estimated at only about 2 ha (Fig. 2.4).

## 2.5  *Brassica campestris*

Two cornfields near Bromptonville, Quebec, were reported in 1977 to be heavily infested with *Brassica campestris* ( = *B. rapa*) following repeated annual applications of atrazine. Further greenhouse and field tests showed that the biotype was resistant to atrazine at rates as high as 30 kg/ha.[20-22] Thus far it has not spread beyond the original 50-ha infestation (Fig. 2.6).

## 2.6  *Kochia scoparia*

As early as 1976, after triazine herbicides had been successfully used for total vegetation control along the Union Pacific Railroad in Idaho for over 13 years, Johnston and Wood[22a] reported that *K. scoparia* had developed resistance to these herbicides. Burnside et al.[23] confirmed a triazine-resistant biotype of normally susceptible *K. scoparia* from several locations on that railroad. They found that cross-resistance extended to all commercial *s*-triazines and metribuzin, and that progeny grown from seed taken from various locations with an atrazine history were less competitive for light and nutrients and were more easily controlled with other types of herbicides (e.g., 2,4-D).[23a] Cowett and Hensley have located and confirmed infestations of this resistant biotype in western Oregon, Oklahoma, Kansas, Utah, Colorado, and Nebraska. Recent reports indicate that infestations of triazine-resistant *K. scoparia* now extend along several railroads and into at least 11 western states from Iowa and Texas to Washington.

## 2.7  Triazine-Resistant Grasses

There have been many observations during recent years that some normally sensitive annual grasses are "gradually developing a tolerance" to triazine herbicides. Among the grasses frequently mentioned are *Panicum dichotomiflorum,* various *Digitaria* spp., and *Setaria faberii* Herrm. Fletchall[24] believes that present-day *P. dichotomiflorum* is generally more robust and grows more vigorously following atrazine treatments than in the "early atrazine era." Skroch[25] has reported that some infestations of *Muhlenbergia schreberi,* a perennial grass, are showing more tolerance to triazine herbicides in perennial crops and on roadsides. Most of these reports have not been confirmed. In fact, there have been few attempts to evaluate other possible explanations for the lack of control. In some cases it is possible that the weed always had a high tolerance to triazine herbicides and took over after their susceptible competitors were controlled. However, such weeds deserve closer scrutiny to determine if truly resistant biotypes have evolved from populations dominated by susceptible biotypes.

Researchers have confirmed the existence of at least three triazine-resistant biotypes of annual grasses that are normally sensitive.[5] *Bromus tectorum* seems to be the most widespread thus far. Isolated infestations have been located on atrazine-treated chemical fallow wheat fields in Montana, southwestern Nebraska, South Dakota, and central Kansas. Triazine-resistant biotypes of this weed have also been confirmed on atrazine-treated roadsides in north-central Oregon and on simazine-treated railroad beds in Washington state, northern Oregon, northeastern Montana, and Nebraska. Only one infestation of a triazine-resistant biotype of *Panicum capillare* has been reported.[5] This was found on simazine-treated railroad beds in southeastern Michigan. There are also confirmed infestations of a resistant *Echinochloa crus-galli* in a northern Maryland cornfield and of a simazine-resistant *Poa annua* in an orchard in the San Joaquin Valley of California.[5] Very recently, a mixture of *S. faberii* and *S. viridis* growing in a cornfield in Nebraska following several years of atrazine use was confirmed to be triazine-resistant.[13] Two triazine-resistant biotypes of normally sensitive annual grasses were reported in October 1981 and confirmed in southern Ontario.[25a] *Echinochloa crus-galli* was discovered in an atrazine-treated corn field near Cambridge, Ontario. Resistant *Panicum capillare* was located near Holstein, Ontario, also in an atrazine-treated corn field. In this latter instance, corn had been grown continuously with only atrazine for chemical weed control for 19 consecutive years. An atrazine-resistant biotype of *Setaria lutescens* has also been reported recently in southern Ontario.[25a] This is currently in the process of being confirmed.

## 3    DIFFERENTIAL TOLERANCE OF WEED BIOTYPES TO VARIOUS HERBICIDES

### 3.1    2,4-D and Other Auxin Herbicides

The evaluation of triazine-resistant weed biotypes is a rather recent phenomenon that was not widely known until at least 10 years after these herbicides were widely used. It may be quite unique in magnitude within some species and also in the mechanisms involved (see Chapter 8), but differential tolerance within a species is certainly not a recent concern.

As early as 1951, Sexsmith[26] noted that morphologically different strains of *Cardaria chalepensis,* which is a common weed in southern Alberta, differed markedly in their sensitivity to 2,4-D (Table 2.2). Although it was not possible to correlate sensitivity to 2,4-D with any particular morphological trait within the 12 *C. chalepensis* collections, he did show that the closely related *C. draba* was markedly more sensitive to 2,4-D. Sexsmith noted that the 2,4-D-tolerant strains were tolerant to TBA as well.

As early as 1950, two recognizable strains of *Commelina diffusa* were known to be important perennial weed problems in Hawaiian sugar cane. One strain was easily controlled with 2,4-D at 1.1 kg/ha, yet 5.6 kg/ha were required to control the other[27,28] (Table 2.2). According to McCall,[27] there have been other cases of weeds in Hawaii with suspected tolerance to 2,4-D that were originally thought to be susceptible. *Physalis peruviana* and *Erechtites hieracifolia* are two such examples, but other explanations, such as large populations of 2,4-D-degrading soil bacteria, have not been excluded.

Morphologically different varieties of *Cirsium arvense* have also been reported. In Montana[29] and Ontario,[30] these different *C. arvense* varieties have exhibited markedly different sensitivities to 2,4-D and related herbicides (Table 2.2). In other similar studies, Whitworth[31] has also documented intra-specific variations in sensitivity to 2,4-D among morphologically different strains of *Convolvulus arvensis* collected in Washington state. More recently there have been reports of similar variations in 2,4-D sensitivity among collections of *Kochia scoparia* and *Sonchus arvensis* from North Dakota.[32] Bell et al.[33] have also found lines of *K. scoparia* that are highly tolerant to 2,4-D and some that are highly tolerant to dicamba. However, they established that responses to 2,4-D or dicamba were independent within *K. scoparia.* The many examples cited here illustrate that it is not uncommon for different biotypes of broad-leaved weed species to vary in their sensitivity to 2,4-D. Possibly unique among the examples documented in the literature is the study of *Daucus carota* on Ontario roadsides by Whitehead and Switzer.[34] This study was promoted by the discovery that after several years of use, 2,4-D had become ineffective for controlling *D. carota* in some isolated roadside locations. Comparative studies confirmed the existence of both 2,4-D-resistant and 2,4-D-susceptible biotypes of *D. carota* which were morphologically indistinguishable from each other. In *D. carota* infestations, previously unexposed to 2,4-D, the 2,4-D-resistant biotype was rarely present and if present, it represented less than 1% of the population.[35] However, if present it could become the dominant *D. carota* biotype if 2,4-D was used repeatedly for at least five seasons. It was quite surprising that the 2,4-D-resistant and susceptible biotypes were both highly sensitive to 2,4,5-T or fenoprop—two very similar herbicides.

In spite of repeated use of rights-of-way and on rangelands, there have been no definitive cases of woody species developing a tolerance to phenoxy herbicides. A possible exception is that reported by Greer.[36] He cited an example of increasing intraspecific tolerance in *Quercus marilandica* to 2,4,5-T. The first time a stand of this species was sprayed, up to 50% root kill was obtained. A few years later the survivors had grown back and retreatment with 2,4,5-T at the same rate caused only 25% root kill. A third treatment of the same area resulted in almost no root kill. He assumed that the survivors were more tolerant to 2,4,5-T than were the original plants.

Table 2.2 Examples of Weeds That Exhibit Intraspecific Variations in Sensitivity to Some Herbicides

| Weed Species | Herbicide(s) | Varying Tolerance (VT) or Resistance (R) | Early Reports | | | Confirmed by Comparative Studies |
| --- | --- | --- | --- | --- | --- | --- |
| | | | Year | References | Location | |
| Agropyron repens (L.) Beauv. | Dalapon | VT | 1958 | 37 | Wisconsin | Yes |
| | Glyphosate | VT | 1978 | 48 | Minnesota | Yes |
| Ambrosia artemisiifolia (L.) | Linuron | VT | 1978 | 51 | New Jersey | No |
| Avena fatua L. | Propham | VT | 1964 | 43 | Idaho | Yes |
| | Diallate, triallate, barban | VT | 1968 | 44 | North Dakota, Minnesota | Yes |
| | Difenzoquat | VT | 1975 | 45 | North Dakota | Yes |
| Chardaria chalepensis (L.) Hand.-Maz. | 2,4-D, TBA | VT | 1951 | 26 | Southern Alberta | Yes |
| Cirsium arvense (L.) Scop. | 2,4-D | VT | 1964 | 29 | Montana | Yes |
| | Amitrole | VT | 1968 | 52 | California | Yes |
| | Dicamba, 2,4-D, 2,4-DB | VT | 1966 | 30 | Ontario | Yes |
| Commelina diffusa Burm. | 2,4-D | R | 1954 | 27, 28 | Hawaii | Yes |
| Convulvulus arvensis L. | 2,4-D | VT | 1964 | 31 | Washington | Yes |
| Cynodon dactylon (L.) Pers. | Dalapon, TCA | VT | 1954 | 27, 28 | Hawaii | Yes |
| Digitaria spp. | Dalapon, TCA | R | 1962 | 42 | Hawaii | No |
| Daucus carota L. | 2,4-D | R | 1957 | 34 | Ontario | Yes |

| Species | Herbicide | | Year | | Location | |
|---|---|---|---|---|---|---|
| *Echinochloa crus-galli* (L.) | Dalapon | VT | 1955 | 40 | Washington | Yes |
| *Eleusine indica* (L.) Gaertn. | Trifluralin | R | 1974 | 53 | South Carolina | No |
| *Hordeum jubatum* L. | Siduron | VT | 1972 | 49 | North-central United States | Yes |
| *Kochia scoparia* | 2,4-D | VT | 1968 | 32 | North Dakota | Yes |
| | Dicamba | VT | 1972 | 33 | North Dakota | Yes |
| *Lolium multiflorum* Lam. | Terbacil | VT | 1967 | 50 | Williamette Valley, Oregon | No |
| *Paspalum dilatatum* Poir. | Dalapon, TCA | VT | 1979 | 28 | Hawaii | No |
| *Setaria lutescens* (Weigel) Hubb. | Dalapon | VT | 1958 | 41 | Maryland, Connecticut | Yes |
| *Setaria faberii* Herrm. | Dalapon | VT | 1958 | 41 | Maryland, Indiana | Yes |
| *Sorghum halepense* (L.) Pers. | Dalapon | VT | 1960 | 39 | Arizona | Yes |
| *Sonchus arvensis* (L.) | 2,4-D | VT | 1968 | 32 | North Dakota | Yes |

## 3.2  Dalapon and TCA

Intraspecific variations in sensitivity of perennial grasses to TCA or dalapon seem to be as common as that observed in broad-leaved weeds for 2,4-D. In 1958, Bucholtz[37] established that different clones of *Agropyron repens* from Wisconsin varied in sensitivity to dalapon. In 1954, there were reports from Hawaii[27,28] (Table 2.2) that different biotypes of *Cynodon dactylon* differed in sensitivity to TCA and dalapon. Later, Rochecouste[38] established that tetraploid biotypes of *C. dactylon* were more tolerant than triploids but that there were significant variations among biotypes with the same chromosome number. There are more recent reports from Hawaii that similar variations in tolerance to dalapon are found among biotypes of perennial *Paspalum dilatatum.*[28] Hamilton and Tucker[39] have found that strains of *Sorghum halepense* can also vary markedly in susceptibility to dalapon.

Biotypes of annual grasses, including *Echinochloa crus-galli* from Washington state,[40] *Setaria lutescens,* and *S. faberii* from Maryland, Connecticut, and Indiana,[41] have also exhibited varying sensitivity to dalapon. *Digitaria* spp. have either an annual or perennial growth habit in Hawaii, and in 1962, Harada reported that dalapon- or TCA-tolerant biotypes were spreading in sugar plantations.[42] However, effective control has been accomplished with diuron.

## 3.3  Carbamates and Thiocarbamates

Rydrych and Seely[43] found that different selections of *Avena fatua* from Idaho varied in tolerance to propham (Table 2.2). Furthermore, it was possible to predict tolerance on the basis of morphological differences. Strains with gray lemmas and nondormant seed tended to be less sensitive to propham than those with brown pubescent lemmas and dormant seed.

Jacobsohn and Anderson[44] made over 200 *A. fatua* collections at over 50 locations in the Red River Valley of North Dakota and Minnesota. They consciously attempted to select as many different seed types as possible at each location. In comparative studies, they observed up to 2.5-fold variations in sensitivity of the lines to diallate or triallate and up to 10-fold differences in tolerance to barban. Furthermore, they observed that some of the lines collected were tolerant to all these herbicides. More recently, in very similar studies, Miller et al.[45] found that in 230 selected lines of *A. fatua,* response to difenzoquat ranged from susceptible to highly tolerant. Although these herbicides are not in the carbamate group, this observation futher emphasizes the genetic diversity in this species.

In the past few years there have been several unconfirmed suggestions that repeated use of EPTC or butylate has led to the appearance of thiocarbamate-tolerant biotypes of *Sorghum bicolor.* However, several studies[46,47] have led to the conclusion that other factors, such as induced populations of degrading microbes, high densities of infestation in some areas, or variations in time of germination or emergence, have been responsible for variations in control. Schuman and Harvey[47a] found that freezing had no effect on the accelerated breakdown of these herbicides, but autoclaving or drying the soil at 55°C almost totally eliminated the accelerated breakdown in predisposed soil. Wilson et al.[47b] reported that microbial inhibitors were effective in prolonging EPTC activity, or persistence, in soil and in reducing the accelerated evolution of $^{14}CO_2$ from $^{14}C$-EPTC-treated soils that had prior exposure to EPTC.

## 3.4  Glyphosate

Glyphosate is a fairly new commercial herbicide that is generally regarded to be toxic to most plants. However, Westra and Wyse[48] studied the variability in glyphosate toxicity to 10 biotypes of *Agropyron repens* and found two of them to be relatively tolerant compared to the others. They suggest that differential tolerance may be one reason that eradication of *A. repens* is difficult to achieve in most infestations.

## 3.5  Urea and Uracil Herbicides

Schooler et al.[49] made 36 collections of *Hordeum jubatum* from Montana east to Wisconsin. They employed a root growth bioassay to confirm their varying sensitivity to siduron. Bioassay data from the $F_2$ seedlings of tolerant X susceptible crosses indicated that inheritance of siduron tolerance was controlled by three complementary dominant factors. They postulated that repeated use of siduron could eliminate potentially susceptible parents and hasten the evolution of siduron-resistant *H. jubatum* by increasing the frequency of tolerant X tolerant crosses. Appleby[50] has indicated that there are many reports of terbacil-tolerant or terbacil-resistant *Lolium multiflorum* in western Oregon and that confirmation studies are required. Similarly, there are reports of a linuron-tolerant *Ambrosia artemisiifolia* in New Jersey.[51]

## 3.6  Amitrole

Smith et al.[52] established that *Cirsium arvense* ecotypes varied in sensitivity to amitrole and that the more tolerant varieties metabolized amitrole more readily.

## 3.7  Trifluralin

There are very recent reports[53] of a trifluralin-resistant biotype of *Eleusine indica* in South Carolina. Information for this is quite limited, but according to Grossett, the weed has been reported in three or four counties in north-central South Carolina, and cross resistance to other dinitroaniline herbicides is suspected. There are no obvious morphological differences from the more common sensitive biotypes.

# 4 CONCLUSIONS

The unique aspect about the many recent discoveries of weeds resistant to triazine herbicides is that they were not recognizable as distinct biotypes until after exposure to the herbicides. In fact, taxonomists have not been able to determine any reliable traits that will differentiate them from the more common susceptible biotypes, except for their repsonse to high rates of a triazine herbicide. It is quite surprising that there have not been more reports of "resistant biotypes" of weeds for other herbicides. The only exceptions seem to be early reports of *Daucus carota* resistance to 2,4-D[34] and the more recent reports of a terbacil-resistant *Lolium multiflorum*,[50] a linuron-tolerant *Ambrosia artemisiifolia*,[51] and trifluralin resistant *Eleusine indica*. However, all of these reports require more documentation. Most of the cases of intraspecific variation in sensitivity to herbicides other than triazines involve biotypes that were visually recognizable prior to the introduction of the herbicide. In many cases, the differential tolerances could be due partially to the observed differences in morphology (e.g., varying cuticle thickness or pubescence, erect vs. prostrate growth habit, time of germination, etc.). The fact that in addition to these more obvious differences, there can be less obvious but even more dramatic physiological differences involving sites of herbicide action is proof of the genetic heterogeneity, environmental flexibility, and ecological adaptability of plants, especially weeds.

In reviewing the information on the discovery and spread of triazine-resistant weed biotypes, several common points can be noted. Resistant biotypes have first appeared as scattered plants in fields or areas where otherwise good weed control had been accomplished with a triazine herbicide. In many instances the fields and areas had received continuous or repeated application of triazine herbicides.

In most instances atrazine or other *s*-triazines had been used for at least 6 to 10 years with little or no inter-row cultivation or soil disturbance. In some cases, a "zero or minimum tillage" production scheme with atrazine for weed control in corn had been employed. The resistant weeds or "escapees" were first noticeable as isolated patches in the field. Their subsequent spread over the entire field was usually very rapid and often occurred within only one or two seasons. Spread within the initial fields may have resulted from harvesting or tillage procedures. Spread from field to field and from farm to farm has been rapid in some cases. This movement was probably a result of moving custom harvesting equipment or the ensiling of ripe seed and subsequent spread to uncontaminated fields in manure. In one case, triazine-resistant *C. album* appeared in a potato field following the purchase of contaminated silage for dairy cattle.[15] In another case contaminated sweet corn wastes had been purchased as feed for cattle.[15] Triazine-resistant weed biotypes have also occurred in orchards or nurseries and on railway or highway rights-of-ways where simazine, atrazine, or other *s*-triazines had been used repeatedly for weed control. However, spread seems to have been most rapid in the corn-producing areas because of the movement of equipment, harvested crops, and spread of manure. Perhaps just as significant is the fact that there are many isolated infestations of triazine-resistant weeds for which no means of movement from one area to the other seems likely, In these cases, the only apparent explanation for the nearly simultaneous appearance of triazine-resistant weeds is parallel evolution of the resistant biotypes in response to the same selection pressure, that is, repeated exposure to high rates of the herbicide.

In nearly all instances where triazine-resistant weed biotypes have been discovered and histories have been developed, it has been concluded that prior to s-triazine use, the resistant biotype made up an infinitesimally small proportion of the "natural" population. Gressel and Segel[54] had postulated that these biotypes were ecologically "less fit" than the sensitive biotypes until the repeated use of s-triazine herbicides was initiated. This has been borne out by the studies of Radosevich and others (Chapter 9).

There is no doubt that there will be many new reports of triazine-resistant weeds in the near future. Undoubtedly, the list of resistant biotypes may become longer for other herbicides as well. When one considers the overall distribution of triazine-resistant weeds in North America, some interesting but speculative points can be made. First, in Washington, Oregon, and Ontario there are confirmed reports of several triazine-resistant biotypes of normally sensitive weeds. One wonders if the resistant biotypes are better adapted to these more temperate areas and thus make up significantly higher percentages of the "pre-triazine" population than is true in warmer areas. Equally plausible is the simple fact that the early discoveries in these areas increased the awareness of weed researchers, who in turn confirmed several other examples. If the latter point is true, we can expect a rapid increase in the documentation of resistant biotypes in the cornfields of most northeastern states, where the existence of at least one or two examples of triazine-resistant weeds is now known. The few examples of triazine-resistant weed biotypes, *Kochia scoparia* and *Bromus tectorum,* in the plains and mountain states have been documented primarily on railroad beds, but there is evidence that they are spreading to adjacent cropland. It is perhaps most significant that there are still vast agricultural areas of North America where there have been no reports of triazine-resistant weeds despite extensive use of triazine herbicides. Most notable are the southeastern states and the midwestern corn belt areas. Fawcett[55] suggests that the lack of triazine-resistant biotypes in Iowa may be due to differences in the prevailing management practices for field crops. In the major corn belt states, most fields are rotated between corn and soybeans and most farmers use herbicide combinations. Possibly even more important is the fact that in addition to the use of herbicides, 97% of all corn and soybean fields in Iowa are cultivated at least once each crop season. These practices may be preventing or delaying the exponential increase in seed populations of triazine-resistant weeds to the extent that although possibly present, they are not yet recognizable weed problems.

# REFERENCES

1   G. F. Ryan, Resistance of common groundsel to simazine and atrazine, *Weed Sci.,* **18**, 614 (1970).

2   S. R. Radosevich and A. P. Appleby, Relative susceptibility of two common groundsel (*Senecio vulgaris* L.) biotypes to six s-triazines, *Agron. J.,* **65**, 553 (1973).

3   G. F. Ryan, Western Washington Research and Extension Center, Puyallup, Wash., personal communication, 1979.

4   J. A. Freeman, Canada Agriculture Research Station, Agassiz, British Columbia, personal communication, 1980.

5   E. Cowett and J. Hensley, CIBA–GEIGY Corporation, Greensboro, N.C. 27419, unpublished data, 1979.

6   A. Bing, Long Island Horticultural Research Laboratory, Riverhead, N.Y. 11901, personal communication, 1980.

7   Anonymous, Aatrex tolerant pigweed in Washington, *Weeds Today,* **4**(2), 17 (1973).

8   D. Peabody, Herbicide tolerant weeds appear in western Washington, *Weeds Today,* **5**(2), 14 (1974).

9   J. Parochetti, U.S. Dept. of Agriculture, Washington, D.C., personal communication, 1980.

10  J. F. Alex and S. J. Weaver, Department of Environmental Biology, University of Guelph and Agriculture Canada, Research Station, Harrow, Ontario, personal communication, 1980.

11  S. J. Weaver and S. I. Warwick, Agriculture Canada, Research Stations in Harrow and Ottawa, Ontario, respectively, personal communication, 1980.

12  V. Souza Machado and J. F. Alex, Department of Horticulture Science, University of Guelph, Guelph, Ontario, personal communication, 1980.

12a W. H. Ahrens, Department of Agronomy, University of Illinois, Urbana, personal communication, 1981.

13  H. M. LeBaron and J. Hensley, CIBA–GEIGY Corporation, Greensboro, N.C. 27419, personal communication, 1980.

14  J. D. Bandeen and R. D. McLaren, Resistance of *Chenopodium album* to triazine herbicides, *Can. J. Plant Sci.,* **56**, 411 (1976).

15  K. I. N. Jensen, Agriculture Canada, Research Station, Kentville, Nova Scotia, personal communication, 1979.

16  J. Doll, University of Wisconsin, Madison, Wis., personal communication, 1979.

17  W. B. Duke, Cornell University, Ithaca, N.Y., personal communication, 1979.

18  A. H. Kates, Virginia Polytechnic Institute and State University, Blackburg, Va., personal communication, 1979.

19  S. I. Warwick, V. Souza Machado, P. B. Marriage, and J. D. Bandeen, Resistance of *Chenopodium strictum* Roth. (late-flowering goosefoot) to atrazine, *Can. J. Plant Sci.,* **59**, 269 (1979).

20  V. Souza Machado, J. D. Bandeen, W. D. Taylor, and P. Lavigne, Atrazine resistant biotypes of common ragweed and bird's rape, *Res. Rep., Can. Weed Comm., East. Sect.,* **306** (1977).

21  B. Maltais and C. B. Bouchard, Une moutarde des oiseaux (*Brassica rapa* L.) résistante à l'atrazine, *Phytoprotection,* **59**, 117 (1978).

22  V. Souza Machado, J. D. Bandeen, G. R. Stephenson, and P. Lavigne, Uniparental inheritance of chloroplast atrazine tolerance in *Brassica campestris, Can. J. Plant Sci.,* **58**, 977 (1978).

22a D. N. Johnston and W. N. Wood, *Kochia scoparia* control on noncropland, *Proc. NCWCC* **31**, 126 (1976).

23 O. C. Burnside, C. R. Salhoff, and A. R. Martin, Kochia resistance to atrazine, *Research Report NCWCC,* 64 (1979).

23a C. R. Salhoff and A. R. Martin, *Kochia* biotype competition and response to herbicides, *Proc. NCWCC,* **35,** 86 (1980).

24 O. H. Fletchall, Control of fall panicum in corn, *Proc. NCWCC,* **33,** 83 (1978).

25 W. A. Skroch, Agricultural Extension Service, North Carolina State University, Raleigh, N.C., personal communication, 1979.

25a R. D. McLaren, H. Martin, R. Upfold, and G. W. Anderson, University of Guelph, Guelph, Ontario, personal communication, 1981.

26 J. J. Sexsmith, Morphological and herbicide susceptibility differences among strains of hoary cress, *Weed Sci.,* **12,** 19 (1964).

27 B. McCall, Are our weeds becoming more resistant to herbicides? *Hawaiian Sugar Technol. Rep.,* **146** (1954).

28 H. W. Hilton, Herbicide tolerant strains of weeds, *Hawaiian Sugar Plant. Assoc. Annu. Rep.,* 69 (1957).

29 J. H. Hodgson, Variations in ecotypes of Canada thistle, *Weeds,* **12,** 167 (1964).

30 W. J. Saidak, Differential reaction of Canada thistle varieties to certain herbicides, *Res. Rep., Nat. Weed Comm. (Can.) East. Sect.,* 212 (1966).

31 J. W. Whitworth, The reactions of strains of field bindweed to 2,4-D, *Weeds,* **12,** 57 (1964).

32 L. W. Mitich, Walster Hall, North Dakota State University, Fargo, N.D., personal communication, 1979.

33 A. R. Bell, J. D. Nalewaja, and A. B. Schooler, Response of *Kochia* selections to 2,4-D, dicamba, and picloram, *Weed Sci.,* **20,** 458 (1972).

34 C. W. Whitehead and C. M. Switzer, The differential response of strains of wild carrot to 2,4-D and related herbicides, *Can. J. Plant Sci.,* **43,** 255 (1963).

35 C. M. Switzer, University of Guelph, Guelph, Ontario, personal communication, 1980.

36 H. A. L. Greer, Cooperative Extension Service, Oklahoma State University, Stillwater, Okla. 74074, personal communication, 1979.

37 K. P. Bucholtz, Variations in the sensitivity of clones of quackgrass to dalapon, *Proc. NCWCC,* **15,** 18 (1958).

38 E. Rochecouste, Observations on the chemical control of chiendent (*Cynodon dactylon*) and Herbe MacKay (*Phalaris arundinacea*), *Rev. Agric. Sucr. Ile Maurice,* **37,** 259 (1958).

39 K. C. Hamilton and H. Tucker, Response of selected and random plantings of Johnsongrass to dalapon, *Weeds,* **12,** 220 (1964).

40 B. F. Roche and T. J. Muzik, Ecological and physiological study of *Echinochloa crus-galli* (L.) Beauv. and response of its biotypes to sodium 2,2-dichloropropionate, *Agron. J.,* **56,** 155 (1964).

41 P. W. Santleman and J. A. Meade, Variations in morphological characteristics and dalapon susceptibility within the species *Setaria lutescens* (Weigel) Hubb., *Setaria faberii* Herrm., *Weeds,* **9,** 406 (1961).

42 K. Harada, Control of grasses with DCMU at Kilanea Sugar Company, *Hawaiian Sugar Technol. Rep.,* 45 (1962).

43 D. J. Rydrych and C. I. Seely, Effect of IPC on selections of wild oats, *Weed Sci.,* **12,** 265 (1964).

44 R. Jacobsohn and R. N. Anderson, Differential response of wild oats lines to diallate, triallate and barban, *Weed Sci.,* **16,** 491 (1968).

45 S. D. Miller, J. D. Nalewaja, and S. Richardson, Variations among wild oat biotypes, *Proc. NCWCC,* **20,** 111 (1975).

46 A. R. Martin and F. W. Roeth, Shattercane control in problem areas, *Proc. NCWCC,* **33,** 108 (1978).

47   R. S. Fawcett and G. G. Guge, Shattercane control in corn with incorporated herbicides at Logan, Iowa, *NCWCC Res. Rep.,* 206 (1978).

47a  D. B. Schuman and R. G. Harvey, Predisposition of soils for rapid thiocarbamate herbicide breakdown, *Proc. NCWCC,* **35,** 19 (1980).

47b  R. G. Wilson, A. R. Martin, and F. W. Roeth, Soil persistence of EPTC with and without a microbial inhibitor, *Proc. NCWCC,* **35,** 77 (1980).

48   P. Westra and D. L. Wyse, Physiology, edaphic factors and control of specific weeds, *Proc. NCWCC,* **33,** 106 (1978).

49   A. B. Schooler, A. R. Bell, and J. D. Nalewaja, Inheritance of siduron tolerance in foxtail barley (*Hordeum jubatum* L.), *Weed Sci.,* **20,** 167 (1972).

50   A. P. Appleby, Crop Science Department, Oregon State University, Corvallis, Oreg. 97331, personal communication, 1971.

51   J. A. Meade, CES, Rutgers University, New Brunswick, N.J., personal communication 1979.

52   L. W. Smith, D. E. Bayer, and C. L. Foy, Metabolism of amitrole in excised leaves of Canada thistle ecotypes and bean, *Weed Sci.,* **16,** 523 (1968).

53   B. J. Gossett, Agronomy Dept., Clemson University, Clemson, S.C. 29631, personal communication, 1980.

54   J. Gressel and L. A. Segel, The paucity of plants evolving genetic resistance to herbicides: Possible reasons and implications, *J. Theor. Biol.,* **75,** 349 (1978).

55   R. S. Fawcett, Dept. of Plant Pathology, Seed and Weed Science, Iowa State University, Ames, Iowa 50011, personal communication, 1979.

# Discovery and Distribution of Herbicide-Resistant Weeds Outside North America

**J. GRESSEL**

Department of Plant Genetics
The Weizmann Institute of Science
Rehovot, Israel

**H. U. AMMON**

Swiss Federal Research Station for Agronomy
Zurich-Reckenholz, Switzerland

**H. FOGELFORS**

Department of Ecology and Environmental Research
Swedish University of Agricultural Sciences
Uppsala, Sweden

**J. GASQUEZ**

Laboratoire de Malherbologie
I.N.R.A.
Dijon, France

**Q. O. N. KAY**

Department of Botany and Microbiology
University College
Swansea, United Kingdom

**H. KEES**

Department of Plant Protection
Bavarian State Institute for Soil and Plant Cultivation
Munich, West Germany

## 1  INTRODUCTION

As in Chapter 2, the sporadic appearance of herbicide resistance outside North America seems to be the compounded outcome of a series of factors; monoculture, heavy reliance on a single cost-effective herbicide, and the availability of weed scientists to verify the cause. Some have suggested that the appearance of resistance is a function of having weed scientists on the scene. Those thinking so forget that the farmers would usually be the first to notice the sudden impotence of an old faithful herbicide and it is doubtful that they would remain quiescent.

Before examining the cases where herbicide resistance has appeared, let us examine where resistance has not occurred to see what we can learn. There is just a single intimation of resistance from an underdeveloped country; herbicide usage is low there. To the best of our knowledge, there are no reports of resistance in the Eastern Bloc countries except for Hungary, possibly due to the extent of herbicide usage (although 2,4-D is used heavily on wheat in the U.S.S.R.). According to Solymosi, research on resistance is being done only in Hungary.[1] Neither have there been documented reports from the Benelux countries, northern Germany, Spain, Ireland, Denmark, South Africa, or Israel. This has been partially attributed to the lack of heavy cultivation of corn, for which the triazines have been the main herbicides of choice and the prime herbicides to which resistance has occurred elsewhere. Japan is a country with a major rice crop and is an intensive user of herbicides; only one case of resistance has been reported there. Ueki attributes this (a) to extensive herbicide development and marketing, which have so far resulted in a short product longevity; (b) most commercial formulations are a combination of more than one active ingredient; and (c) it is always recommended to follow preemergence herbicide treatments with postemergence treatments.[2] In parts of Japan, rice is grown in summer followed by barley in winter, both under very different conditions, which may also delay appearance of resistance. Thus despite what must be very heavy herbicide usage, resistance has not appeared. A theoretical analysis of these practices is given in Chapter 17.

As with North America (Chapter 2), the most significant type of resistance appearing in other countries to date has been to the s-triazine herbicides. It is also interesting and significant that most of the resistant biotypes belong to the same species as those which have occurred in North America. This relationship could not be by chance, nor is it likely that their appearance in other countries could be by distribution from North America or from a common source. Reported below are also occurrences of resistance to bipyridylium and even phenoxy herbicides. The case histories presented are important, as together with those reported in Chapter 2, they may be portents of the future.

## 2  TRIAZINE HERBICIDES

### 2.1  Genera Where Resistance Has Also Appeared in North America

#### 2.1.1  *Senecio*

Holliday and Putwain[3] screened 46 populations of *Senecio vulgaris* from various sites in England for simazine resistance. Many of the sites were in horticultural crops where simazine had been used for up to 10 years continuously. Still, only just recently did they find full resistance in any of the populations of this weed; much more partial tolerance was found. The percentage mortality varied from 44 to 95% at 0.7 kg/ha simazine, with the degree of tolerance positively correlated with the number of years used. In a second series of tests, the percentage mortality of seven populations of *Senecio vulgaris* ranged from 91 to 99%. Plants from these seven populations were subject to a cycle of artificial selection for simazine resistance; six did not respond, but one of the most susceptible populations (Malpas, Cheshire, with 98.5% mortality) yielded plants that had full simazine resistance (0.0% mortality at 2.8 kg/ha simazine).[4] Putwain[5] found that the Malpas population of *S. vulgaris* has increased in abundance substantially and appears to be no longer controllable by simazine. In early 1981 a population of *S. vulgaris* that was resistant to triazines was found in a tree nursery in Staffordshire, U.K., which had received 14 annual triazine treatments. In laboratory experiments it withstood 5 kg/ha atrazine or simazine.[5]

#### 2.1.2  *Amaranthus*

This is a widespread weed genus in Europe and one of the major triazine-resistant genera in North America. A triazine-resistant biotype of *Amaranthus retroflexus* sensu lato* is rapidly spreading in several areas of France[6] (Table 3.1, Fig. 3.1).

*A. retroflexus* and *A. lividus* have generally been easily controlled with simazine in Switzerland. According to field observations, control is no longer possible in some vineyards in the Valais, Switzerland[13] (Fig. 3.2). The affected area is about 1000 ha. The resistance has been confirmed in the laboratory.[15] The results of very recent studies in Switzerland (Table 3.2) show that in addition to *A. retroflexus* and *A. lividus*, resistant populations of *A. hybridus* have been found. *A. graecizans* and *A. bouchonii* populations were found which had varying degrees of tolerance, and *A. blitoides* has remained susceptible.[17]

There is a report of the appearance of triazine-"resistant" *A. retroflexus* in Austria.[18] Unfortunately, quantitative data on doses or on the extent of the problem are unavailable. *A. retroflexus* sensu lato growing in a 1200-ha atrazine-treated cornfield in Kalsdorf near Graz, Austria, was found to possess plastid resistance to triazines (J. Gasquez, unpublished data).

---

*The North American resistant *Amaranthus* was originally designated *A. retroflexus*, but more recent expert taxonomic analysis has redesignated it as *A. hybridus* and/or *A. powellii* (see Chapters 2 and 5).

A: Chenopodium album
B: Solanum nigrum
C: Amaranthus retroflexus
D: Polygonum lapathifolium
E: Polygonum persicaria
F: Poa annua
G: Chenopodium polyspermum

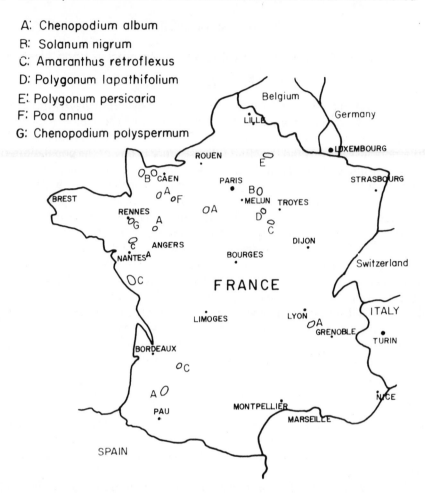

**Figure 3.1**    Distribution of triazine-resistant weeds in France in early 1979. (J. Gasquez, previously unpublished.)

There are considerable areas in Hungary which have been in a corn monoculture with atrazine as the sole herbicide.[1] Within 10 years of cultivation, in the 1970s, atrazine-resistant *Amaranthus retroflexus* was found[19] and its spread has been followed.[20,21] Seed from 246 cornfields was tested and half "exhibited total resistance" to 2 kg/ha atrazine.[20,21] Hartmann[20] followed the relationship between years in corn monoculture with atrazine, and the resistance of the *A. retroflexus* to atrazine. Seventy percent of seed from 74 sample locations where atrazine was used for 7 to 10 years was "totally resistant", and all seed from the six sample locations with 10 years of continuous treatment was resistant.[20] Reciprocal crosses were made between 10 resistant isolates and 10 sensitive isolates and resistance to 3 kg/ha was shown to be inherited maternally in both the $F_1$ and $F_2$ generations.[21a] This indicates cytoplasmic inheritance, as

## Table 3.1  Distribution and Properties of Triazine-Resistant Weeds Found in France

| Species | Locale: Town, Department | Reference[a] | First Observed | In Crop | Treatment | Estimated Distribution in 1979 (ha)[b] | ID$_{50}$ in Chloroplast[c] (M atrazine) Wild Type | Resistant Biotypes |
|---|---|---|---|---|---|---|---|---|
| *Amarathus retroflexus* (sensu lato) | Pannecé (near Nantes), Loire Atlantique | 6 | 1977 | Corn | Atrazine | 10,000 | $2.5 \times 10^{-7}$ | $2 \times 10^{-4}$ |
| *Chenopodium album* | LeLion D'Angers (near Angers), Marne et Loire | 7–9 | 1973 | Corn | Atrazine | 30,000 | $2 \times 10^{-7}$ | $10^{-4}$ |
| *Chenopodium polyspermum* | Pacé (near Rennes), Ille et Vilaine | 6 | 1974 | Corn | Atrazine | 1 locale | $2 \times 10^{-7}$ | $8 \times 10^{-5}$ |
| *Echinochloa crus-galli*[d] | Pau, Pyrénées Atlantiques | 10, 11 | | Corn | Atrazine | | Plastids not affected[d] | |
| *Poa annua* | La Ferté Macé (near Alençon), Orne | 8 | 1976 | Roadside | Simazine | 1 locale | $2 \times 10^{-7}$ | $1.2 \times 10^{-4}$ |
| *Polygonum lapathifolium* | Villiers Bonneux (near Sens), Yonne | 12, 12a | 1973 | Corn/wheat | Atrazine | 2,000 | $3.5 \times 10^{-7}$ | $2 \times 10^{-4}$ |
| *Polygonum persicaria* | Berry au Bac (near Laon), Aisne | 6 | 1978 | Corn | Atrazine | 1 locale | $2.5 \times 10^{-7}$ | $10^{-4}$ |
| *Solanum nigrum* | Isigny (near Bayeux), Calvados | 12 | 1976 | Corn | Atrazine | 50,000 | $2.5 \times 10^{-7}$ | $3 \times 10^{-4}$ |

*Note:* The populations observed were tested by treating seedlings with 3 kg/ha of atrazine and the fluorescence of whole leaves having absorbed atrazine. All were resistant by the first test, and all except *Echinochloa* were resistant by the fluorescence test.

[a] Part of the material on observations and distribution is previously unpublished.

[b] In 1981 it is estimated that >200,000 ha of corn-growing areas have trazine-resistant weed infestations.

[c] The concentration of atrazine required for 50% inhibition of activity in isolated chloroplasts was determined as outlined in ref. 8. The chloroplast inhibition data are from J. Gasquez, previously unpublished.

[d] A new population near Pau has been found in 1980 which shows chloroplastic resistance.

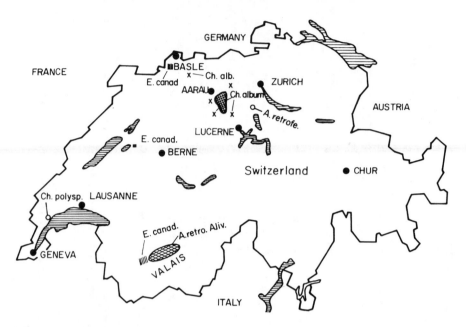

**Figure 3.2** Distribution of triazine-resistant weeds in Switzerland. (H. U. Ammon, previously unpublished.)

described in Chapter 13. The resistant biotype was classified as *A. retroflexus f. Aquinci* Soo.[1] The resistant biotype is being controlled with ethalfluralin and atrazine, alachlor, and chlorbromuron, among others.

### 2.1.3 *Chenopodiaceae*

Instead of cultivating crops such as potatoes, sugar beets, and cereals as has been done traditionally in Switzerland for years, a farmer changed in 1968 to cultivating corn as a monoculture, using 2 kg/ha atrazine with good effect. In 1973, he observed in an area of about 10 m² a dense stand of *Chenopodium album*. He explained this lack of control as inadequate spraying. The fact that only *Chenopodium* remained was not taken into consideration. In 1974 on about 1 ha of the same Swiss field, *C. album* again resisted the atrazine treatment. The farmer was worried; he spent "several weeks with his family in the cornfield cutting the remaining *Chenopodium* by hand." He was told to use higher dosages, up to 5 kg/ha. In 1976, trials with simazine at 3 kg preemergence + atrazine 1.5 kg postemergence had no effect on the *C. album*. The unusually dry weather of this year resulted in poor weed control in many regions; therefore, the resistance was still in doubt.

In 1977, field trials showed that normal field rates of triazine herbicides had no effect on this *Chenopodium* biotype, even though conditions for herbicide activity were ideal and control of *Chenopodium* from other regions of Switzerland was normal.[22] In these experiments it was shown that this biotype was not

Table 3.2  Degree of Resistance and Tolerance to Simazine
among Some Swiss Populations of *Amaranthus*, spp.

| Species[a] | Province[b] | Dosage of simazine (kg/ha) (% of Control) | | | |
|---|---|---|---|---|---|
| | | 0.1 | 1.0 | 2.0 | 5.0 |
| *A. retroflexus* | Changins (VD) | 66 | 6 | 0 | 0 |
| *A. retroflexus* | Miège (VS) | 100 | 82 | 0 | 0 |
| *A. retroflexus* | Vétroz (VS) | 95 | 105 | 90 | 75 |
| *A. retroflexus* | Chamoson (VS) | 103 | 110 | 87 | 63 |
| *A. retroflexus* | Vuisse (VS) | 89 | 83 | 89 | 71 |
| *A. retroflexus* | Ardon (VS) | 110 | 95 | 95 | 75 |
| *A. hybridus* | Ardon I (VS) | 98 | 80 | 80 | 0 |
| *A. hybridus* | Ardon II (VS) | 116 | 116 | 80 | 72 |
| *A. lividus* | Changins (VD) | 90 | 8 | 0 | 0 |
| *A. lividus* | Vétroz (VS) | 105 | 108 | 100 | 40 |
| *A. lividus* | Chamoson (VS) | 99 | 85 | 86 | 41 |
| *A. graecizans* | Vuisse (VS) | 125 | 33 | 17 | 0 |
| *A. graecizans* | Ardon (VS) | 116 | 38 | 32 | 0 |
| *A. bouchonii* | Changins (VD) | 104 | 53 | 0 | 0 |
| *A. blitoides* | Leytron (VS) | 98 | 0 | 0 | 0 |

*Note:* Laboratory results. Means of three replicates, 15 plants per pot, treated pre-emergence with the dry weight measured when the controls reached the five leaf stage.
*Source:* Unpublished data of E. Beuret.[17]
[a]Taxonomy according to Tutin.[16]
[b]VS, Valais; VD, near Lausanne.

only resistant to triazine herbicides, but also to other compounds.[23] Problems with *C. album* arose in neighboring fields. Potatoes treated with metribuzin and sugar beets treated with pyrazon resulted in poor control of the *Chenopodium*.[22,23] The resistance of this *C. album* to triazines and the sensitivity to other herbicides is quite similar to the resistant *C. album* found in Canada,[24] as shown in Table 3.3. In 1978, resistant *C. album* was predominant in some fields where no corn had been grown and no atrazine used for many years, but poor control of *C. album* had been observed in potatoes the year before. These fields were a few kilometers downwind from the infested cornfields.

The maximum spread of *C. album* observed to date in Switzerland was reported in 1977 near Aarau. There was a 15-km-long strip along a 20-km² valley, with about 1100 ha infested. Since 1977 farmers have been taking severe control measures; control is by atrazine plus bromophenoxim postemergence, and crop rotation has been reintroduced on most farms. Susceptible types have begun to predominate again, but new localities outside the known area have been reported recently (Fig. 3.2).

Seeds from resistant *C. album* plants where atrazine had been applied, and seeds from untreated plots near these fields in Switzerland where atrazine was effective, were collected. Plants from these seeds were treated in the laboratory.

Table 3.3 Comparison of the Canadian and Swiss Chenopodium album Biotypes[a]

| Treatment | Application Method | Sensitive Biotypes (Dose in kg/ha) | | Resistant Biotypes (Dose in kg/ha) | | Highest Tolerated Dosage Rate for Corn in Switzerland (kg/ha) |
| --- | --- | --- | --- | --- | --- | --- |
| | | Canadian | Swiss | Canadian | Swiss (Aarau) | |
| Atrazine | Pre/post | + (1.1) | + (1.5) | − (1.4) | − (1.5) | 0.5 |
| Simazine | Pre/post | + (2.2) | (+) | − (2.2) | (−) | |
| Metolachlor + atrazine | Pre | | + (1.7) | | − (1.7) | 3.5 |
| Pendimethalin + atrazine | Pre | + (2.2) | + 1.9 | + (2.2) | + (1.9) | 2.5 |
| Dicamba | Pre | + (0.6) | | + (0.6) | | |
| Methabenzthiazuron + atrazine | Pre | | + (3.1) | | − (3.5) | 3.4 |
| Linuron | Pre | + (0.8) | | + (0.8) | | |
| Metobromuron | Pre | + (0.8) | + (1.5) | + (0.8) | − (1.5) | |
| Metribuzin | Pre | + (0.8) | + (1.5) | + (0.84) | − (1.5) | |
| Chlorbromuron | Pre | | + (1.5) | | + (1.5) | |
| 2,4-D | Post | + (0.6) | + (0.8) | + (0.6) | + (0.8) | |
| Bentazon | Post | | + (1.4) | | + (1.4) | 1.9 |
| Dinoterb | Post | | + (1.0) | | + (1.0) | 1.0 |
| Ioxynil | Post | | + (0.4) | | + (0.4) | |
| Bromoxynil | Post | | + (0.5) | | + (0.5) | |
| Bromofenoxim | Post | | + (1.5) | | + (1.5) | 3.3 |

*Source:* Results with Canadian biotypes according to Bandeen and McClaren.[24] Swiss—field results in Switzerland in 1977.[23]

[a] +, good control; −, no control with the given dosage (kg/ha). With herbicide mixtures the atrazine was given at 1.5 kg/ha and the second herbicide at the level in parentheses.

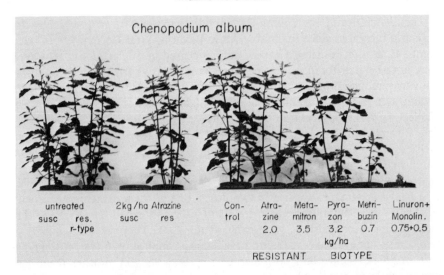

**Figure 3.3**   Effects of various herbicides on the triazine-resistant biotype of *Chenopodium album* from Switzerland. Five pots per treatment containing 4-week-old *Chenopodium* plants at the four- to six-leaf stage were sprayed in the laboratory and grown at 20°C. There was a minimum of variability within treatments. Representative pots are shown 14 days after treatment. (H. U. Ammon, previously unpublished data.)

It was demonstrated that atrazine at 2 kg/ha postemergence had no effect on the resistant type, but resulted in immediate control of the susceptible plants (Table 3.4, Fig. 3.3). Nearly the same reaction occurred with pyrazon at 3.2 kg/ha (Fig. 3.3). Partial resistance can also be observed for metribuzin (0.7 kg/ha) and metamitron (3.5 kg/ha) for the resistant type, whereas the susceptible type was controlled by these dosages (not shown). A mixture of linuron and monolinuron (0.75 + 0.5 kg/ha) controlled both types.

Triazine resistance in *C. album* has appeared in many localities throughout France with an estimated 300,000 ha affected (Fig. 3.1, Table 3.1). Two years of atrazine treatment were sufficient for an invasion of resistant *C. album* to appear in Eure et Loir. A genetic study of different populations has shown that they vary from one region to the next and that consequently their origins probably vary.[9] All cases appeared in corn and it was first noticed as early as 1973 in Marne et Loire.[8] In a susceptible population near a resistant one, plants were found to resist only 0.5 kg/ha,[7] but after continuous triazine treatment, the population became resistant, like those which had been selected in cornfields. One case of resistance was also found in *C. polyspermum* in Ille et Vilaine near Rennes[6] (Fig. 3.1, Table 3.1).

A few hectares of resistant *C. polyspermum* have also been found near Geneva, Switzerland.[15] Szith and Furlan[27b] recently reported that a *C. polyspermum* biotype has developed in southern Austria having apparent resistance to atrazine.

The difficulties of controlling Chenopodiaceae weeds in corn with atrazine have also increased since the mid-1970s in Bavaria, West Germany.[25] In addition to *C. album*, resistance has steadily appeared in *C. ficifolium*. However, there have been differences in experience with intensive corn cultivation, especially with monoculture, where it was no longer possible to control the *Chenopodium* spp. in spite of using considerably higher than normal rates of atrazine (Table 3.4). In 1978, five cases of this kind were observed in Bavaria near Freising, Rosenheim, Fürstenfeldbruck, Krumbach, and Kitzingen (Fig. 3.4). Near Freising, after 5 to 6 year years of continuous corn cultivation on a mineral soil, a weed population consisting mostly of *C. ficifolium* and a small amount of *Atriplex patula* and *C. album* developed. None of them could be controlled by 3 kg/ha of atrazine applied in two doses, one of them postemergence with oil. Field and greenhouse tests in 1978 and 1979 confirmed the suspicion that these strains were resistant to atrazine (Table 3.5). They proved to be completely resistant even at 5 kg/ha of atrazine when applied on different soil types both in preemergence and postemergence treatments. On another farm near Freising with peat soil, where intensive fertilization with liquid manure from pigs had been applied for years, *C. ficifolium* also formed the main weed population after atrazine application. Similar observations with this species have been made near Rosenheim. In both cases, resistance to atrazine has been confirmed. In all other cases, resistance was observed in the related members of the Chenopodiaceae (i.e., *Atriplex patula* and *C. album*); resistance occurred on slightly humic soils after more than 5 years of permanent corn monoculture. There is also evidence that a resistant biotype of *A. patula* was already present in corn fields of northern Switzerland when atrazine was first used in the early 1970's.[25a]

Atrazine-resistant *C. album f. corymbose paniculatum* Soo has been found in two places in Hungary, at frequencies of above 60%, having 15-year monocultures of corn.[1]

Holliday and Putwain[3] screened *C. album* populations in England and found variations in tolerance to simazine (0.7 kg/ha), with the percentage mortality varying between 23 and 88% at the various sites.

## 2.2 Cases of Resistance Indigenous to Europe

Cases of triazine resistance have occurred in *Brassica* and *Ambrosia* in North America (Chapter 2) but not yet in Europe. The reasons must be intriguing, as there is worldwide distribution of these genera of weeds.[26] With *Ambrosia* spp. it may be due to a lack of genetic variability in Europe. *Ambrosia* is native to the western hemisphere and was introduced to Europe. This theory is somewhat supported in reverse by the appearance of triazine resistance in *Atriplex patula* in Bavaria (Section 2.1), but not in America. *Atriplex* is found in America but it was introduced from Europe. Thus the same degree of genetic variability is not likely to be present in North America. The following species are also pests in North America, but to date they have developed triazine resistance only outside North America.

Table 3.4 Swiss *Chenopodium album* Biotypes Treated with Various Herbicides

| Biotypes (Seeds Collected in 1978) | Location | Herbicide (kg/ha) | | | | | |
|---|---|---|---|---|---|---|---|
| | | Control | Atrazine (2.0) | Metamitron (3.5) | Pyrazon (3.2) | Metribuzin (0.7) | Linuron + Monolinuron: (0.75 + 0.5) |
| | | | (Necrotic/Chlorotic Leaf Surface, as %) | | | | |
| *Sensitive* | | | | | | | |
| Reckenholz[a] | Near Zurich | 0/0 | 100/0 | 100/0 | 80/20 | 100/0 | 100/0 |
| *Resistant* | | | | | | | |
| Liebegg[b] | Near Aarau | 0/0 | 0/0 | 10/0 | 5/5 | 55/25 | 100/0 |
| Unterkulm[c] | Near Aarau | 0/0 | 0/0 | 10/35 | 5/10 | 55/35 | 100/0 |
| Unterkulm[d] | Near Aarau | 0/0 | 0/0 | 10/45 | 5/5 | 55/45 | 100/0 |

*Note:* Five pots with two plants each were measured 7 days after treatment. (Data of Ammon and Siegrist.[14]) In addition, the resistant Swiss *Chenopodium album* was controlled with pendimethalin, chlorbromuron, 2,4-D, bentazon, dinoterb, ioxynil, bromoxynil, and bromophenoxim[23] (data of Aeschlimann[15]).

[a] "Sensitive" type from Reckenholz.

[b] Seed from farmer's fields, atrazine treated.

[c] Seeds from methabenzthiazuron-treated experimental plots.

[d] Seeds from atrazine-treated plots.

○ Chenopodium sp.     □ Solanum nigrum     ▽ Stellaria media

**Figure 3.4** Distribution of triazine-resistant weeds in Bavaria. Herbicide-resistant weeds have not been reported in other West German federal states. (H. Kees, previously unpublished.)

### 2.2.1 Stellaria

A biotype of *Stellaria media* has appeared in Bavaria (West Germany) which is resistant to agricultural levels of triazines and is tolerant to higher levels. Its appearance has a temporal correlation with the increase in the amount of corn grown.[25,27] The production of corn in Bavaria has rapidly and continuously increased from about 40,000 ha in the beginning of the 1960s to 360,000 ha in 1979. In most cases corn is in the rotation once in 3 years and is in monoculture or continuous culture only in exceptional cases. Weed control has usually been performed with atrazine alone.

Table 3.5 Cases of Resistance and Tolerance of Bavarian (Germany) Biotypes to Atrazine[a]

| Species | Location | Area Infested (ha) | Tested Soil | Atrazine (kg/ha) | | | | | |
| | | | | Preemergence | | | Postemergence | | |
| | | | | 1.25 | 2.5 | 5.0 | 1.0 | 2.0 | 5.0 |
| | | | | (% of untreated sensitive control, fw) | | | | | |
| *Chenopodium ficifolium* (res) | Freising | 20 | Light sandy loam | 80 | 60 | 11 | 97 | 63 | 48 |
| | Freising | | Humus sand | 100 | 87 | 93 | 90 | 87 | 80 |
| *Solanum nigrum* | Deggendorf (res.) | 50 | Light sandy loam | 100 | — | 100 | 100 | — | 100 |
| | Deggendorf (sens.) | | Light sandy loam | 0 | — | 0 | 0 | — | 0 |
| | Deggendorf (res.) | | Humus sand | 100 | — | 100 | 100 | — | 100 |
| | Deggendorf (sens.) | | Humus sand | 10 | — | 0 | 0 | — | 0 |
| *Stellaria media* | Vilsbiburg (res.) | 5 | Light sandy loam | — | — | 56 | 100 | 100 | 100 |
| | Vilsbiburg (sens.) | | Light sandy loam | — | — | 0 | 0 | 0 | 0 |

*Source:* Unpublished data of H. Kees from 1979 experiments.

[a]Seed was gathered from plants from the locations listed and germinated in the greenhouse in the soil types designated. Preemergence treatments were measured 12 weeks after spraying; postemergence treatments were measured 5 weeks after treatment.

43

The first incidence of atrazine-resistant biotypes among dicotyledonous weeds in Bavaria was found in 1974 in an area of some 5 ha near Landshut (Lower Bavaria) (Fig. 3.4) where corn had been continuously grown since 1969 and where *Stellaria* could no longer be satisfactorily controlled with normal rates of atrazine. In 1975, the sixth year of continuous corn, even 4.5 kg/ha of atrazine (applied in three doses) was no longer sufficient to control *Stellaria*. In the following years, field trials and greenhouse experiments confirmed the *Stellaria* biotype to be almost resistant to atrazine.[27] Seedlings from other locations which were used as controls in these experiments were fully susceptible to normal rates of the herbicide. However, there were differences in susceptibility to atrazine at rates of 5 kg/ha, depending on the development of the weed. Applications of atrazine before emergence and at the two-leaf stage of growth resulted in a stronger temporary growth depression than at the four-to-six-leaf stage. This resistance was not diminished during the next 6 years. Later, weed control in this field was successfully accomplished with atrazine in combination with bromophenoxim. Some weeks after treatment new seedlings of resistant *Stellaria* grew up and formed a carpet a few centimeters tall. This weed infestation had no effect on the development of the corn; it even seemed to be an advantage in preventing soil erosion. Since that time no further incidence of resistant *Stellaria* has been observed in Bavaria.

### 2.2.2 *Solanum*

Since the mid-1970s a considerable increase of atrazine-resistant *Solanum nigrum* in corn has been observed in a limited area near Deggendorf in Lower Bavaria (West Germany) (see Fig. 3.4). Subsequent experiments in 1979 have shown that seedlings from this location were completely resistant to high doses of atrazine (Fig. 3.5, Table 3.5), both to preemergence and postemergence treatments on different soils. Seedlings from other areas were killed by normal rates of atrazine. Although corn in this region is grown only once every 3 years, the atrazine-resistant *Solanum* biotype has become the main weed population as a result of gradual selection since 1963. In this case the problem was solved by using contact herbicides in postemergence treatments.

*S. nigrum* having resistant plastids was found in an atrazine-treated cornfield near Vivero in northern Italy (J. Gasquez, unpublished data).

Triazine-resistant *S. nigrum* was found in one locale of France[12] in 1976. It has since spread to 50,000 ha (Fig. 3.1, Table 3.1). Although this biotype has the same type of plastid resistance as most other triazine-resistant weed species studied so far (Table 3.1), the inheritance seems to be only partially maternal (i.e., maternal with possible nuclear modifiers).[27a]

### 2.2.3 *Polygonum*

Beginning in 1976, resistant populations of *Polygonum lapathifolium* rapidly invaded one village in France[12,12a] (Fig. 3.1, Table 3.1). Except when accidentally carried by human beings, it has not seemed to have spread over the neighboring area of corn culture. Three years of atrazine treatments out of 6 years

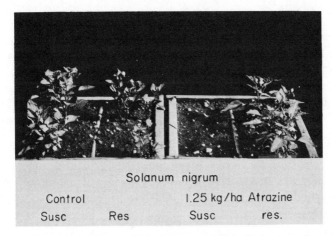

**Figure 3.5** Triazine-resistant and triazine-susceptible *Solanum nigrum* biotypes from Bavaria. (H. Kees, previously unpublished.)

of corn cultivation were sufficient to allow the first major invasion of *P. lapathifolium*.

Resistant *P. persicaria* has remained circumscribed to one valley on a few hundred hectares in France since the first report in 1978[6] (Fig. 3.1, Table 3.1). Szith and Furlan[27b] recently reported that a resistant population of *P. convolvulus* has developed after three years of atrazine treatments in Austrian corn fields.

### 2.2.4 *Gramineae*

Triazine-resistant *Poa annua* has appeared on one simazine-treated roadside in France where the herbicide was used once yearly for several years at 10 kg/ha[8] (Table 3.1, Fig. 3.1).

Since 1970, some populations of *Echinochloa crus-galli* previously treated with triazine herbicides in the Pau area of France were found to consist of plants that were much more tolerant than those which grew in locales which were not treated with triazines. This high tolerance borders on complete resistance. With preemergence treatments of atrazine, 80% of the plants survived at 3 kg/ha and 50% of the plants survived at 6 kg/ha.[10,11] This species has invaded more than 100,000 ha in the southwest of France and in Alsace (Fig. 3.1).

It is interesting that within the Gramineae two major types of triazine resistance have been reported. *Bromus tectorum* (Chapter 2 and ref. 28) and *Poa annua* (Table 3.1) exhibit resistance at the plastid level, whereas the French population of *Echinochloa crus-galli* failed to show resistance to atrazine at the plastid level (Table 3.1). This very high tolerance in *Echinochloa* is apparently due to an enhanced detoxification of the herbicide, as has been shown by Jensen for panicoid grasses (see Chapter 8). A resistant *E. crus-galli* in Maryland (Chapter 2), on the other hand, seems to have the plastid resistance.[28] A lower level of tolerance to atrazine by *Echinochloa* is discussed in Section 2.3.

### 2.2.5 *Erigeron*

Triazine-resistant *Erigeron canadensis* has been found near Basle and Martigny (Valais) in Switzerland, but it is limited to a few hectares at each location. Field observations showed that the *Erigeron* could no longer be controlled with 9 kg/ha atrazine applied preemergence. In recent years, this resistant biotype has spread along railroad beds and into some cropland of Switzerland.

### 2.2.6 *Bidens*

A rather extensive area near Gross-Meinbach in Steinmark, Austria, is covered by *Bidens tripartita*, which is resistant to 5 kg/ha atrazine.[29] Plastid triazine resistance was found in *B. tripartita* grown in a 100-ha atrazine-treated field in nearby Gliesdorf, Austria (J. Gasquez, unpublished data).

### 2.2.7 *Galinsoga*

*Galinsoga ciliata* has been identified as becoming resistant to 5 kg/ha of atrazine in pre- and postemergence applications. This species was found resistant in an area of 25 ha of slightly humic soil at Eichenreid near Munich in West Germany. There had been 5 years of continuous corn cultivation.[29a]

## 2.3 Differential Tolerance in *Echinochloa*, *Setaria*, *Digitaria*, *Veronica*, and *Capsella*

Increases in tolerance (but not resistance) to triazines were noted following repeated treatments in cornfields (including in rotation with wheat) and in vineyards near Montpellier, France[30] and in cornfields in northern Italy.[31] Some of the data for *Setaria viridis*, *Echinochloa crus-galli*, *Digitaria sanguinalis*, and *Veronica persica* are presented in Table 3.6. The problem in France seems to be localized and has not spread. Seeds were not available for further laboratory testing.[32]

Holliday and Putwain[3] screened populations of *Capsella bursa-pastoris* from various sites in England for simazine resistance. Although many of the sites had been planted to horticultural crops on which simazine had been used continuously for up to 10 years, they did not find full simazine resistance in any population in the field. When exposed to simazine at 0.7 kg/ha, the percentage mortality of the *C. bursa-pastoris* populations ranged from 25% (West Peckham, Kent) to 86% elsewhere.

## 3 PHENOXY HERBICIDES

### 3.1 *Chenopodium*

Seventeen years ago Beauge[33] reported a population of *Chenopodium album* that was resistant to 2,4-D. Only recently, with the enhanced interest in the

Table 3.6  Increase in Differential Tolerance to Atrazine
by Some Weeds in France and Italy

| Species | Rotation or Crop | France[a] | | Italy[c] | |
|---|---|---|---|---|---|
| | | Number of Previous Atrazine Treatments | Percent Mortality at 5 kg/ha Atrazine[b] | Number of Previous Atrazine Treatments | Percent Mortality at 5 kg/ha Atrazine[b] |
| *Echinochloa crus-galli* | Untreated | 0 | 96[a] | 0 | 97[a] |
| | Grapes | 10 | 86[b] | | |
| | Corn | 14 | 69[c] | 10 | 66[b] |
| *Setaria viridis* | Untreated | 0 | 92[a] | 0 | 93[a] |
| | Grapes | 5 | 70[b] | | |
| | Corn, wheat | 9 | 68[b] | | |
| | Corn | | | 10 | 62[b] |
| *Digitaria sanguinalis* | Untreated | 0 | 79[a] | 0 | 96[a] |
| | Grapes | 10 | 67[b] | | |
| | Corn | | | 10 | 68[b] |
| *Veronica persica* | Untreated | 0 | 79[a] | | |
| | Corn | 14 | 55[b] | | |
| *Sorghum halepense* | Untreated | 0 | | 0 | 95[a] |
| | Corn | 10 | | 10 | 60[b] |

[a]Data condensed from Grignac.[30]

[b]For a given species, results followed by different letters differ significantly at the 0.05 level in a multiple-range test.

[c]Data condensed from Miele and Vannozzi.[31]

topic, was it thought to reconfirm this observation and to further study the phenomenon, but viable seed was no longer available.[34] It is, therefore, questionable to say whether that observation was truly a herbicide resistance phenomenon.

## 3.2  Cirsium

In the 1950s, observations indicated an increasing tolerance of *Cirsium arvense* to MCPA in certain parts of Sweden.[35] The common explanation was that the spraying was being done increasingly earlier, which resulted in unsatisfactory control of *C. arvense*; the best time of treatment being when most of the shoots have emerged. As a result, an investigation was started in 1976 which was planned to include both a survey of the literature as well as green-

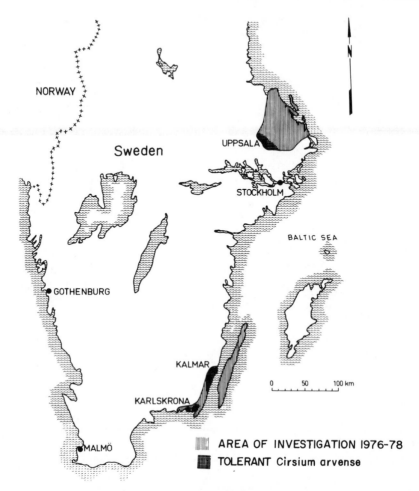

**Figure 3.6**  Distribution of *Cirsium arvense* tolerant to MCPA in Sweden. The criteria for tolerance and appearance of tolerant plants are illustrated in Figs. 3.7 and 3.8. (H. Fogelfors, previously unpublished.)

house experiments and field studies with regard to changes in the flora of farmland (arable land) with special emphasis on chemical weed control.[36] Both inter- and intraspecific changes were included, but primarily the latter. This naturally led to *C. arvense*, as well as other species, being included in the investigation. About 60 clones were tested in greenhouse experiments for susceptibility to MCPA. These clones were collected in eastern and southeastern parts of Sweden (Fig. 3.6), both from places where there had been an intensive and fairly regular use of MCPA, as well as from similar areas where it was highly unlikely that herbicides had ever been used. Almost all clones from the larger and more intensive agricultural districts within both areas were found to be more tolerant, whereas with clones from the areas where MCPA had not been used there was a fairly even distribution between susceptible and more tolerant biotypes. The differential response of the tolerant and susceptible biotypes is illustrated in Figs. 3.7 and 3.8. Hodgson[37] had noticed ecotype differences in tolerance to the related 2,4-D in the United States, but this seems to be the first correlation with the treatment history of the area.

## 3.3 Polygonum

Hammerton[38] found considerable differences in susceptibility to dichlorprop among population samples of the inbreeding species *Polygonum lapathifolium* from four British localities in Yorkshire (having a median lethal dose of 3.2 kg/ha), Shropshire (3.0 kg/ha), Oxford (2.9 kg/ha) and Aberystwyth (1.6 kg/ha), and one locality in the Netherlands (1.5 kg/ha) (Fig. 3.9). It seems most likely that these differences were due to differences in the ability of the weeds to metabolize and detoxify dichlorprop. However, it was unlikely that these variations resulted from differential exposure of the parent populations to dichlorprop. Dichlorprop was introduced only in 1961 and the populations were collected in 1962. There may have been some earlier differential exposure to other phenoxy acid herbicides which could have brought about the differential susceptibility to dichlorprop.

## 3.4 Matricaria

Ellis and Kay[39,40] screened farmland and roadside populations of the outbreeding species *Matricaria perforata (Tripleurospermum maritimum* ssp. *inodorum)* from 43 sites in England and Wales and five sites in France for tolerance to MCPA and ioxynil (see below). The mean fresh weight per plant of the samples exposed to MCPA at 1.75 kg/ha ranged from 10 to 60% of those of the controls. Tolerance to MCPA varied geographically and was significantly greater in populations with a history of heavy spraying with herbicides. The least tolerant populations (about 20% of the samples) originated from wasteland or roadsides, and the most tolerant ones were found in heavily sprayed cereal crops. Populations showing high or relatively high tolerance to MCPA occurred in all the areas of Britain that were sampled (Glamorgan, Somerset, Wiltshire, Berkshire, Hampshire, Sussex, Oxfordshire, Essex, Suffolk, Norfolk, Staffordshire, and Warwickshire). There was no correlation between spray retention and herbicide resistance, and it seemed likely that general physiological resistance was involved.

## 3.5 Taraxacum, Ranunculus, and Trifolium

In probably the first documented case of enhanced tolerance, Stryckers[40a] in 1950 found uncontrolled *Taraxacum officinale, Ranunculus* spp., and *Trifolium repens* in Belgian pastures following up to nine 2,4-D or MCPA treatments. From 20 to 100% tolerance was reported for treatments of 250 g/ha. Unfortunately, it appears that this work was not followed up.

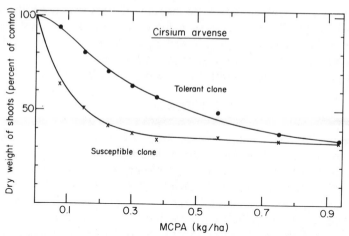

**Figure 3.7** Dose-response curve for MCPA-tolerant and MCPA-susceptible *Cirsium arvense*. Equal root sections of two clones (Fig. 3.6) were planted in pots and grown in the greenhouse. Twenty-eight days later they were sprayed with the dimethylamine salt of MCPA. At the time of spraying the dry weights of the tolerant and susceptible clones were 1.15 and 0.95 g per pot, respectively. When the plants were harvested 15 days later, the control values were 2.74 and 2.71 g, respectively. Doses are presented as MCPA free-acid equivalents. (H. Fogelfors, previously unpublished.)

**Figure 3.8** Tolerant and susceptible *Cirsium arvense* 15 days after spraying with 0.225 kg/ha MCPA. All other particulars as in Fig. 3.7. (H. Fogelfors, previously unpublished.)

**Figure 3.9** Sites in the United Kingdom where *Polygonum* and *Matricaria* were assayed for tolerance to herbicides (see Sections 3.3, 3.4, 4, and 7 in the text; by Q. O. N. Kay.)

## 4  BENZONITRILES

### 4.1  *Matricaria*

In the same research[39,40] on resistance to MCPA, resistance to ioxynil in *Matricaria perforata* was also studied. Plants from the seed collected at the different locations were treated with 0.28 kg/ha ioxynil. These had fresh weights varying between 2 and 40% of the untreated controls. Tolerance to ioxynil showed a correlation with the geographic origin of the population but no overall correlation with spray history. There was a significant cross tolerance between MCPA and ioxynil, even though few of the populations had ever been exposed to the latter herbicide.

# 5  UREAS

## 5.1  Poa

Grignac[30] reported experiments in which he subjected three ecotypes of *Poa annua* to three annual treatments of 3 kg/ha metoxuron for 8 years. Mortality of one of the three ecotypes dropped from 94% to 16% over the period, although the "vigor" of the viable seedlings was not as great as that of the control. This shows the relative ease of obtaining partial tolerance.

# 6  BIPYRIDYLIUMS

## 6.1  Poa

A market gardener near Reading, England, had been applying paraquat as the sole herbicide treatment, two or three times each year for 10 years, to clear annual weeds from his pathways and seedbeds. In 1978, he was no longer able to control *Poa annua,* even with greatly increased doses of the herbicide. Plant samples were collected from the site, which he claimed had been sprayed with, and had survived, paraquat. These plants were grown in a greenhouse at Jealott's Hill, and propagated to obtain several clones of the supposed "resistant" types. Samples of typical *P. annua* from fields that had not been exposed to such abnormal herbicide pressure were propagated in the same way. Both types were tested for response to paraquat and there was a marked increased tolerance in the clones having a history of paraquat treatment. Normal *Poa* was killed by applications of 0.1 to 0.2 kg/ha. The purported paraquat-resistant material required in excess of 0.8 kg. Superficially, the resistant strains appeared to have acquired very much the same mechanisms of tolerance as the resistant ryegrass (Chapter 11). Localized damage occurred on leaves beneath individual spray droplets, but the effects hardly spread. In particular, juvenile tillers shielded from the spray grew normally, and did not become chlorotic, even when sufficient paraquat had been applied to severely damage the more mature foliage. Hence the resistant type eventually recovered from high levels of early damage. As far as the grower was concerned, the "problem" was solved very simply by adopting a rotation of different herbicides.[41]

## 6.2  Conyza (Erigeron)

Resistant *Conyza linifolia* (possibly synonymous with *C. bonariensis*) has appeared in vine and citrus treated five times annually with paraquat in Egypt.[41] A rate of 10 kg/ha paraquat is now required to control this composite weed (see also Chapter 11, Section 2.2.1, and ref. 42).

*Erigeron philadelphicus* has become resistant to paraquat in Japan.[43] Resistance appeared in mulberry plantations in the Saitama prefecture, 60 km northwest of Tokyo, and covers most of the 100 mulberry patches in a 50-ha area. The farmers treated this with paraquat about 2 to 4 times each year since 1969. The selected biotypes can withstand levels of paraquat well beyond those necessary to kill the susceptible biotypes.[43] Both the *Erigeron*[43] and the *Conyza*[41] are only moderately tolerant to diquat. Many taxonomists classify *Conyza* and *Erigeron* as the same genus.

## 7 CONCLUDING REMARKS

Most of the conclusions to be derived from this short history of the appearance of herbicide resistance are to be found in the following chapters.

The case histories all have a common thread which must be borne in mind when reading on. Why is it that some species seem more apt to become resistant? *Poa annua* is a case in point. Why is it that the use of triazines has brought about resistance, whereas with the phenoxy acids there seem to be more cases of increasing tolerance and changes in weed distribution (Chapter 4)? Most cases occurred where the crops were grown in continuous culture with the repeated use of a single herbicide. Does this mean that with rotation resistance will not happen? These are all points to be touched upon in later chapters. The history should help us develop strategies to prevent recurrence of the phenomena described in Chapters 2 and 3. It is clear that resistance can be overcome at present—at a cost. The farmer can no longer routinely use the most cost-effective herbicide; more expensive ones must now be considered to replace them where resistance has become a problem.

## REFERENCES

1 P. Solymosi, Institute of Plant Protection, Budapest, Hungary, personal communications, 1980.

2 K. Ueki, Weed Science Laboratory, Kyoto University, Kyoto, Japan, personal communication, 1979.

3 R. J. Holliday and P. D. Putwain, Variation in the susceptibility to simazine in three species of annual weeds, *Proc. 12th Br. Weed Control Conf.*, 649 (1974).

4 R. J. Holliday and P. D. Putwain, Evolution of resistance to simazine in *Senecio vulgaris* L., *Weed Res.*, **17**, 291 (1977).

5 P. D. Putwain, Dept. of Botany, University of Liverpool, England, personal communications, 1980 and 1981.

6 J. Gasquez and J. P. Compoint, Trois nouvelles mauvaises herbes résistantes aux triazines en France; *Amaranthus retroflexus* s.l., *Chenopodium polyspermum* L., *Polygonum persicaria* L., *Chemosphere*, **9**, 39 (1980).

7   J. Gasquez and G. Barralis, Observation et sélection chez *Chenopodium album* L. d'individus résistants aux triazines, *Chemosphere,* 7, 911 (1978).

8   J. M. Ducruet and J. Gasquez, Observation de la fluorescence sur feuille entière et mise en évidence de la résistance chloroplastique à l'atrazine chez *Chenopodium album* L. et *Poa annua* L., *Chemosphere,* 7, 691 (1978).

9   J. Gasquez and J. P. Compoint, Isoenzymatic variations in populations of *Chenopodium album* L. resistant and susceptible to triazines, *Agro-ecosystems,* 7, 1 (1981).

10   J. Gasquez and J. P. Compoint, Apport de l'électrophorèse en courant pulsé à la taxonomie d'*Echinochloa crus-galli* (L.) P. B., *Ann. Amelior. Plant.,* 26, 345 (1976).

11   J. Gasquez and J. P. Compoint, Mise en évidence de la variabilité génétique infra-population par l'utilisation d'isoenzymes foliaires chez *Echinochloa crus-galli* (L.) P.B., *Ann. Amelior. Plant.,* 27, 267 (1977).

12   J. Gasquez and G. Barralis, Mise en évidence de la résistance aux triazines chez *Solanum nigrum* L. et *Polygonum lapathifolium* L. par observation de la fluorescence de feuilles isolées, *C. R. Acad. Sci. (Paris) Sec. D,* 288, 1391 (1979).

12a   H. Darmency, J. P. Compoint and J. Gasquez, La résistance aux triazines chez *Polygonum lapathifolium* L. *Acad. Agric. France* pp. 231-238 (1981).

13   L. Stalder, Der Rauhhaarige Amarant *(Amaranthus retroflexus)* wird resistent. *Schweiz. Z. Obst.-Weinbau,* 115, 386 (1979).

14   H. U. Ammon and A. Siegrist, Swiss Federal Research Station for Agronomy, Zurich-Reckenholz, unpublished results, 1979.

15   J. Aeschlimann, CIBA-GEIGY Ltd., CH-4002, Basle, Switzerland, personal communication, 1979.

16   T. G. Tutin, *Florae Europeae,* Vol. 1, Cambridge University Press, Cambridge, 1964.

17   E. Beuret, Swiss Federal Research Station for Agronomy, Changin, Switzerland, personal communication, 1980.

18   H. Neururer, Verstärktes auftreten von resistenten zurückgekrümmten fachsswanz in mais, *Der Pflanzenartz,* 30, 104 (1977).

19   G. Csala and F. Hartmann, Investigation of expansion of *Amaranthus retroflexus* L. in maize mono-cultures of Agricultural Kombinat in Babolna, Hungary, *Növényvedelem,* 15(1) 28 (1979).

20   F. Hartmann, The atrazine resistance of *Amaranthus retroflexus* L. and the expansion of resistant biotype in Hungary, *Növényvedelem,* 15(11), 491 (1979).

21   M. Ötvös and F. Hartman, The resistance of *Amaranthus retroflexus* to atrazine in Hungary, *Proc. 9th Int. Congr. Plant Prot.,* Abstr. 936 (1979).

21a   P. Solymosi, Inheritance of herbicide resistance in *Amaranthus retroflexus, Növénytermelés,* 30, 57 (1981).

22   H. U. Ammon, Kombination chemisch-, mechanisch- und biologischer Methoden zur Unkräutbekämfung im mehrjährigen Maisbau und erste Resultate über die Beeinflussung bodenphysikalischer Kenwerte, *Proc. EWRS Symp. Methods of Weed Control and Their Integration,* 243 (1977).

23   H. U. Ammon, Praxiserfahrungen mit Praparaten zur Hirsebekampfung und neue Resistents Unkräuter im Mais, *Mitt. Schweiz. Landwirtsch.,* 26, 33 (1978).

24   J. D. Bandeen and R. D. McClaren, Resistance of *Chenopodium album* to triazine herbicides, *Can. J. Plant Sci.,* 56, 411 (1976).

25   H. Kees, Beobachtungen über Selektion und Resistenzbildung bei Unkräutern durch Herbizide und Fruchtfolgevereinfachung im Bayern, *Proc. EWRS Symp. The Influence of Different Factors on the Development and Control of Weeds,* 225 (1977).

25a   H. M. LeBaron, CIBA-GEIGY Corporation, Greensboro, N.C. 27419, personal communication, 1981.

26  L. G. Holm, D. L. Plucknett, J. V. Pancho, and J. P. Herberger, *The World's Worst Weeds: Distribution and Biology,* University Press of Hawaii, Honolulu, 1977, p. 609.

27  H. Kees, Beobachtungen über Resistenzerscheinungen bei der Vogelmiere (*Stellaria media*) gegen Atrazin im Mais, *Gesunde Pflanz.,* **30**(1), 137 (1978).

27a  J. Gasquez, H. Darmency and J. P. Compoint. Etude de la transmission de la résistance chloroplastique aux triazines chez *Solanum nigrum, C.R. Acad. Sci. (Paris) Sec. D.,* **292,** 847 (1981).

27b  R. Szith and H. Furlan, Der windenknöterich (*Polygonum convolvulus* L.), ein "neues" atrazinresistentes unkraut im maisbau, *Der Pflanzenarzt,* **33,** 95 (1980).

28  J. R. Hensley, CIBA-GEIGY Corporation, Agricultural Chem., Greensboro, N.C., personal communications, 1980.

29  R. Szith and H. Furlan, Der dreiteilige zweizahn (*Bidens tripartita,* L.) ein neues atrazinresistentes unkraut im mais, *Der Pflanzenarzt,* **32,** 6 (1979).

29a  H. Kees, Zum. Auftreten Atrazinresistenter Samenunkraeuter in Bayern. *Bayer. Anw. Jahrbuch* 1981. In press.

30  P. Grignac, The evolution of resistance to herbicides in weedy species, *Agro-ecosystems,* **4,** 377 (1978).

31  S. Miele and G. P. Vannozzi, Induzione di resistenza all'atrazina in alcuna infestanti del mais, *Agric. It.,* **106** (32 n.s.), 179 (1977).

32  P. Grignac, École Nationale Superieure Agronomique, 9 Place Viala, 34060 Montpellier Cédex, France, personal communications, 1979.

33  A. Beauge, *Chenopodium album et espèces affinés,* SEDES, Paris (1974), 409 p.

34  A. Beauge, Laboratoire de Botanique Historique et Palynologie, Faculté des Sciences et Techniques de St-Jérôme, 13397 Marseille Cédex 4, France, personal communication, 1978.

35  A. L. Abel, The rotation of weed killers, *Proc. Br. Weed Control Conf.,* **1,** 249 (1954).

36  H. Fogelfors, Changes in the flora of farmland, *Swed. Univ. Agric. Sci. Dept. Ecol. Eviron. Res. Rep. No. 5, Upps.* (1979), 66 p.

37  J. M. Hodgson, Variations in ecotypes of Canada thistle, *Weeds,* **12,** 167 (1964).

38  J. L. Hammerton, Studies on weed species of the genus *Polyonum* L.: III. Variation in susceptibility to 2-(2,4 dichlorophenoxy)propionic acid within *P. lapathifolium* L., *Weed Res.,* **6,** 132 (1966).

39  M. Ellis and Q. O. N. Kay, Genetic variation in herbicide resistance in scentless mayweed (*Tripleurospermum inodorum* L., Schultz Bip.): I. Difference between populations in response to MCPA, *Weed Res.,* **15,** 285 (1975).

40  M. Ellis and Q. O. N. Kay, Genetic variation in herbicide resistance in  scentless mayweed (*Tripleurospermum inodorum* L., Schultz Bip.): II. Intraspecific variation in response to MCPA and ioxynil, and the role of spray retention characteristics, *Weed Res.,* **15,** 295 (1975).

40a  J. Stryckers, Onderzoekingen naar de toepassings-mogelijkheden van synthetische groeistoffen als selektieve herbiciden im grasland en akkerbouwgewassen, *Gent Rijkslandbouwhogsch. Rep.,* **100** (1958).

41  A. F. Hawkins and M. Parham, ICI, Jealott's Hill, England, personal communications, 1979.

42  R. J. Youngman and A. D. Dodge, On the mechanism of paraquat resistance in *Conyza* sp., in *Proc. 5th Int. Congr. Photosynth. 1980,* G. Akoyunoglou (Ed.), Balaban International Science Services, 2242 Mt. Carmel Ave., Glenside, Pa. 19038 (1981).

43  Y. Watanabe, Central Agricultural Experiment Station, Konosu, Saitama, Japan, personal communication, 1981.

# CHAPTER FOUR

# Changing Patterns of Weed Distribution as a Result of Herbicide Use and Other Agronomic Factors

H. HAAS and J. C. STREIBIG

Department of Crop Husbandry and Plant Breeding
The Royal Veterinary and Agricultural University
Copenhagen, Denmark

## 1  INTRODUCTION

Human beings have thoroughly changed the original vegetation by cultivating particular plant species in monoculture for about 7000 years, and throughout this period weeds have been an everlasting menace to our ability to grow crops. Weeding and hoeing by hand have been the most prevalent methods of weed control. Only for the last 200 to 300 years have agricultural implements been used in the struggle against weeds. The first efforts to use inorganic chemical means of weed control are barely 100 years old, and organic herbicides were introduced only 40 years ago.

In retrospect, the history of weed control had been relatively static for more than six millennia and has undergone profound changes only in the last 30 years, during which time about 200 chemical compounds have been introduced to agriculture. In return, the accelerating development of herbicides has certainly had a great impact on the weed flora and has given the farmer a safety or selection factor in cultivating crops, which was previously unknown.

In 1950, about 5% of the total consumption of pesticides consisted of herbicides. In 1976, herbicides accounted for 45%.[1] At the present time, herbicide use is even higher in many developed countries (e.g., in Denmark 76% of the total pesticides sold in 1978 were herbicides).[2] By summarizing the global statistics on herbicides,[3] the total amount of herbicide use in arable land was equal to approximately 1.0 to 1.5 kg/ha in many developed countries. During the period 1973 to 1977 there has been an annual increase in the total herbicide consumption of about 4%. Consequently, herbicide use has become the dominant weed control method in many countries.

The use of herbicides exhibiting similar modes of action and sprayed on the same area year after year has imposed new factors of selection, either by radical changes in the mutual competition between present weed species[4] or by increasing tolerance or even resistance of species or strains within species that had been susceptible in the past. Most of this book is about intraspecific selection for the rare herbicide-resistant plants in a susceptible species. This chapter, however, will try to describe how various factors interrelate to change species distribution in arable land.

The evaluation of the changing pattern of weed distribution, described in this chapter, is based on existing botanical data covering a 60-year period.[5-9] Hence the period covers the time before and after introduction of modern herbicides. In order to explain the apparent causality seen in the data, we have used the Raunkiaer's life forms[10] (therophytes, geophytes, etc.) as being more descriptive than the commonly used life duration terms (annual, biennials, etc.). These life forms are defined in Table 4.1. Using this form of classification, some of the changes brought about by herbicides and other agronomic factors become more readily apparent.


## 2 THE WEED FLORA: ECOLOGICAL AND GENETIC CONSIDERATIONS

Weeds are the pioneers in an ecological succession which results in a natural vegetation based upon the prevailing local conditions. Weed species increase the diversity of agricultural ecosystems by utilizing the environmental potential especially developed by human beings for crop production.[11] The weed flora is made up of native species recruited from labile natural habitats as well as nonnative species introduced and spread by human beings or by other means. The evolution of weeds has been dealt with by Baker[12] and McNeill,[13] and the migration history, global distribution, and occurrence of important weed species are given by Holm et al.[14,15]


### 2.1 Diversity of the Weed Flora

About 206 weed species important to human beings have been recorded on a worldwide scale,[16] and 43% of these belong to only four families. The Gramineae and Compositae alone contain no fewer than 76 weed species. Of 59 families altogether, 12 dominate the population and cover 68% of the total number of species. In Europe, about 650 weed species belonging to 50 different families have been recorded, half of which are confined to only five families, particularly Compositae, Cruciferae and Gramineae.[17] In Denmark several investigations on the importance and distribution of weed species throughout the last 60 years have recorded a total of 207 weed species in arable land.[5,6,9] Although only 32 species accounted for 75% of the total area covered by weeds around 1960,[6] the arsenal of potential noxious weeds is high.

Table 4.1 Rank Distribution of Common Weed Species in Arable Land in Denmark

| Rank | Species | Life Form | Occurrence | Nitrophilic | Shade Tolerant | Combine harvest | Harrowing |
|------|---------|-----------|-----------|-------------|----------------|-----------------|-----------|
| 1 | *Stellaria media* | Th | 85 | × | × | × | × |
| 2 | *Poa annua* | Th | 72 | × | | × | |
| 3 | *Plantago major* | H | 71 | | | | |
| 4 | *Polygonum convolvulus* | Th | 64 | × | ×[c] | | |
| 5 | *Viola arvensis* | Th | 64 | | × | | × |
| 6 | *Polygonum aviculare* | Th | 62 | | | × | × |
| 7 | *Agropyron repens* | G | 61 | × | | | |
| 8 | *Chenopodium album* | Th | 55 | × | × | × | × |
| 9 | *Taraxacum* spp. | H | 54 | | | | |
| 10 | *Myosotis arvensis* | Th | 52 | | | × | |
| 11 | *Capsella bursa-pastoris* | Th | 48 | × | | × | |
| 12 | *Matricaria inodora* | Th | 47 | × | | × | × |
| 13 | *Cerastium caespitosum* | Ch | 40 | | | | |
| 14 | *Anagallis arvensis* | Th | 39 | | | | × |
| 15 | *Veronica persica* | Th | 39 | | | × | × |
| 16 | *Polygonum persicaria* | Th | 37 | | | × | |
| 17 | *Veronica arvensis* | Th | 30 | | | | |
| 18 | *Cirsium arvense* | G | 28 | | | | |
| 19 | *Ranunculus repens* | H | 26 | | | | |
| 20 | *Matricaria matricarioides* | Th | 25 | | | × | |
| 21 | *Arenaria serpyllifolia* | Th | 23 | × | | | |
| 22 | *Aphanes arvensis* | Th | 23 | | | | |
| 23 | *Geranium pusillum* | Th | 20 | | | | |
| 24 | *Atriplex patula* | Th | 20 | × | | | |
| 25 | *Lapsana communis* | Th | 18 | | × | | |
| 26 | *Lamium purpureum* | Th | 16 | | × | | × |
| 27 | *Polygonum lapathifolium* | Th | 12 | | × | × | |

*Note:* The species are ranked from most common (e.g., percent fields in which the species occur) to least common. The data were sampled during 1967–1970 and comprise 17 different crops in 466 fields. (Partly from Haas.[8]) The weed characters are defined as follows:

*Life form*
Th: *Therophytes.* Survive unfavorable climatic seasons as seeds.
Ch: *Chamaephytes.* Survive unfavorable climatic seasons as buds placed above soil level.
H: *Hemicryptophytes.* Survive unfavorable climatic seasons as buds placed on soil level.
G: *Geophytes.* Survive unfavorable climatic seasons as buds placed below soil surface.

*Occurrence:* Percent of investigated fields, in which the species occur.

*Nitrophilous species:* Species that favor soils high in nitrogen. (Partly from Mikkelsen.[54])

*Shade-tolerant species:* Species that can endure low light intensities at ground level or (×[c]) by being a climber. (Partly from Rademacher[72] and Fogelfors.[73])

*Combine harvest:* Species that profit by harvesting with a combine. (Partly from Aamisepp et al.[57] and Petzoldt.[56])

*Harrowing:* Species susceptible to harrowing procedures. (Partly from Habel,[58,59] Kees,[60] Koch,[61,62] and Neururer.[63])

A characteristic of many successful weed species is that they are inbreeders[11,12] which produce stable duplicates of readily adapted genotypes. The advantages of self-pollination in weeds could be the capability to establish a seed-producing colony from a single immigrant. The inbreeding system may be linked with occasional environmentally controlled outcrossing, permitting adaptation to new or changing ecosystems.[11] In many weed species, polyploidy provides a means of adaptation to less favorable environments. For example, in Germany the proportion of polyploids within species classified as weeds was 62% compared with 48% within the total angiosperm flora. *Stellaria media* growing in natural habitats in central and northern Europe is chiefly diploid, whereas the species is tetraploid in arable land. Another successful weed, *Galium aparine*, appears to be hexa- and octaploid, whereas other nonweed species within the genus *Galium* are diploid or tetraploid.[18] The degree of polyploidy seems also to be dependent upon latitude. The flora of Sicily has about 37% polyploidy, whereas that of Denmark and Spitzbergen has 52% and 80%, respectively.[18] From an evolutionary point of view, the advantages of polyploidy should be that the possibility of a mutation is increased due to the multiplied loci. Moreover, these mutations of polyploid species may not inevitably lead to lethality, as do most mutations in diploid species, because the unmutated alleles remain active. Consequently, the polyploidy enhanced the adaptation of species to environmental changes more rapidly than that of diploid species.[18]

Another important factor contributing to the maintenance of diversity among weed floras is the "buffering capacity" of the soil seed bank, which will, at least for a period of time, tend to offset the effects of radical environmental changes brought about by human beings.[19] This will be particularly important in arable land, where the seed population in the soil plays a major part in determining the weed problems that must be faced each time a crop is grown.[20] In an analysis of cultivated fields in Denmark, Jensen[21] found about 50,000 viable seeds per square meter at a depth of 0 to 20 cm, 80% of which were of the seven most common species, including *Chenopodium album, Plantago major, Poa annua*, and *Stellaria media*. A correlation between the observed weed vegetation in the field and the seed reservoir in the soil, however, appeared only for a few of the more common weeds. Under the conditions prevailing in Denmark, the seed bank is a source of potential weed problems of the future. Species such as *C. album, S. media*, and other weeds that are less noxious at present remain viable for many years in soil,[22] and under certain circumstances, seeds produced before the introduction of herbicides are still able to germinate.

Investigations on the weed seed content of the soil have been dealt with by numerous workers, for example, in England,[23,24] in Germany[25,26] and on the American continent.[27,28] Korsmo[29] has carried out a countrywide investigation on the weed seed content of Norwegian soils, and Pawlowski[30] has published data for five different soil types in Poland. To evaluate the long-term effect of weed control methods, however, those factors that determine the relative abundance of viable seeds, their dormancy, emergence, and longevity will be of importance. Several publications deal with those problems.[31-36] It seems that among the species or strains of species occurring in unstable environments, the "seed survivor" type of life form obviously exists, and may have developed as a physiologically adapted ecotype that guards against local extinction for decades or even centuries.[37] The significance of the soil seed reservoir and the probable appearance of herbicide resistance is dealt with in Chapter 17.

Another major aspect of the weed flora to be considered when evaluating the effect of weed control measures is the number of seeds per plant and the fact that sheer numbers of weed plants can be a rather misleading measure of the actual amount of a species present in an area.[38-40] The very same species grown under different conditions may show a great variation in seed production,[38] and the same number of weed plants per unit area may, in the same crop on two different locations, result in different dry matter production.[40] These examples demonstrate the great plasticity of weed species.

## 2.2  The Changing Weed Flora of This Century

Our findings on the changes in the weed flora during the last 60 years in Denmark will be presented as an example. We have analyzed botanical data obtained from 4600 sample plots, on which 19 different crops were grown in 466 fields, unsprayed in the sampling year. The data from 1960–1970 illustrate the present situation in Denmark.[6,7,41,42] As the data represent botanical analyses in unsprayed fields, they are comparable to two earlier investigations carried out in the same manner at the turn of this century and around 1945.[5,9] In light of this, the data do not tell us what actually happens to the weed communities in the year they were treated with herbicides, but it *does* tell us something about the changing distribution of weed species following the use of herbicides since the late 1940s. The distribution of some of the more common Danish weeds are ranked in Table 4.1 according to their occurrence as the percentage of fields in which the species were present. The life forms of the species chiefly describe how their reproductive parts survive unfavorable climatic seasons.[10] It is noted that the most frequent life form is that of therophyte weeds (annual species), which survive unfavorable seasons as seeds in the soil. The ranking of species agrees very well with data from other European countries.[43-45] The relative importance of some of the widely distributed species in Table 4.1, however, differs from region to region. For example, *C. album* is considered to be most frequent in Sweden,[43] whereas *Sinapis arvensis* and *C. album* are considered more frequent and widespread on arable land in Britain.[46]

The relative occurrence of weed species shown in Table 4.1, however, does not itself explain the factors governing the distribution pattern of the weed species. In order to reveal a distribution pattern of weeds, an ordination analysis on the basis of the average frequency of 71 common species in 17 crops is illustrated in Fig. 4.1. For the average frequency,[*] the Raunkiaer's technique[10]

---

[*]The term "frequency" is defined as the percentage of sample plots, in which the weed species occurred within a field (10 sample plots per field, about 30 fields per crop type).

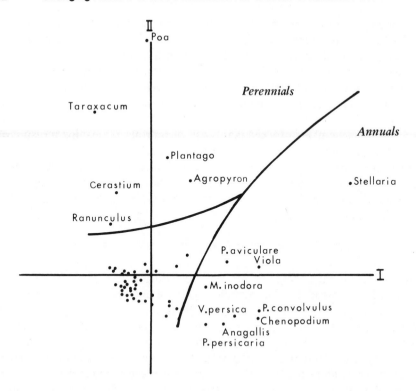

**Figure 4.1**    Ordination analysis of the average frequency of 71 weed species in 17 different crops. The names of some of the common species are listed near their points. The average frequencies of the 71 weed species are plotted in a 17-dimensional space, in which each axis represents a crop. The space is reduced by projecting it into two dimensions in such a way that variation of the 71 points is maximized. The axes thus represent certain biological factors, the interpretation of which requires a knowledge of the particular weed species involved. In this analysis the axes appear to be associated with the natural growth period of the weed species, the first axis (I) describing an "annual factor" and the second axis (II) a "perennial factor" [42,47] This analysis was performed using a factor analysis procedure from Biomedical Computer Programs.[74] Kershaw[75] has described the use of ordination analysis in plant ecology.

has been used. To visualize the ordination analysis presented in Fig. 4.1, we have to imagine a space of 17 dimensions, each axis representing a crop. In this space, the 71 weed species are plotted according to their frequencies in the 17 crops. The space is reduced by projecting it into two dimensions in such a way that the variation of the 71 points is maximized. This space reduction, which minimizes the loss of information in terms of variation in the distribution of weed species, enables us to get an idea of the distribution pattern of the weed species. By virtue of the separation of the more frequent weed species in Fig. 4.1, we can interpret the first axis as one describing the distribution of annual weed species. Similarly, the second axis describes the distribution of the perennial weed species. Further details of the ordination analysis are given elsewhere.[47]

Figure 4.1 shows that the distribution pattern can be explained in terms of recognizable biological features.[42] A rather dense group of weed species is distributed around the origin; they represent the relatively less frequent weeds. A loose group of annual species is located along the horizontal axis, and on the vertical axis we find the perennial weeds. *Poa annua* is an exception because its average frequencies within crops are more similar to the perennial weeds than they are to the annual weeds (see Section 3.1). The pattern of weed species warrants the conclusion that their distribution pattern seems to be a function of their life cycle. It is evident from this relationship between distribution and life cycle (Fig. 4.1) that changes in agricultural techniques and weed control measures will impose a strong influence on the occurrence of particular weeds.

Comparisons of vegetation analyses of weeds were carried out in the same fields with the same crop type in the 1940s and again about 1970.[7,8] They represent the situation before and after the advent of herbicides; similar changes in the distribution of weed species have also been found in Germany[44] and probably elsewhere. The total number of weed species has declined, particularly geophytes (weed species surviving unfavorable seasons as buds below the soil surface) and some therophytes, such as *Agrostemma githago* and *Lolium temulentum*,[44,48] which contaminate crop seeds. Generally, the number of therophytes has been rather constant during this period, but there have been changes in composition. For example, *Centaurea cyanus*, *Galeopsis ladanum*, and *Myosurus minimus* have been replaced in part by *Euphorbia* spp., *Matricaria* spp., *Polygonum persicaria*, and *Silene noctiflora*.[8] Conversely, an extensive survey of the displacement of the weed flora during a period of 15 years *without* major agronomic changes showed that the total number of weed species in 11 different crops was virtually unaffected.[49]

We can ascertain *when* changes in the distribution of weed life forms first become apparent by summarizing the life form pattern of the weed flora in certain crop types (Table 4.2). From the data it can be concluded that changes of weed communities were already in progress well before the introduction of herbicides. During that period the proportion of therophytes increased with a decrease in geophytes. This was probably brought about by a more intensive soil cultivation when the tractor replaced the horse. The long-term effects of herbicide treatments after 1944–1945 have probably enhanced this tendency by further increasing the number of therophytes and diminishing the number of geophytes. In some regions with a warmer climate, noxious geophytes such as *Cyperus rotundus* and *Cynodon dactylon* have become severe pests,[14] probably because of the removal of competition from annual weeds by herbicides.

Table 4.2  Distribution Changes of Weed Life Forms in Crops of Different Longevity

| Crop Type | Time of Analysis | Percent Occurrence of | | |
|-----------|-----------------|-------------|------------------|-----------|
| | | Therophytes | Hemicryptophytes | Geophytes |
| Spring cereals | 1911–1915[a] | 76 | 8 | 16 |
| | 1944–1945[b] | 88 | 1 | 11 |
| | 1964–1969[c] | 92 | 1 | 7 |
| Winter cereals | 1911–1915[d] | 77 | 13 | 10 |
| | 1967–1968[e] | 83 | 10 | 7 |
| Grass leys | 1911–1915[f] | 45 | 33 | 23 |
| | 1969[g] | 53 | 38 | 10 |

Source: Modified from Laursen and Haas.[7]

Note: The occurrence is based on the percent of investigated fields, in which the weed species occur.

[a] 11 fields (Ferdinandsen[5]).

[b] 75 fields (Thorup and Petersen[9]).

[c] 90 fields (Laursen and Haas[7]).

[d] 15 fields (Ferdinandsen[5]).

[e] 60 fields (Laursen and Haas[7]).

[f] 32 fields (Ferdinandsen[5]).

[g] 91 fields (Laursen and Haas[7]).


## 3  AGRONOMIC FACTORS AFFECTING THE WEED FLORA

In examining the influence of herbicides on the distribution of weed species on arable land in the last decades, we have to keep in mind that several other agronomic factors also operate, and some of the major ones even vary concurrently with the development and use of herbicides.

The data in Table 4.3 clearly illustrate a development found in many developed countries over the last 30 years. The crop rotation pattern has changed in favor of the continuous growing of cereals. The application of nitrogen and herbicides has increased simultaneously with the decrease in crop rotation. The greatest changes have occurred between 1960 and 1970. Furthermore, liming and draining have severely reduced the importance of calcifuge weeds favoring acid soils (e.g., *Rumex acetosella*, *Scleranthus annuus*, *Erodium cicutarium*, and *Spergula arvensis*) and species favoring wet soils (e.g., *Juncus bufonius* and *Mentha arvensis*).[48,50] Improved techniques of crop seed cleaning have almost eradicated some weeds that were noxious in the past. The rapid growth of mechanization has favored certain weeds or groups of weeds and has suppressed others. All these agronomic measures have to be taken into account when evaluating which factors impose major changes on the weed flora.

Table 4.3  Increase in Cereal Crops, N Fertilization, and use of Herbicides

|  | 1950 | 1960 | 1970 | 1975 | 1979 |
|---|---|---|---|---|---|
| Cereals (percent of arable land) | 48 | 53 | 65 | 65 | 70[a] |
| Average nitrogen fertilizer use on arable land (kg/ha) | 63 | 84 | 145 | 161 | 187[a] |
| Herbicides purchased (thousands of tons) | — | 1.2 | 3.8 | 3.9 | 5.1[b] |

*Source:* Modified from Haas.[2]

[a]Updated figures for 1979 from: Danmarks Statistik, *Statistiske Meddelelser 1980:9 Agricultural Statistics 1979.* Danmarks Statistik, Copenhagen 1980.

[b]Updated figure for 1979 from: Kemikaliekontrollen 1981, *Statistiske oplysninger vedrørende forbrug og salg af bekæmpelsesmidler for årene 1978, 1979 og 1980.* Kemikaliekontrollen, Lyngby 1981.

## 3.1  Crop Rotation

In the past, the rotation of a number of crop types has been an important weed control method. The chances of any one weed species becoming predominant were minimized as cultural practices and competitive effect of crops varied, favoring some weed species and discouraging others.[50] The result was a highly diverse weed flora. The introduction of the phenoxy acid herbicides in the late 1940s, the reduced availability of labor, and the rapid mechanization of agriculture all constituted drastic changes in agricultural management practices.

As pointed out earlier, the proportion of land under cereal crops has increased, with a parallel increase in the stress imposed by herbicides upon the weed flora, in the last 30 years. This favored a simplification of the weed flora. As a result, a few major weed species, well adapted to monoculture, became dominant.[50,51] It is of interest that the phenoxy acid herbicides changed the distribution of weed species but did not select for resistant strains (Chapter 17). From Table 4.2 we can see that a higher proportion of therophyte weed species was found in spring cereals than was found in winter cereals, while the opposite trend applies to the hemicryptophyte weed species, which survive unfavorable seasons as buds at soil level. By arranging the weed life forms according to 17 different crops, it could be shown that the proportion of therophyte weeds was reduced with increasing growth period of the crop, while the amount of hemicryptophyte and geophyte weeds increased.[42] The actual number of weed species present, however, did not appear to be changed.[42,49]

These results, however, give no indication of the influence of crop type upon the weed flora compared with other agronomic factors. A hierarchic classification (cluster analysis) of 17 crops based on the average frequency of the 71 most common weed species illustrates the effect of crop type on the weed flora. This method is based on two independent procedures. The first procedure calculates the degree of similarity between crops using the squared Euclidian distance:

$$D_{ik} = \frac{1}{n} \sum_{j}^{n} (x_{ij} - x_{kj})^2 \qquad (4.1)$$

where $D_{ik}$ denotes the distance between crop $i$ and $k$, $x_{ij}$ the frequency of weed species $j$ in crop $i$, $x_{kj}$ the frequency of species $j$ in crop $k$, and $n$ the number of weed species used in the analyses. This procedure results in a table consisting of the mutual distances between the 17 crops, with little similarity between weed floras of the crops having large distances. The second procedure sorts the table of distances (Ward's method)[52] and arranges crops in hierarchical clusters according to the similarity of their weed flora. In Fig. 4.2, two broad groups of exclusively annual crops (II) and mainly biennial and perennial crops (I) are recognized. By studying these two broad groups in more detail it is seen that the perennial grass leys and the winter annual cereals are classified in small subgroups. Even though winter wheat and winter rye are grown on different soil types and geographic locations in Denmark, cluster analysis indicates that the growth period of the crop seems to be of greater importance for the distribution of the weeds than does geographic location and soil type (Fig. 4.2). The classification of oats, winter rape, and swedes (rutabagas) in the perennial group can, to a certain extent, be partially explained by the joint influence of the particular competitive effect of these crops, the weed flora, and by the limited geographical region in which the crops are grown. Further details are given elsewhere.[41,42] The classification shows that the weed flora of unsprayed crops is closely related to the growth period of crops, and further analyses have demonstrated that the distribution of annual weeds was especially sensitive to crop type.[41]

The frequency pattern of five common weeds also demonstrated the influence of the growth period of crops (Fig. 4.3). It is noted that the frequency patterns of *Poa annua* and *Stellaria media* were essentially opposite. This illustrates their preference to crop life duration. *P. annua* dominated (frequency > 80) in grass leys, whereas *S. media* dominated in annual crop types. *Polygonum convolvulus* and *Chenopodium album* almost disappeared in biennial and perennial crops. The frequency distribution of the noxious geophyte *Agropyron repens* was largely unaffected by the growth period of the crop.

In assessing the results of Figs. 4.2 and 4.3 we have to remember that any one crop has its own "history" in terms of different soil preparation, fertilization, and so on, each affecting the development of particular weed floras. These agronomic measures cannot be distinguished separately, and consequently the conclusions must to a certain extent be based on circumstantial evidence.

### 3.2 Nitrogen and Other Fertilization

The prevailing practice in several countries to switch to cereal monoculture has changed not only the proportions of life forms of the weed flora (Table 4.2),

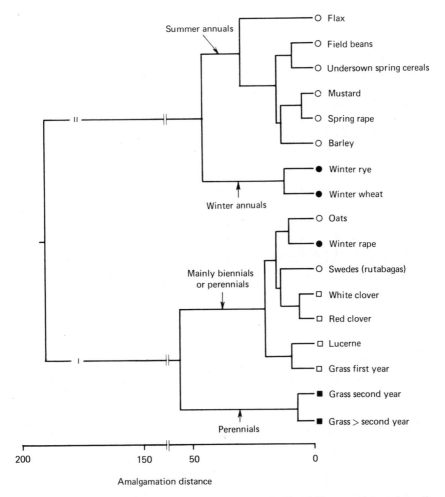

**Figure 4.2** Hierarchic classification (cluster analysis) of 17 crops based on the average frequency of the 71 most common weed species. Summer annuals, O; winter annuals, ●; perennial that can be considered either biennials or perennials, □; perennials, ■. I and II represent two broad groups of exclusively annual crops (II) and mainly biennial and perennial crops (I). The degree of similarity between weed floras of the 17 crops, expressed as a squared Euclidian distance measurement (see the text), has been used in a sorting algorithm to arrange the crops in a numerical classification system. High amalgamation distances (e.g., distances to the points where two crops and/or clusters of crops merge) indicate that their weed floras differ to a greater extent than do the weed floras of crop at smaller distances. (From Streibig and Haas.[42])

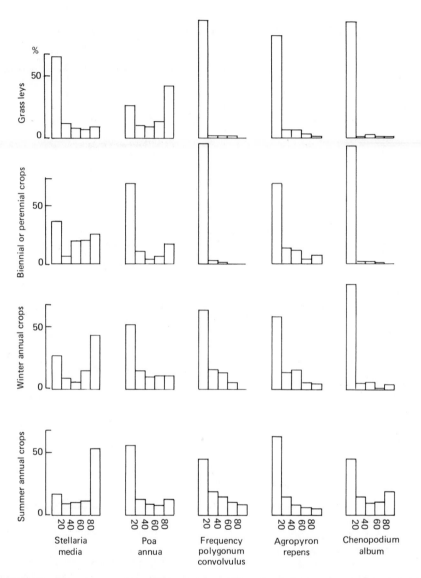

**Figure 4.3** Frequency of five important weed species in four crop types of different longevity. The bars represent the proportion of fields within the crop types in which the species was found at frequencies between 0 and 20, 21 and 40, 41 and 60, 61 and 80, 81 and 100. (From Streibig and Haas.[42])

Table 4.4  Displacement of Weed Species in Danish Spring Cereal Fields

| | Occurrence in Fields (%) | | | Susceptibility[d] | |
|---|---|---|---|---|---|
| | 1918[a] | 1944–1945[b] | 1966–1970[c] | MCPA | Dichlorprop |
| *Species on the increase* | | | | | |
| *Matricaria inodora* | 18 | 34 | 27 | + | + + |
| *Matricaria matricarioides* | 0 | 1 | 18 | + | + + |
| *Plantago major* | 36 | 49 | 75 | + + | + + |
| *Poa annua* | 18 | 34 | 46 | 0 | 0 |
| *Stellaria media* | 73 | 94 | 92 | + | + + |
| *Species in retreat* | | | | | |
| *Capsella bursa-pastoris* | 46 | 53 | 22 | + + + | + + |
| *Rumex acetosella* | 55 | 20 | 15 | + | + + |
| *Scleranthus annuus* | 55 | 23 | 21 | | |
| *Sinapis arvensis* | 75 | 83 | 12 | + + + | + + + |
| *Sonchus arvensis* | 73 | 60 | 19 | + + | + + |
| *Taraxacum* spp. | 73 | 35 | 27 | + + + | + + + |
| *Unaffected species* | | | | | |
| *Agropyron repens* | 55 | 47 | 60 | 0 | 0 |
| *Atriplex patula* | 27 | 60 | 21 | + + + | + + + |
| *Chenopodium album* | 64 | 85 | 68 | + + + | + + + |
| *Myosotis arvensis* | 36 | 45 | 50 | + + | + + |
| *Polygonum convolvulus* | 91 | 93 | 82 | + + | + + + |
| *Polygonum aviculare* | 82 | 85 | 64 | + | + + |
| Number of fields examined | 11 | 73 | 90 | | |

*Note:* The numbers in the first three columns are percent of fields examined in which the species were present.

[a]From Ferdinandsen.[5]

[b]From Thorup and Petersen.[9]

[c]From Laursen and Haas[7] and Haas.[8]

[d]From Anonymous.[65] The susceptibility of species to MCPA and dichlorprop are denoted: + + +, very susceptible: + +, susceptible: +, weakly susceptible: 0, tolerant.

but also the distribution of weeds within similar life forms.[8,43,44,48,51,53] This cannot be linked entirely with rotation practices as discussed in the preceding section, or with herbicide use (as discussed in Section 3.4). *Chenopodium album* and *Atriplex patula* have remained common weeds despite their susceptibility to certain commonly used herbicides (Table 4.4). Conversely, *Scleranthus annuus* and *Spergula arvensis* are retreating species, although they are not sensitive to most phenoxy acids.

As shown in Table 4.3, the nitrogen fertilization per unit area has tripled in Denmark since 1950. Similar trends are also found in other countries, which have similar agricultural production systems. Nitrogen may cause a partial increase in nitrophilous weeds [e.g., *Agropyron repens*, *Poa annua*, *Chenopodium album*, and *Stellaria media* (Table 4.1)]. As nitrogen also enhances the shading ability of crops, this will favor weed species which either possess physiological shade tolerance [e.g., *Chenopodium album*, *Viola arvensis*, and *Stellaria media* (Table 4.1)] or are able to climb into more favorable light conditions (e.g., *Polygonum convolvulus* and *Galium aparine*).

The Danish consumption of phosphorus and potassium fertilizers per unit area has increased about 40% and 10%, respectively, during the last 20 years. Some weed species (e.g., *Solanum nigrum* and *Lamium purpureum*) seem to be favored by high phosphorus and potassium levels in the soil.[54] Another notable factor is that the use of lime has tripled during the last two decades in Danish agriculture, which has had an impact on those weed species favoring basic soils (e.g., *Veronica persica*, *Anagallis arvensis*, and *Lamium purpureum*). This impact on the weed flora brought about by the maintenance and improvement of soil fertility is rather difficult to evaluate because the long-term effects are interrelated with numerous other factors governing the nutritive status of the soil. Another significant outcome of soil improvement is that certain crops, such as sugar beets, previously restricted to particular soil types are now more widely grown in Denmark.

## 3.3  Mechanization and Changes in Mechanical Control Methods

During the period outlined in Table 4.3, the tractor as well as the combine have become common agricultural implements in most developed countries. The use of heavy machinery in the fields will probably lead to an increase of certain weed species. Thorup[55] has reported that this seems valid for therophytes, such as *Polygonum aviculare* and *Matricaria inodora*. The effect of agricultural implements on the weed flora, however, are multifarious and rather complex. Thus we only wish to point out some trends, by means of a few illustrative examples.

The combine has provided certain therophyte weeds with better possibilities for seed dispersal. Some of these species shed their seeds later than the time at which reaping would previously have been carried out, but earlier than the present combine harvest. Moreover, 35% of the total amount of combine-harvested seeds of certain weeds are dispersed, with the straw and chaff left in the field.[56,57] Some of the more frequent weed species that have profited by the combine are recorded in Table 4.1. Furthermore, the delay of the harvest period introduced with the combine compared with the reaper-binder harvest has partly replaced stubble cultivation with rotary hoeing and so on in the fall. This has in turn favored the increase of weeds such as *Tussilago farfara*.

Experiments in Germany[58-62] and Austria[63] have all shown that particular weed species are more sensitive to harrowing than others. Some of these species are listed in Table 4.1. The practice of spring harrowing to control annual weeds is no longer widespread, largely because the tractor has reduced the interval between seedbed preparation and sowing, and also because harrowing increases soil evaporation. Finally, spring harrowing as a weed control method was displaced by the introduction of herbicides in the late 1940s.

## 3.4 Herbicide Use

It is necessary to look at some statistics concerning the pattern of herbicide consumption to understand some of the changes in weed distribution caused by herbicide use. An analysis of the consumption of different groups of herbicides used in different crops, however, seems to be more pertinent to the subject than the total consumption shown in Table 4.3.

The consumption of herbicides in Denmark, for example, has increased four-to fivefold during the last two decades (Table 4.3). At present, a total of 2.6 million ha of arable land is annually treated with 4000 to 5000 tons of herbicides, in which three large groups of compounds dominate. About 64% of the total amount of herbicides are used exclusively in cereals. The phenoxy acids, MCPA and/or dichlorprop, account for more than 80% of this group. Similar trends have also been reported from Sweden.[43] Another large part (24%) is used primarily against weeds in root crops. The last 12% are sprayed against a single weed species, *Agropyron repens*. If we bear in mind that the area occupied by cereals has increased considerably (Table 4.3), it is obvious that the same field is now exposed to the same herbicide group more frequently than before. Consequently, perpetual spraying with the phenoxy acid herbicides must clearly have an impact on the existing weed flora.

The general tendency in spring cereals since the introduction of MCPA is shown in Fig. 4.4. The amount of weeds in control plots, not treated in the experimental year, has markedly decreased, whereas the amount of weeds in treated plots has been more or less unaffected. Another interesting feature is the obvious increase in crop yield with time in both treated and untreated plots. Among other things, the yield increase may have been caused by increased nitrogen fertilization (Table 4.3). The decrease of weeds in the untreated control plots may partly be explained by depletion of the soil seed reservoir brought about by a preceding history of continuous herbicide use or by increased competition from the cereals in response to nitrogen fertilization.

The impact of uncritical spraying of MCPA on weed productivity can be illustrated by data on *Galium aparine*[64] (Table 4.5). Only one year of spraying with an inappropriate herbicide (MCPA) resulted in a rapid increase in the biomass of the tolerant *G. aparine*; MCPA eradicates the potential competitors of the *Galium*. Conversely, a herbicide, wisely chosen on the basis of the indigenous weed species, may easily overcome the noxious *Galium aparine*. Analogous results have been reported for DNOC,[62] which controlled broad-leaved weed species and decreased the total amount of weeds, but considerably increased the problems caused by grass weeds.

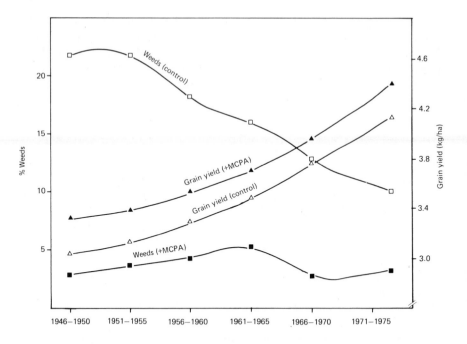

**Figure 4.4** Effect of MCPA on grain yield and weed-infestation level. Weeds untreated (□) and MCPA treated (■). Yield of grain in areas untreated (△) and in MCPA treated (▲). Data from all Danish field trials (more than 4000 experiments) with MCPA in spring cereals during the period 1946–1975. Weed infestation is expressed as percent weed fresh weight in relation to total vegetation (fresh weight of crop plus weed) in several subsamples of 0.25 m² per plot. The experiments have been placed within the normal crop rotation systems. The control plots were not sprayed in the experimental year but had been sprayed in previously years. (From Thorup.[55])

**Table 4.5  Changes in Dry Matter Production of *Galium aparine* after MCPA and Mecoprop Treatments**

|  | Control | (kg/ha) | | |
|  |  | 0.5 | 1.0 | 2.0 |
| --- | --- | --- | --- | --- |
|  |  | Percent of Untreated Control | | |
| MCPA | 100 | 170 | 185 | 162 |
| Mecoprop | 100 | 52 | 35 | 6 |

*Source:* From Petersen.[64]

*Note:* Average of several field trials in spring barley.

The distribution of common weed species in cereal crops before and after the introduction of herbicides may help us in understanding the displacements of weed species with special reference to herbicide spraying. Our investigations in Denmark have included only fields that were untreated in the year of analysis. Thus the distribution pattern does not reflect the immediate impact of herbicides during the year of sampling, but it does give us a possibility of interpreting the long-term effect of herbicide use, which is probably more important.

The occurrence of species considered under Danish conditions to be either increasing, declining, or unaffected during the last 60 years are listed in Table 4.4. A semiquantitative assessment of their susceptibility to MCPA and dichlorprop[65] has also been included in the table.

None of the species that are on the increase appear to be very susceptible to MCPA and dichlorprop. *Poa annua* is tolerant to MCPA and dichlorprop, and has profited by the eradication of its susceptible competitors. *Matricaria matricarioides*, a species first appearing in Denmark in the middle of the last century,[66] has also increased, as has *Plantago major*. *P. major* has two subspecies, *pleiosperma*, which is common in annual crops, and *major*, which is more frequent in biennial and perennial crops.[67] Ecologically, *Plantago* spp. and *Matricaria matricarioides* are very common in "Trittpflanzen-Gesellschaft"[68] (trampled communities) and hence can endure heavy trampling and can stand heavy agricultural implements in the field. The distribution of *Stellaria media* has not changed to any appreciable extent since 1945, probably due to its tolerance to the herbicides and to increased nitrogen fertilization (Tables 4.1 and 4.4).

In the group of weed species in retreat, the decrease in *Rumex acetosella* is due to liming and not to herbicides; this species is rather tolerant of phenoxy acids. *Sinapis arvensis*, on the other hand, represents a species that is susceptible to phenoxy acids, which explains its rapid decrease since 1945. As the occurrence is shown in fields not treated in the experimental year, the decline in *S. arvensis* may be attributed to a decline in viable seeds in the soil brought about primarily by effective weed control.[43] The continued persistence of *S. arvensis* in some regions of the United Kingdom, despite its susceptibility to many commonly used herbicides, has been explained by the longevity of the seeds in the soil.[48] *Sonchus arvensis*, a geophyte, and *Taraxacum* spp., a hemicryptophyte, have also declined. *S. arvensis* has largely declined after 1945, probably due to phenoxy acid herbicides.

In the group of unaffected species, we have *Chenopodium album* and *Atriplex patula*, both susceptible to MCPA and dichlorprop. Nevertheless, the soil seed reserves of these species do not appear to be greatly affected over the last 30 years. This is very similar to the situation in Sweden.[43]

Another important aspect that must be considered when interpreting the changing weed flora after the introduction of herbicides is that these chemical weed control methods have changed other agronomic practices—both the pattern of spring harrowing and stubble cultivation in fall, as discussed in Section 3.3. In the period between the first botanical investigations on Danish arable land[5] and the analyses carried out in 1945,[9] the most important weed control method was that of fallow and/or half fallow. Later, the increase in root crops in which mechanical weed control was possible, partly replaced the role of the fallow. At the present time, weed control in root crops is achieved by chemical means, which has its limitations and gives poor control of some noxious rhizomatous geophytes. Moreover, the control of some therophytes, such as *Polygonum aviculare,* which is tolerant to most herbicides used in root crops, and the late-germinating *Solanum nigrum,* is insufficient. This may result in local increases in these species.

Another notable feature is the status quo of *Agropyron repens.* Despite the intensive chemical weed control of this noxious rhizomatous geophyte, it has endured the herbicide pressure and kept its position as one of the most serious weed problems in agriculture in Denmark, as elsewhere. This situation is probably due to the decrease of fall stubble cultivation and to increased nitrogen fertilization (Table 4.1).

### 3.5 Cause and Effect

Assessing the evidence discussed in previous sections, several factors have evidently operated together to influence the changes in weed distribution during the last decades. Some of the factors seem to be interrelated so that the cause-and-effect relationship between any two factors must be based upon circumstantial evidence. The decline of weeds in untreated fields (Fig. 4.4) may, for instance, be linked with a concomitant increase in crop yield during the period, because of the increased competitive effect of the crop and with a consequent suppression of weeds.

Attempting an overall assessment of the available data, we have to admit that herbicide use probably plays a major part in the changing patterns of the weed flora. Yet herbicides cannot bear the full responsibility. The effect of other agricultural measures, such as crop rotation, increased fertilizer level, mechanization, and changed mechanical weed control methods, all seem to have worked in the same direction or at least collectively (Table 4.1). This conclusion is also supported by the fact that some displacements of the weed flora began with changes in tillage practices and cropping sequences before the introduction of herbicides (Table 4.2).

### 4  OUTLOOK FOR FUTURE CHANGES IN THE WEED FLORA

The present trend toward continuous cereal cultivation appears to be a general pattern throughout the world. This trend will lead to the type of changes in weed floras discussed earlier, but may also give rise to distribution changes in weed species that are very sensitive to even minor changes in agricultural practice, thereby shifting even further from the climax vegetation with which human beings started. Different varieties of barley have shown considerable variation in their ability to suppress weeds by shading[69] and thus indicate that even a choice of varieties can lead to different weed population patterns. Moreover, competition experiments with some weed species having a different growth habit have revealed that *Viola arvensis,* for example, the fifth most frequent weed species in Denmark, was capable of enduring high competitive stress. On the contrary, *Chenopodium album* and *Matricaria inodora* were very sensitive to such a competitive stress.[70]

The sensitivity of weed species to changes in agricultural practices can be further illustrated by means of two weed species and the introduction of no-till systems. The advantages of no-till are obvious; this method of soil cultivation preserves soil moisture and saves fuel as well as labor. *Taraxacum* spp. growing in a wayside along a field with a rapidly growing spring barley crop cannot be considered a weed, because seed production begins in the second year of growth and because appropriate plowing does not permit its establishment in a summer annual crop. A different situation exists, however, when a summer annual crop has been established by a no-till system. With no-till systems, *Taraxacum* may easily become a prominent pest because of the lack of this cultivation. Under no-till it also might be expected that grass weeds (e.g., *Apera spica-venti*), a winter annual under Danish conditions, will attain a foothold in summer annual barley crops. Thus grass weeds might become even a greater problem than they are today under no-till conditions,[71] unless new herbicides become available which are more effective on grasses, to counteract this trend.

The examples and the discussion in this chapter show that the kind of changes in the weed flora that can be expected in the future will also depend on many future developments in agricultural chemicals and management.

## 5 CONCLUDING REMARKS

Interpretation of the data and the literature presented in the preceding sections indicates that the changes in the Danish weed flora during the last 60 years are affected by changes in crop rotation, fertilization, mechanization, and last but not least, by herbicide use. We do not think that in our or other agricultural situations it will be possible to quantify unambiguously the changes in distribution of weed species brought about by herbicides alone. Herbicides seem to be a major force, but not the only one. It is our opinion that the use of plant sociology and synecology can provide us with an all-round picture of the dynamic weed flora. This may contribute to an understanding of the distribution changes in specific weed species in specific crops under particular agronomic situations.

The data also warrant the conclusion that no weed species appears to have been eradicated by herbicides. *Lolium temulentum* and *Agrostemma githago* have been almost eradicated, but this has been achieved *physically* by improved seed cleaning, not by herbicides. However, bearing in mind that herbicide use is a very new weed control method compared to the control methods employed for millennia, changes in weed distribution brought about by herbicides are probably just beginning to emerge.

# REFERENCES

1  M. Hanf, Einführung zu Methoden der Unkrautbekämpfung und deren Integration, *Proc. EWRS Symp. Methods of Weed Control and Their Integration,* 1 (1977).

2  H. Haas, Økologiske aspekter ved kemisk ukrudtsbekaempelse, *Miljøforvaltning,* DSR Forlag, The Royal Veterinary and Agricultural University, Copenhagen, 1979, p. 285.

3  FAO, *1978 FAO Production Yearbook,* FAO Stat. Ser. No. 15, 1979, p.32.

4  D. J. Tomkins and W. F. Grant, Effects of herbicides on species diversity of two plant communities, *Ecology,* **58**, 398 (1977).

5  C. Ferdinandsen, Undersøgelser over danske ukrudtsformationer på mineraljorder, *Tidsskr. Planteavl,* **25**, 629 (1918).

6  V. M. Mikkelsen and F. Laursen, Markukrudtet i Danmark omkring 1960, *Bot. Tidsskr.,* **62**, 1 (1966).

7  F. Laursen and H. Haas, Faktorer af betydning for ukrudtsarternes forekomst i danske afgrøder: I. Livsvarighedens betydning, *NJF-kongr. 1971: Fortrykk av foredrag Seksjon II Plantedyrkning,* 29 (1971).

8  H. Haas, Herbicidernes uønskede virkninger på floraen, *Pesticider, Forureningsrådets-Sekretariat, København,* **17**, 99 (1971).

9  S. Thorup and H. I. Petersen, National Research Center for Plant Protection, Institute for Weed Control, Flakkebjerg, unpublished data from 1944/45, 1970.

10  C. Raunkiær, *The Life Forms of Plants and Statistical Plant Geography,* Clarendon Press, Oxford, 1934.

11  J. A. Young and R. A. Evans, Responses of weed populations to human manipulations of the natural environment, *Weed Sci.,* **24**, 186 (1976).

12  H. G. Baker, The evolution of weeds, *Annu. Rev. Ecol. Syst.,* **5**, 1 (1974).

13  J. McNeill, The taxonomy and evolution of weeds, *Weed Res.,* **16**, 399 (1976).

14  L. G. Holm, D. L. Plucknett, J. V. Pancho, and J. P. Herberger, *The World's Worst Weeds: Distribution and Biology,* University Press of Hawaii, Honolulu, 1977.

15  L. G. Holm, J. V. Pancho, J. P. Herberger, and D. L. Plucknett, *A Geographical Atlas of World Weeds,* Wiley, New York, 1979, p. 391.

16  L. G. Holm, The importance of weeds in world food production, *1976 Br. Crop Prot. Conf. Weeds,* 753 (1977).

17  M. Hanf, Einführung zu Einfluss verschiedener Faktoren auf Entwicklung und Bekämpfung von Unkräutern, *Proc. EWRS Symp. The Influence of Different Factors on the Development and Control of Weeds,* 5 (1979).

18  R. Rauber, Evolution von Unkräutern, *Z. Pflanzenkr. Pflanzenschutz,* Spec. issue 8, 37 (1977).

19  D. E. Schafer and D. O. Chilcote, Factors influencing persistence and depletion in buried seed populations: I. A model for analysis of parameters of buried seed persistence and depletion, *Crop Sci.,* **9**, 417 (1969).

20  H. A. Roberts, Viable weed seeds in cultivated soils, *Rep. Natl. Veg. Res. Stn. for 1969,* 25 (1970).

21  H. A. Jensen, Content of buried seeds in arable soil in Denmark and its relation to the weed population, *Dan. Bot. Ark.,* **27**, 1 (1969).

22  S. Ødum, Germination of ancient seeds, floristic observations and experiments with archaeologically dated soil samples, *Dan. Bot. Ark.,* **24**, 2 (1965).

23  W. E. Brenchley, Buried weed seeds, *J. Agric. Sci.,* **9**, 1 (1918).

24  W. E. Brenchley and K. Warington, The weed seed population of arable soil: I. Numerical estimation of viable seeds and observations on their natural dormancy, *J. Ecol.,* **18**, 235 (1930).

25 K. Snell, Über das Vorkommen von keimfähigen Unkrautsamen im Boden, *Landwirtsch. Jahrb.,* **43**, 323 (1912).

26 O. Wehsarg, Das Unkraut im Ackerboden, *Dtsch. Landwirtsch. Ges.,* **226**, 1 (1912).

27 R. G. Robinson, Annual weeds, their viable seed population in the soil, and their effect on yields of oats, wheat, and flax, *Agric. J.,* **41**, 513 (1949).

28 A. C. Budd, W. S. Chepil, and J. L. Doughty, Germination of weed seeds: III. The influence of crops and fallow on the weed seed population of the soil, *Can. J. Agric. Sci.,* **34**, 18 (1954).

29 E. Korsmo, Undersøkelser over: Innhold av ugressfrø i melle, agner, høimo, husdyrgjødsel og kulturjord, 1900–1925, *Meld. Nor. Landbrukshoegsk. XV,* **1**, 1 (1935).

30 F. Pavtowski, Content and specific composition of weed seeds in more important soils of the Lublin Districts, *Ann. Univ. Mariae Curie-sktodowska, Sect. E,* **18**, 125 (1963).

31 W. E. Brenchley and K. Warington, The weed seed population of arable soil: II. Influence of crop, soil and methods of cultivation upon the relative abundance of viable seeds, *J. Ecol.,* **21**, 103 (1933).

32 J. L. Harper, Ecological aspects of weed control, *Outlook Agric.,* **1**(6), 197 (1957).

33 H. A. Roberts, Emergence and longevity in cultivated soil of seeds of some annual weeds, *Weed Res.,* **4**, 296 (1964).

34 H. A. Roberts and P. A. Dawkins, Effect of cultivation on the numbers of viable weed seeds in soil, *Weed Res.,* **7**, 290 (1967).

35 H. A. Roberts and P. M. Feast, Emergence and longevity of seeds of annual weeds in cultivated and undisturbed soil, *J. Appl. Ecol.,* **10**, 133 (1973).

36 H. A. Roberts and J. E. Neilson, Seed survival and periodicity of seedling emergence in some species of *Atriplex, Chenopodium, Polygonum* and *Rumex, Ann. Appl. Biol.,* **94**, 111 (1980).

37 S. Ødum, *Dormant Seeds in Danish Ruderal Soils: An Experimental Study of Relations between Seed Bank and Pioneer Flora,* The Royal Veterinary and Agricultural University, Hørsholm Arboretum, Denmark, 1978, p. 247.

38 J. L. Harper, Factors controlling plant numbers, in J. L. Harper (Ed.), *The Biology of Weeds,* Blackwell, Oxford, 1960, p. 119.

39 J. L. Harper, *Population Biology of Plants,* Academic Press, London, 1977, p. 852.

40 F. Laursen, Studies of weed competition in barley, *Royal Vet. and Agric. Univ., Copenhagen, Denmark, Yearbook 1971,* 201 (1971).

41 J. C. Streibig, Numerical methods illustrating the phytosociology of crops in relation to weed flora, *J. Appl Ecol.,* **16**, 577 (1979).

42 J. C. Streibig and H. Haas, Zusammensetzung der dänischen Unkrautflora und deren Veränderung in den letzten 60 Jahren, *Proc. EWRS Symp. The Influence of Different Factors on the Development and Control of Weeds,* 273 (1979).

43 H. Fogelfors, Changes in the flora of Farmland, *Rep. 5 Swed. Univ. Agric. Sci. Dept. Ecol. Environ. Res.,* 65 (1979).

44 A. Mittnacht, C. Eberhardt, and W. Koch, Wandel in der Getreideunkrautflora seit 1948, untersucht an einem Beispiel in Südwestdeutschland, *Proc. EWRS Symp. The Influence of Different Factors on the Development and Control of Weeds,* 209 (1979).

45 R. R. Schmidt, Änderungen der Unkrautflora auf drei Bodenarten durch unterschiedlichen Fruchtwechsel und durch Herbizide, *Proc. EWRS Symp. The Influence of Different Factors on the Development and Control of Weeds,* 293 (1979).

46 R. J. Chancellor, A preliminary survey of arable weeds in Britain, *Weed Res.,* **17**, 283 (1977).

47 J. C. Streibig, Classification of agricultural crops based on their weed flora, *Symposium i anvendt statistik: Teknometri og Biometri,* Polyteknisk Boghandel, Lyngby, Denmark, 1979, p. 109.

48  J. M. Way and R. J. Chancellor, Herbicides and higher plant ecology, in L. J. Audus (Ed.),
    *Herbicides, Physiology, Biochemistry, Ecology*, Vol. 2, 2nd ed., Academic Press, London,
    1976, p. 345.

49  B. Rademacher and W. Koch, Kulturartbedingte Veränderungen in der Unkrautflora
    eines Feldes von 1956–1971, *Z. Pflanzenkr. Pflanzenschutz*, Spec. issue 6, 149 (1972).

50  J. D. Fryer, Key factors affecting important weed problems and their control, *Proc.
    EWRS Symp. The Influence of Different Factors on the Development and Control of Weeds*,
    13 (1979).

51  G. Bachthaler, Entwicklung der Unkrautflora in Deutschland in Abhängigkeit von den
    veränderten Kulturmethoden, *Angew. Bot.*, **43**, 59 (1969).

52  L. Orloci, An agglomerative method for classification of plant communities, *J. Ecol.*, **54**,
    129 (1966).

53  J. D. Fryer and R. J. Chancellor, Evidence of changing weed populations in arable land,
    *Proc. 10th Br. Weed Control Conf.*, **3**, 958 (1970).

54  V. M. Mikkelsen, Agerlandets vilde Flora, in A. Nørrevang and T. J. Meyer (Eds.), *Danmarks
    Natur (Agerlandet*, Vol. 8), Politikens Forlag, Copenhagen, 179 (1970).

55  S. Thorup, Ukrudt og ukrudtsbekaempelse i økologisk perspektiv, *Ugeskr. Jordbrug*, **125**,
    107 (1980).

56  K. Petzoldt, Wirkung des Mähdruschverfahrens auf die Verunkrautung, *Z. Acker- Pflanzenbau*,
    **109**, 49 (1959).

57  A. Aamisepp, V. Stecko, and E. Åberg, Ogräs frö spridning vid bindarskörd och skördetrö-
    skning, *Lantbrukshögsk. Medd. Ser. A*, **81**, 31 (1967).

58  W. Habel, Über die Wirkungsweise der Eggen gegen Samenunkräuter sowie die Empfind-
    lichkeit der Unkrautarten und ihrer Altersstadien gegen den Eggvorgang, Dissertation, Hohen-
    heim, 1954.

59  W. Habel, Über die Wirkungsweise der Eggen gegen Samenunkräuter sowie deren Empfind-
    lichkeit gegen den Eggvorgang, *Z. Acker- Pflanzenbau*, **104**, 39 (1957).

60  H. Kees, Untersuchungen zur Unkrautbekämpfung durch Netzegge und Stoppelbearbeitung-
    smassnahmen unter besonderer Berücksichtigung des leichten Bodens, Dissertation, Hohen-
    heim, 1962, p. 102.

61  W. Koch, Untersuchungen zur Unkrautbekämpfung durch Saatpflege und Stoppelbearbeitung-
    smassnahmen, Dissertation, Hohenheim, 1959, p. 123.

62  W. Koch, Einige Beobachtungen zur Veränderung der Verunkrautung während mehrjährigen
    Getreidebaus und verschiedenartiger Unkrautbekämpfung, *Weed Res.*, **4**, 350 (1964).

63  H. Neururer, Mechanische Unkrautbekämpfung mit modernen Hackeggen, *Proc, EWRS
    Symp. Methods of Weed Control and Their Integration*, 65 (1977).

64  H. I. Petersen, *Ukrudtsplanter og ukrudtsbekaempelse*, Copenhagen, 1960, p. 144.

65  Anonymous, Kemiske midler til bekaempelse af ukrudt, *Statens Forsøgsvirksomhed
    Plantekult.*, 785, medd. nr. 21 (1966).

66  K. Jessen and J. Lind, Det danske markukrudts historie, *Det kongelige Danske Videnskabernes
    Selskabs Skrifter, Naturvidenskab og Matematik*, Vol. 8, Part VIII, Copenhagen, 1922.

67  P. Mølgaard, *Plantago major* ssp. *major* and spp. *pleiosperma*. morphology, biology and
    ecology in Denmark, *Bot. Tidsskr.*, **71**, 35 (1976).

68  Oberdorfer, Zur Syntaxonomie der Trittpflanzen-Gesellschaften, *Beitr. Naturk. Forsch.
    Südwestdtschl.*, **30**, 95 (1971).

69  B. Dennis and G. Mortensen, Varietal differences in weed suppression in barley, *Nord.
    Jordbrugsforsk.*, **62**, 433 (1980).

70  B. Dennis, Dept. of Crop Husbandry and Plant Breeding, The Royal Veterinary and
    Agricultural University, Copenhagen, personal communication, 1980.

71  G. W. Cussans, S. R. Moss, F. Pollard, and B. J. Wilson, Studies of the effects of tillage on annual weed populations, *Proc. EWRS Symp. The Influence of Different Factors on the Development and Control of Weeds,* 115 (1979).

72  B. Rademacher, Über die Lichtverhältnisse in Kulturpflanzenbeständen insbesondere in Hinblick auf Unkrautwuchs, *Z. Acker- Pflanzenbau,* **92**, 129 (1950).

73  H. Fogelfors, *Några ogräsarters utveckling under skilda ljusförhållanden och koncurrensförmåga i kornbestånd,* Dissertation, Swedish University of Agricultural Sciences, Department of Plant Husbandry, Uppsala, 1973.

74  W. J. Dixon, *BMDP—Biomedical Computer Programs,* University of California Press, London, 1975, p. 790.

75  K. S. Kershaw, *Quantitative and Dynamic Ecology,* 2nd ed., Edward Arnold, London, 1973, p. 308.

# CHAPTER FIVE

# Taxonomy and Biological Considerations of Herbicide-Resistant and Herbicide-Tolerant Biotypes

ROBERT J. HILL

Department of Biology
York College of Pennsylvania
York, Pennsylvania

## 1 INTRODUCTION

The evolutionary history, or phylogeny, of weeds has been the subject of some attention in recent years.[1,2] The role of herbicides in this evolution is only now becoming apparent. Herbicide resistance and tolerance is reported in natural field populations of normally susceptible plants (Chapters 2,3), laboratory (Chapter 14) and field selections (Chapters 12,13), weeds that exhibit intraspecific variation, cell culture, and crop plant varieties. This chapter deals with weeds where there has been the appearance of field resistance or tolerance.

The plant populations that show resistance or tolerance are species in which individuals have a highly variable expression of characters. Individual polymorphism is seen to be an important factor in population responses to herbicides. This variability has been a source of taxonomic confusion. Tables of diagnostic characters are provided to distinguish the closely allied weed taxa where field tolerance or resistance occurs. Numerous nomenclatural problems in the closely related *Chenopodium album* complex are presented.

Previous North American reports of resistant *Amaranthus retroflexus* are in part referable to *A. powellii* and *A. hybridus*. With the exception of one Canadian population (Chapter 2), no resistant *A. retroflexus* biotypes have been seen by the author. Nomenclatural differences in European and American taxonomy are discussed in an attempt to clarify the identity of tolerant and resistant species of *Amaranthus.* In addition to the taxa listed above, others treated are *Polygonum lapathifolium* and *P. persicaria,* for which taxonomic or nomenclatural lucidity is needed.

Tolerant and resistant biotypes have appeared in species that display a constellation of characters, including a preponderance of herbaceous annuals, autogamy, association with the agricultural habitat, r-selection, evolutionary strategies selecting for fitness, and high fecundity. Each of these is defined and described in detail in the following sections in relation to acquired herbicide resistance or tolerance in weeds.

## 2  PHYLOGENY: A BRIEF DISCUSSION

It is important that no herbicide-resistant or herbicide-tolerant weeds are known from among Thorne's superorders[3]: Annoniflorae, Theiflorae, Santaflorae, Hamamelidiflorae, Myrtiflorae, and Lamiiflorae within the Dicotyledonae, and are found in only one order of the Monocotyledonae. The superorders where resistance or tolerance is found (Commeliniflorae, Chenopodiiflorae, Asteriflorae, Cistiflorae, Corniflorae, Malviflorae, Gentianiflorae, Geraniiflorae, Rutiflorae, and Rosiflorae) have a common ancestry with the Annoniflorae and Theiflorae. The superorders where it is not known are presently predominantly woody, tropical groups. No members from among them are utilized worldwide as major crop plants or grown in monoculture. Some of them do contain weed taxa of temperate distribution, especially the Lamiiflorae, but most are not weeds of temperate-zone agriculture. A few weeds of tropical and subtropical distribution are found in these superorders. Many of the groups contain either wind-pollinated (e.g., the amentiferous Hamelidae) or entomophilous plants; they are outcrossing taxa. The groups in which herbicide resistance and tolerance are found are, on the contrary, predominantly associated with temperate agriculture or mass monoculture and are wholly or partially self-fertile.

Although there are some trends in which plants of major lines of evolution do share herbicide resistance, the lack of resistant biotypes in other major lines of evolution, correlated with few weeds (temperate agriculture) in these superorders, suggests that this phenomenon may not be a phylogenetic character. Resistance and tolerance is derived (or at least detected) from the application of herbicides in agricultural context. The occurrence of resistant or tolerant populations must result from natural gene mutations which are artificially selected for by herbicides. Certainly naturally occurring variation to chemicals exists in populations of temperate weeds. These variations could supply the material upon which selection acts. The field herbicide-resistant and herbicide-tolerant weeds have several attributes in common, including having a variety of morphologically different biotypes in each taxon (polymorphism).

## 3  VARIABILITY AND BREEDING SYSTEM

Of the weeds acquiring field resistance or tolerance, only *Cirsium arvense* is a perennial; the others are annuals or winter annuals. Some do survive for longer periods and act as facultative perennials. A notable feature of temperate annual weeds is the ability to produce viable seeds by autogamy (wholly or partially self-fertile) or agamospermy (no sexual fusion hence embryos arise without fertilization).[1] The advantage of this type of reproductive strategy is obvious; one seed or one plant can result in an entire population, usually rapidly. *Cirsium arvense* produces lateral roots which are brittle and easily broken. Small portions of these roots bear buds, which can result in the scatter of the plant by cultivation. The product is an increase of individuals of the same genotype. This type of evolutionary strategy is defined as selection for fitness; an optimum genotype exploits a given habitat. This is in contrast to heterogamous plants, which evolve under a flexible regime. Most species are apt to exhibit a compromise between the evolutionary strategies of fitness or flexibility. Plants that possess closed recombination systems (i.e., autogamous taxa) predominately have a fitness strategy for survival. Examples of autogamous plants, especially weeds, are replete; several have shown the herbicide-response shift from susceptible to tolerant/resistant and include field-resistant *Stellaria media* and field-tolerant *Capsella bursa-pastoris*. *Capsella* is pollinated before the flower opens. *Poa annua,* field-resistant, is also self-fertile and may be apomictic.[4] In fact, *Poa* and *Capsella* are both highly heterozygous species complexes with different ''microspecies.'' Each genus contains numerous morphologically similar taxa, making satisfactory taxonomic treatment unlikely. In *Poa,* morphometric investigations discern a pattern of a greatly magnified hybrid amalgam. *Arabidopsis thaliana,* regularly self-pollinated, has been selected for herbicide resistance in the laboratory. Many of the most successful weeds are primarily self-fertile. Those plants that have open recombination systems (i.e., outcrossing taxa) possess a flexible strategy. An open recombination system means that new gene combinations as well as the parental ones will appear in each generation.[5] Genetic variability is generally associated with open recombination systems, in which panmixis ensures random distribution of characters. At first glance it seems paradoxical that autogamous weeds have genetic variability. However, self-fertile plants have variable phenotypic expressions. This polymorphism is discussed in the following section. The pattern of variation is different from that seen in heterogamous taxa. Regardless of breeding system, the source of variation remains the same: mutation (spontaneous and induced).

There are specific genodemes, genetic lines of descent in autogamous populations, that are distinguished by phenotypic characters such as vestiture (trichome covering) and leaf and fruit-capsule outline. The genodemes tend to be lines of a single genotype resulting from habitual selfing.[6] Yet true-breeding, pure lines are probably seldomly found in nature, as autogamy is usually not obligate. Occasional outcrossing between biotypes in a predominantly self-breeding population will introduce new variability at periodic intervals.

## 4 INDIVIDUAL VARIATION WITHIN HERBICIDE-RESISTANT AND HERBICIDE-TOLERANT SPECIES

Populations of weeds are known to be especially prone to both phenotypic and chemical variation. The appearance of herbicide-resistant biotypes undistinguishable from sensitive plants is merely an expression of variation at the chemical level. It is of value to examine briefly both the phenotypic and genotypic variations that exist in natural populations.

A question that always concerns population biologists is: What is the "normally distributed variation" in biological material? Variation in genotype, with associated or correlated variations in phenotype, provide the materials upon which selection operates within a population. Selection allows for both immediate adaptation, and ultimately for evolutionary change.[7] Phenotypic plasticity of plants is reflected in the confused taxonomy and confused naming of some economically important groups. A list of highly variable susceptible weeds would include many known to have herbicide-resistant and herbicide-tolerant biotypes (e.g., *Stellaria media*, a genetically complex cosmopolitan weed, much subdivided by European authors, excessively variable in number, size, and development of petals, and number of stamens).[8,9] *Capsella bursa-pastoris*, a taxon that has limited field herbicide tolerance, is extremely variable in foliage and silicle outline.

*Solanum nigrum*, with acquired field-resistant populations, is polymorphic in foliage outline, amount of pubescence, size of corolla, shape and size of calyx lobes, and size and color of berries.[9] Herbicide-resistant or herbicide-tolerant species that are highly variable in phenotype include *Chenopodium album*, *Amaranthus hybridus*, and *Atriplex patula*. Many of the species developing resistant biotypes have intergrading races, or are taxa that have been subdivided by taxonomists because of the polymorphism seen in the group. The presence or absence of ray flowers in the capitulum of the Compositae is diagnostic of tribal affinities. Yet inflorescences of *Senecio vulgaris* can be radiate, half-radiate, or nonradiate. In England there exists a gradient in *Senecio;* races in the south bear five to eight ray florets per capitulum but become nonradiate northward.[4,6] Many varietal names have been proposed under *Bidens tripartita*, *Matricaria perforata*, and *Cirsium arvense*, which are attributable to variability.

An examination of the nomenclature in any floristic work will indicate the natural variability found in weed taxa. Specific epithets in the Latin binomials indicating this include *mutabilis* (changeable), *intermedia* (intermediate), *hybridus* (mongrel), and *spurium* (previously confused with another taxon), among others. Imprecise understanding of the taxonomy and the nomenclature of highly polymorphic species results in confusion. Several herbicide-resistant or herbicide-tolerant plants that need clarification are found in the genera *Chenopodium, Amaranthus,* and *Polygonum.* Within these genera are species that present taxonomic or nomenclatural difficulty.

## 5 POPULATION ECOLOGY

The herbicide-resistant biotypes are all colonizing species playing a pioneering role in the open community of the cultivated field. A demand is placed on the reproductive potential of such a population. Maximum population expansion is a result of few favorable gene combinations being maintained by restricting recombination (autogamous reproductive strategy). Present adaptations are maintained and variability is reduced, but as we have seen, not absent. Population ecologists describe the regulation of population size with the concept of r- and K-selection. Colonizing taxa display r-selection (reproductive selection), which results in an intrinsic rate of increase (from many propagules), periodic mass death of individuals (at times of germination, establishment, cultivation, herbicide selection, winter, etc.), high fecundity, and rapid development. Colonizing species evolving in an open habitat produce large numbers of propagules. The known herbicide-resistant biotypes are r-selections; they can produce large quantities of seed. High reproductive yield is seen in *Chenopodium album* (averaging 3000 seeds per plant), *C. ficifolium* (130,000 seeds per plant), *C. polyspermum* (over 80,000 seeds per plant), *Amaranthus retroflexus* (some plants are known to yield 500,000 seeds per plant), *Polygonum persicaria* (up to 1200 seeds per plant), *P. lapatifolium* (up to 1500 seeds per plant), *Erigeron canadensis* (over 60,000 achenes per plant), and *Stellaria media* (over 2000 seeds per plant).[4] Many of the field-tolerant or herbicide-resistant biotypes have several generations maturing in a single season. Within 5 weeks of being shed, seeds of both *Senecio vulgaris* and *Stellaria media* will produce ripe seeds. *Stellaria media* can produce three generations in one year. Plants of several resistant or tolerant taxa (including *Poa annua, Senecio vulgaris, Veronica persica, Capsella bursa-pastoris,* and *Bidens* spp.) periodically survive the winter in temperate areas, producing seeds during this time or maturing for seed production in advance of other plants. It is estimated that over a 12-month period, if most survive, 15 billion plants could be produced from one *Stellaria media* plant.[4]

## 6 TAXONOMY AND NOMENCLATURE

The continuing need for clear, simple methods permitting identification of weed species at all stages of development has been recognized.[2,10] Taxonomic information about weeds should include lucid descriptions of seed, seedling, vegetative plant body, and reproductive structures of the flower if it is to be of practical value to all members of the plant science community. It is hoped that this section will aid in the proper identification of some of the presently herbicide-resistant and herbicide-tolerant biotypes to correct species. Proper identification is obviously important in recommending control measures.

## 6.1  Chenopodiiflorae

The following genera are known to contain herbicide-resistant or herbicide-tolerant populations: *Chenopodium, Amaranthus,* and *Polygonum.*

### 6.1.1  Chenopodium

Several species of *Chenopodium* are competitive weed problems in agriculture. The genus is large, containing many species that are confusing in taxonomy and nomenclature.

The phenology of *Chenopodium album* is complex. Anthesis can occur under low light intensities (shade). Short days (spring, autumn) increase maturity and during long days (summer) flowering is less rapid but vegetative growth results.[11] High temperatures (summer) during growth reduce the time to flowering; thus the vegetative/reproductive equilibrium is skewed toward maturity and seed production. This perhaps contributes to the success of the plant. Flowering plants can thus vary between 5 cm and 1.5 m tall, with implications that are discussed in Chapter 17.

The expression of polymorphism in the very variable and complex *Chenopodium album* has caused much confusion in the nomenclature of this plant over its entire range in both North America and Europe. As the species has been much subdivided, published names, both synonymous and varietal, are legion.

In the western United States, *C. berlandieri* (not known to be herbicide-tolerant), also a variable taxon, is frequently interpreted as *C. album*. In the eastern United States, *C. album* is confused by many with *C. bushianum* (also not yet tolerant) and *C. missouriense* (which has resistant populations). With the exception of *C. album*, these species are not distributed in Europe or Asia. The ranges of these taxa are given in Table 5.1. In North America small plants of *C. strictum* var. *glaucophyllum* have the profuse branching habit of *C. album*; in the shade it appears like shade forms of *C. missouriense*.[12] Herbicide-resistant taxa from Europe include *C. album, C. ficifolium,* and *C. polyspermum* (Chapter 3). The latter two species are highly distinctive and present few identification problems. They are rare adventive species in North America; *C. polyspermum* has been collected along the St. Lawrence River, Quebec, and Ontario, Canada, and as waifs at ports in Pennsylvania. *Chenopodium ficifolium* has been recorded at Philadelphia, Pennsylvania, and Pensacola, Florida. Young plants of *C. ficifolium* can be mistaken for *C. glaucum*, owing to tapering leaf bases and farinose glands beneath the leaves.

It is significant that evolution in the genus has resulted in regionally distributed taxa in North America. *Chenopodium album* is cosmopolitan, occurring in 40 different crops from 47 countries.[13] *Chenopodium missouriense*, a closely related species, is found in the Missouri Basin but not in southern Canada, a region where *C. strictum* var. *glaucophyllum* occurs. *Chenopodium missouriense* and *C. strictum* var. *glaucophyllum* are closely allied taxa in the *C. album* complex. Table 5.1 provides a complete complement of characters differentiating the species.

The texture of the pericarp surrounding the seed and the seed coat (testa) are diagnostic characters for separation of many confused taxa in the genus *Chenopodium*. Critical examination of the pericarp is imperative for proper identification of all chenopods. The pericarp is smooth, or merely pitted in *C. ficifolium*, *C. album*, *C. missouriense*, *C. polyspermum*, and *C. strictum* var. *glaucophyllum*. It is honeycombed (rugose to alveolate) with a reticulum in *C. berlandieri* and *C. bushianum*.

Chromosome data have been reported for all taxa[14-19]; *C. album* has a basic chromosome number of 18 in some European populations but is usually hexaploid along with *C. missouriense,* (54 chromosomes); *C. berlandieri, C. strictum* var. *glaucophyllum*, and *C. bushianum* are tetraploid, (36 chromosomes); and *C. ficifolium* and *C. polyspermum* are diploid, (18 chromosomes). According to Cole[17] and Uotila,[18] the tetraploid counts for *C. album* are misidentifications, the correct chromosome complement being 54 (hexaploid). Al Mouemar[19] has reported 36 plus a variable number of additional chromosomes for *C. album* from Cote d'Or.

*C. album* seedlings can generally be distinguished[20] by a threadlike tap root and laterals, hypocotyl often red in color, at first succulent; seed leaves 0.75 to 4.5 mm by 2.0 to 12.0 mm, leaving distinct stubs when shed. The first leaves are paired, later leaves alternating, covered beneath with (mealy) glistening glands that may be shed as the leaf matures. Stems are covered with similar granules. The biology and taxonomy of *C. album* in comparison to *C. strictum* and *C. berlandieri* are discussed at length by Bassett and Crompton.[21]

### 6.1.2 *Amaranthus*

The genus *Amaranthus* is a predominately native American group which has spread worldwide and contains primary crop competitors.[22]

Although *A. retroflexus* has been reported to have become herbicide resistant[23-26] in Washington state, the plants were later identified as an aberrant, indehiscent-capsule form of *A. powellii*. Plants from Pennsylvania and others provided by various researchers working with herbicide resistance (original seed source unknown) are typical *A. hybridus*. Verification of previous reports of resistance in North American *A. retroflexus* is impossible in the absence of voucher specimens. The need for proper taxonomic procedures in weed studies has been expressed recently.[10] All North American material studied by this author proved to be *A. hybridus*, with the exceptions of one population of *A. retroflexus* from Ayr, Ontario, and the *A. powellii* from Washington state (see Chapter 2). It seems likely that *A. hybridus* is, in fact, the more com-

Table 5.1 Useful Characteristics of the *Chenopodium* Species Involved in Herbicide Resistance Studies

| Character | C. album | C. bushianum | C. missouriense | C. berlandieri |
|---|---|---|---|---|
| Basic Chromosome number | 54 (hexaploid) | 36 (tetraploid) | 54 (hexaploid) | 36 (tetraploid) |
| Habitat | Cultivated sites, disturbed soil | Most often cultivated sites; wastel | Wastel, ruderal, undisturbed soil | Disturbed soil, ruderal |
| Phenology (anthesis) | June–October | Late August, September | Mid-September | Not restricted |
| Seeds | 1.1–1.5 mm broad, smooth surface | 1.5–2.5 mm broad, alveolate | 0.9–1.2 mm broad, smooth surface | 1.0–1.5 mm broad, alveolate |
| Leaves | Ovate, rhombic to lanceolate; some leaves trilobate, irregularly toothed | Large, thin, broadly rhombic; not trilobed, irregularly toothed; light green color | Oblong, rhombic; slightly trilobed, median lobe coarsely toothed | Thin to coriaceous; ovate, rhombic, oblong to elliptical; trilobed, median lobe obtuse or acuminate |
| Inflorescence | Stiff, erect glomerules; narrow, compact inflorescence; grayish green; sparsely leafy | Heavy, drooping irregular inflorescence; lead gray in color; leafy | Delicate, (nodding) flexuous, arching inflorescence of small glomerules; green, sparsely leafy | Glomerules small, slender or stout, dense; branches moniliform; inflorescence sometimes spreading; sparsely leafy |
| Plasticity and range | Very variable, many subspecific taxa named; widely distributed in world: Asia, Europe, and northern Africa; may be in part native to North America; distributed in Americas from Yukon, southern Canada, to northern Mexico, sporadic in South America | Uniform, but with several forms and varieties formerly recognized, widely distributed in eastern North America from Massachusetts to North Dakota, south to Missouri and Virginia; cool regions | Uniform, widely distributed eastern North American, rare in cool regions (e.g., New England) | Very variable, many subspecific taxa named; widely distributed in western North America, but adventive eastward, typically frequents warmer regions; adventive in northeastern U.S. and Europe |
| Confused taxa | Polymorphic, many species placed here | Resembles C. album and C. missouriense | Plants from late seeds produced in fall; resembles C. berlandieri, especially in shade | |

| Character | C. ficifolium (C. serotinum) | C. polyspermum | C. strictum var. glaucophyllum |
|---|---|---|---|
| Basic Chromosome number | 18 (diploid) | 18 (diploid) | 36 (tetraploid) |
| Habitat | Fields | Wastel | Plants respond dramatically to habitat differences; wastel, ruderal |
| Phenology (anthesis) | | August–October, not restricted | Not restricted in flowering time |
| Seeds | 0.8–1.0 mm broad; smooth to regularly minutely reticulate (scarcely discernible at 10X) with radially elongate pits | 0.9–1.1 mm broad; exposed; smooth surface with round margin | 0.9–1.2 mm broad; exposed; smooth surface |
| Leaves | Narrowly ovate to linear, median lobe linear, long with nearly parallel sides above the basal lobes, leaves trilobed; conspicuously toothed | Thin, ovate, entire leaves, plant glabrous | Ovate-laceolate, mostly entire; lower leaves shallowly serrate, the median oblong, entire |
| Inflorescence | | Glomerules of flowers in spikes or cymes from the axils of lowermost to uppermost branches in var. *acutifolium* (Sm.) Gaudin pyramidal, var. *obtusifolium* Gaudin cymose, axillary, ebracteate | Glomerulose, cymose, axillary inflor. spicate and stiff but terminally becoming paniculate and spreading |
| Plasticity and range | Uniform and distinctive; found in Eurasia but is very infrequent in North America as an adventive | Two varieties which in extreme forms are strikingly different in appearance. Of Old World origin (Europe and northern Asia) but adventive in the New World; frequents cool regions | The species C. *strictum* is Asiatic in origin and several varieties are recognized. American material (central U.S.) referable to the species have lower leaves only 2 times as long as wide with margins abundantly shallowly serrate; cool regions of North America (southern Canada, Mississippi Basin, New England) may be adventive and not native in America |
| Confused taxa | Young plants confused with C. *glaucum* | Phylogenetic relationship uncertain yet easily recognized by apiculate sepals, thin entire leaves | Small plants may have the profuse branching habit of C. *album*; in shade it simulates shade forms of C. *missouriense* |

89

mon herbicide-resistant biotype. However, Sauer[26a] has recently identified a triazine-resistant biotype from Connecticut to be a hybrid between *A. retroflexus* and *A. hybridus*.

In Europe *A. bouchonii* is reported to have acquired field tolerance. The specific epithet *A. bouchonii* used by European taxonomists is based on indehiscent fruit. In describing this species, Thellung[27] expressed uncertainty as to whether it was a new species or merely a form of ordinary *A. hybridus*. Indehiscence is the rule in some sections of the genus, but is anomalous in others. Indehiscent mutants spontaneously arise in circumsessile-fruited taxa, especially in *A. powellii*. Interspecific hybrids can also have indehiscent capsules. Although *A. powellii* is a plant native to the western Cordilleran system of North and South America, it is found in disturbed habitats elsewhere. This taxon began appearing as a rare adventive in eastern North America about 1900, becoming a widespread, troublesome weed about 1931, although still frequently misidentified as *A. retroflexus*. This species arrived in Germany as an adventive before 1900, becoming an abundant weed presently undergoing range extension in north and central Europe,[28] and is not yet reported in *Flora Europaea*.[29] In Europe it has been misidentified as *A. chlorostachys*, a synonym of *A. hybridus*. Sauer[28] states that the European *A. bouchonii* is an indehiscent form of *A. powellii*; an examination of the fragment of the type of *A. bouchonii* in the U.S. National Herbarium supports this.

Red forms of *Amaranthus hybridus* have been called *A. paniculatus* by some American authors. The type specimen of *A. paniculatus* in the Linnaean herbarium is not *A. hybridus* but *A. cruentus*, a cultivated plant.[26a] The name *A. chlorostachys* has been applied to green forms, and as stated previously, is synonomous with *A. hybridus*.

Taxonomic and nomenclatural problems arise with the use of the names *A. graecizans* and *A. blitoides*. *A. blitoides* was described by S. Watson[30] in 1877 as having a prostate growth habit. The Linnaean species *A. graecizans* has an erect growth. Fernald,[8] in *Gray's* 8th edition, named *A. blitoides* as *A. graecizans,* erroneously concluding that it matched the type of a Gronovian name, cited by Linnaeus as a synonym of *A. graecizans*. Thellung, and Dandy and Melderis, postulated that Linnaeus' material of *A. graecizans* is different from material of *A. blitoides,* and that the name *A. graecizans* properly belongs to an Old World species otherwise known as *A. augustifolius*.[31] Recent American authors treat *A. graecizans* and *A. blitoides* as conspecific, accepting the earlier Linnaean name.[32] The appellation *A. graecizans* has been misapplied by some taxonomists for that which is referable to *A. albus*. Table 5.2 lists all key characters used to differentiate these taxa.

Seedlings of some *Amaranthus* spp. have been grown in the greenhouse of the Pennsylvania Department of Agriculture. The seedlings vary from glabrous to sparingly pubescent, cotyledons are all lanceolate, attenuated at the base into a petiole (2.0 mm) with a thick blade (8.0 mm). The cotyledons are maculate with a green reticulum, inter-reticular spaces, white, lucid, large in *A. retroflexus* but small in *A. hybridus* and *A. powellii*. The primary leaves are initiated unevenly, resulting in alternating leaves, one large and one small. The first leaves are red-purple beneath, glistening, green above, notched at the apex to varying degrees, obcordate in *A. hybridus* and *A. powellii*, retuse in *A. retroflexus*. Initially, the leaf veins of *A. retroflexus* bear a few hairs, whereas *A. hybridus* is smooth beneath, although hairs may be present on the basal margins. Hypocotyls of all are silverish to green or red. The biology and taxonomy of *A. powellii*, *A. retroflexus* and *A. hybridus* are also discussed at length in a recent review by Weaver and McWilliams.[33]

### 6.1.3 *Polygonum*

The genus *Polygonum* is a cosmopolitan, highly variable group of plants. Resistance is known in populations of *P. persicaria* and *P. lapathifolium* in France. Resistant populations of *P. convolvulus* have developed in Austria. The first two species are readily distinguishable with proper taxonomic discernment (Table 5.3) but might be confused by an untrained eye. Seedlings[20] of both species early develop a brown root and hypocotyl, both suberized. Cotyledons of *P. lapathifolium* are smaller (oblong-ovate) (1 by 3.5 to 3.5 by 12 mm) than *P. persicaria* (ovate) (2 by 8.5 to 3 by 12 mm) and are vested with hairs on the upper surface. In *P. persicaria*, glandular hairs may be present on the basal margins. Primary leaves of *P. lapathifolium* are with short soft hairs beneath (arachnoid) the ocrea, transparent, and margin entire. In *P. persicaria*, the primary leaves are punctate, the margin ciliate, and the ocrea fringed on the upper end. *P. convolvulus* is very distinctive and poses no problem in separating it from the other species based on its vining characteristic.

### 6.2 Conclusions

The plants that have acquired field resistance or tolerance to herbicides can be placed in 10 of the possible 21 superorders of angiosperms. These superorders contribute many, but not all of the weeds and cultivated plants of the more temperate flora. Members of the superorders where acquired resistance and tolerance are not known are predominately woody tropical species; some are weeds of the region where the shift in herbicide response has been discovered.

The following correlation of characters is seen in resistant or tolerant species: herbaceous annuals, at least partially self-fertile, association with agricultural synusium, high fecundity, complex genetic variability expressed as a polymorphic phenotype, evolutionary strategies selecting for fitness, periodic mass death of individuals, and rapid development of plants to maturity.

The importance of taxonomic verification of resistant species before reporting is stressed. There has been much confusion in the literature in the absence of proper specific taxonomic identification of specimens.

Reports of resistance in *Chenopodium album* are corrected to include *C. missouriense* from Pennsylvania. *Chenopodium strictum* var. *glaucophyllum* (Canada) and *C. missouriense* are part of the *C. album* species complex. Resistance is now known in all three taxa.

Table 5.2 Useful Characteristics of the *Amaranthus* Species Involved in Herbicide Resistance Studies

| Character | A. retroflexus | A. powellii | A. hybridus |
|---|---|---|---|
| Basic Chromosome number | 32 (and 34 in Europe)[28] | 34 | 32 |
| Habitat | Disturbed sites, native riverbank plant, cultivated areas, temperate to warmer regions | Disturbed sites, open, dry habitats, cultivated areas | Disturbed sites, native riverbank plant, mesic, milder regions, waste and cultivated areas |
| Habit | Stems stout 3–30 (+) dm tall, erect or ascending, much branched or simple | Stems stout 3–20 (+) dm tall; usually much branched | Erect or ascending 3–25 dm, usually branched; stems may be red or red tinged |
| Phenology | Seed mature by mid-August; plant brown by early October | | Seed mature by mid-September; plants green at time of frost |
| Hairs | Plants abundantly villous above, often densely so in inflorescence, finely villous to scurfy villous below; hairs flat, crisped, multicellular | Plants never villous, glabrous to pubescent | Plants finely villous above, rough puberulent below or glabrous, hairs flat, crisped, multicellular |
| Leaves | Hairy beneath along veins; petioles to 8 cm; blades 12 cm; ovate or rhombic-ovate | Glabrous to sparsely pubescent on veins; petioles to 5 cm; blades to 8 cm; ovate or rhombic ovate | Petioles may be pubescent; leaves pubescent to glabrous beneath; petioles to 9 cm; blades to 15 cm; ovate or rhombic-ovate, rarely lanceolate; may be red |
| Leaf apex/base | Acute to obtuse, often emarginate/acute or obtuse | Acute to obtuse, sometimes emarginate/cuneate to obtuse | Acute or rarely rounded at apex/acute |
| Perianth | (2.5) 3–4 mm; greater than mature fruit | 2.0–3.8 mm; equaling, rarely greater than mature fruit | 1.0–2.5 mm; equaling or less than mature fruit, only occasionally exceeding the fruit |
| Style | Branches: 3, erect short | Branches: 3, recurved | Branches: 3, erect |
| Confused taxa | A. powellii is subglabous with sharply acute sepals and may be confused with C. retroflexus | May have indehiscent capsules; may be the same as A. bouchonii in Europe, which has 32 chromosomes[28] | Erroneously called A. retroflexus |

| Character | A. lividus | A. graecizans | A. blitoides |
|---|---|---|---|
| Basic chromosome number | 32 | 32 | 32 |
| Habitat | Disturbed sites, ruderal, wastel | Ruderal, wastel and cultivated areas | Dry waste or cultivated ground |
| Habit | Branched, succulent; 3–10 dm tall, *prostrate* matt plant | Stems stout, *erect* 3–7 (+) dm; divaricate or ascending branches | Stems stout, *prostrate*, much branched to 10 dm long, may be red tinged |
| Phenology | | | |
| Hairs | Glabrous | Glabrous or sparingly puberlent or villose | Glabrous or sparingly pubescent |
| Leaves | Glabrous, green or red; petioles 1.5–7 cm long, slender; broadly ovate, or *orbicular-ovate*, spotted, *emarginate* | Petioles slender 3–5 cm long; leaves elliptical to oblong, *elliptical-rhombic*, glabrous prominently veined | Glabrous, petioles stout 2.0–20 mm long; leaves obovate to *oval*, *spatulate*, or elliptic 0.8–4 cm; veins white beneath |
| Leaf apex/base | *Rounded, emarginate*/cuneate to rounded; margin undulating | *Acute*/cuneate | Rounded/cuneate to attenuate, obtuse with mucronate tip |
| Perianth | | Oblong to linear, acute, 1 nerved nerved thin, may be red tinged | 2.5–3.0 mm long acuminate, 1 nerved, oblong |
| Style | 2 or 3 branched | 3 | 3, short recurved |

Table 5.2 (Continued)

| Character | A. retroflexus | A. powellii | A. hybridus |
|---|---|---|---|
| Stamens | 5(4) | 3(4-5) | 5 |
| Utricle/seed | Circumsessile/dark reddish brown, shining | Circumsessile/black, shining | Circumsessile/shining, reddish or black |
| Distribution | Common in waste and cultivated grounds, southern Canada throughout U.S. to northern Mexico; naturalized in Europe, Asia and Africa | Waste or cultivated ground in U.S. and Canada, northern Mexico; adventive eastern North America and Europe | Southern Canada, U.S., adventive warmer parts of South America, Europe, Asia, Africa |
| Sepals | Recurved, retuse, or obtuse, mucronate; membranous, white with green nerve, nerve short, excurrent, 5 | Erect, lanceolate to oblong, acute, bristle tipped; scarious mid-nerve excurrent as spinous tip 3(5) | Erect oblong to linear, inner rarely obtuse, aristate, may be recurved; green nerve excurrent as a pungent tip, 5 |
| Panicle | Terminal and axillary; entire "panicle" 5-2 cm, lateral panicles 1-5 cm thick; short dense, crowded; lateral branches numerous | Terminal and axillary; entire "panicle" 2.5-4 cm long; central spike 1-2.5 cm, lateral spike 4-12 cm; lateral branches few, widely spaced | Terminal and axillary; entire "panicle" 5-15 cm, lateral panicles 1-8 cm, 1-1.2 cm thick; slender, cylindric; lateral branches of inflorescence numerous, crowded |
| Bracts | Ovate, tapering to tip, 4-8 mm long, rigid; may be 2 (3) times as long as sepals | Lanceolate to ovate, 4-8 mm long, attenuate to a rigid spinose tip, very thick midnerve; 2 to 3 times as long as sepals | Lanceolate to ovate 2-4 mm, tapering to slender, stout tip; 2 times as long as sepals or less |

94

| Character | A. lividus | A. graecizans | A. blitoides |
|---|---|---|---|
| Stamens | 3 | 3 | 3 |
| Utricle/seed | Indehiscent/reddish brown or black shining; fruits slightly rugose | Circumsessile/dark reddish brown shining; fruits veined | Circumsessile/dull black to glossy; fruits smooth |
| Distribution | Europe | Europe | North America; adventive in Europe |
| Sepals | Linear-oblong or oblong, obtuse or acute, 3(−5) | Staminate flowers ovate, lanceolate, acute, 3 | Staminate flowers, scarious, oblong, acute; pistillate flowers, oblong, 5(4) |
| Panicle | Terminal or axillary, erect or drooping spikes, 2–7 cm long X 5–8 mm thick, leafless | Dense or loose wholly axillary clusters; usually shorter (rarely longer) than petioles | Dense wholly axillary clusters, usually shorter, sometimes longer than petioles |
| Bracts | Scarious, ovate to oblong obtuse or acute, 1-nerved; not pungent and not exceeding the sepals | Ovate, mucronate rigid, pointed, spreading | Oblong to lanceolate, erect, short spinous tip, slightly exceeding the sepals |

**Table 5.3 Diagnostic Characters Useful in Distinguishing Herbicide-Resistant *Polygonum* Species**

| Character | *P. lapathifolium* | *P. persicaria* |
|---|---|---|
| Ocreolae | Ovate, oblique | Obliquely truncate, entire or with short cilia to 1 mm |
| Ocreae | Entire, eciliate, cylindric (lowest may bear very short inconspicuous cilia) | Fringed with bristles minutely strigose, short ciliate |
| Stems | Up to 25 dm tall, erect or occasionally prostrate | Up to 10 dm tall, erect or ascending |
| Leaves | Linear—lanceolate to elliptic, acuminate; tomentose beneath when immature (some permanently so), leaves may be glandular or incanous | Narrowly lanceolate with a dark purple-brown blotch |
| Inflorescence | Numerous racemes, thin erect (to pendulous) nodding 1–8 cm long X 5–9 mm thick on glabrous to glandular peduncles | Numerous dense, thick usually erect, leading 1–4.5 cm long X 7–12 mm thick *eglandular* |
| Calyx | 3–4 mm at maturity; rose (purplish), white or green at base | Pink to rose (green or very rarely white) |
| Mature calyx | Ovoid to rhomboid, constricted toward summit forming beak (anchor-shaped fork) overtopping the achene | |
| Achene | 1.7–3.2 mm long X 1.5–2 mm broad | 2.5–3 mm |
| Phenotype | Very variable, probably native to both the Old and New World, distributed widely in temperate North America on mesic soil; widespread in Europe | Somewhat more uniform phenotype, native of Europe and adventive in the New World as a common ubiquitous weed in mesic areas, cultivated or ruderal, wastel |

Reports of resistant *Amaranthus retroflexus* in North America are questioned, as only a Canadian population (Ayr, Ontario) could be confirmed by the author; it is also reported in Connecticut and Switzerland. Most North American reports are referable to *A. hybridus* or *A. powellii*. Questions exist as to whether *A. bouchonii* should be accepted as a valid taxon. This indehiscent-fruited plant, tolerant in Switzerland, parallels resistant populations of an abberant, indehiscent capsuled form of *A. powellii* from North America. In fact, previous taxonomic authorities have argued that the European *A. bouchonii* is adventive *A. powellii*. *A. powellii* commonly gives rise to indehiscent-capsuled populations in its native North American range.

## REFERENCES

1  H. G. Baker, The evolution of weeds, *Annu. Rev. Ecol. Syst.,* **5**, 14 (1974).

2  J. McNeil, The taxonomy and evolution of weeds, *Weed Res.,* **16**, 399 (1976).

3  R. F. Thorne, A phylogenetic classification of the Angiospermae, *Evol. Biol.* 9, 1976.

4  E. Salisbury, *Weeds and Aliens,* Collins, London, 1961, p. 384.

5  V. Grant, *Plant Speciation,* Columbia University Press, New York, 1971, p. 435.

6  D. Briggs and S. M. Walters, *Plant Variation and Evolution,* World University Library, New York, 1969, p. 256.

7  T. H. Hamilton, *Process and Pattern in Evolution,* Macmillan, Toronto, 1967, p. 118.

8  M. L. Fernald, *Gray's Manual of Botany,* 8th ed., Van Nostrand Reinhold, New York, 1970, p. 1632.

9  H. A. Gleason, *The New Britton and Brown Illustrated Flora of the Northeastern United States and Adjacent Canada,* Vol. 2, Hafner, New York, 1952, p. 654.

10  R. J. Hill, The necessity for proper taxonomic procedures in weed studies, *WSSA Newslett.,* **8**(3), 5 (1980).

11  J. T. Williams, Biological flora of the British Isles: *Chenopodium album* L., *J. Ecol.,* **51**, 711 (1963).

12  H. A. Wahl, A preliminary study of the genus *Chenopodium* in North America, *Bartonia,* **27**, 1 (1954).

13  L. G. Holm, D. L. Plucknett, J. V. Pancho, and J. P. Herberger, *The World's Worst Weeds: Distribution and Biology,* University Press of Hawaii, Honolulu, 1977, p. 609.

14  C. S. Keener, Documented plant chromosome numbers 70:1, *Sida,* **3**, 533 (1970).

15  C. S. Keener, Documented plant chromosome numbers 74:1, *Sida,* **5**, 290 (1974).

16  P. Aellen, and T. Just, Key and synopsis of the American species of the genus *Chenopodium* L., *Am. Mid. Nat.,* **30**, 47 (1943).

17  M. Cole, Interspecific relationships and intraspecific variation of *Chenopodium album* L. in Britain: II. The chromosome numbers of *C. album* L. and others species, *Watsonia,* **5**, 117 (1962).

18  P. Uotila, Chromosome counts on the *Chenopodium album* aggregate in Finland and NE Sweden, *Ann. Bot. Fenn.,* **1**, 29 (1972).

19  A. Al Mouemar, Variations caryologiques et isoenzymatiques chez *Chenopodium album* L., *C.R. Acad. Sci. (Paris) Sec. D,* **288**, 677 (1979).

20  A. P. Kummer, *Weed Seedlings,* University of Chicago Press, Chicago, 1951, p. 435.

21   I. J. Bassett and C. W. Crompton, The biology of Canadian weeds, 32: *Chenopodium album,* *Can. J. Plant Sci.,* **58,** 1061 (1978).

22   M. K. Moolani, *Amaranthus hybridus* competition with crops, *Weeds,* **12,** 126 (1964).

23   S. R. Radosevich, Mechanism of atrazine resistance in lambsquarters and pigweed, *Weed Sci.,* **24,** 68 (1976).

24   D. Peabody, Herbicide tolerance weeds appear in western Washington, *Weeds Today,* **5,** 14 (1974).

25   L. D. West, T. J. Muzik, and R. E. Witters, Differential gas exchange response of two biotypes of redroot pigweed to atrazine, *Weed Sci.,* **24,** 68 (1976).

26   C. J. Arntzen, C. L. Ditto, and P. E. Brewer, Chloroplast membrane alterations in triazine-resistant *Amaranthus retroflexus* biotypes, *Proc. Natl. Acad. Sci. USA,* **76,** 278 (1979).

26a   J. Sauer, Department of Geography, UCLA, Los Angeles, California, personnal communication, 1981.

27   A. Thellung, *Amaranthus Bouchonii* Thll. spec. (?) nov., *Monde Plant.,* **27**(160), 4 (1926).

28   J. Sauer, The grain amaranthus and their relatives; a revised taxonomy and geographic survey, *Ann. Mo. Bot. Gard.,* **54,** 103 (1967).

29   T. G. Tutin, et al., *Flora Europaea,* Vol. 1, Cambridge University Press, New York, 1964.

30   S. Watson, Contributions to American Botany, no. 7; Descriptions of new species of plants, *Proc. Am. Acad.,* **12,** 273 (1877).

31   J. Sauer, and D. Davidson, Wisconsin Flora No. 45, *Amaranthus, Wis. Acad. Sci., Arts Lett.,* **50,** 76 (1961).

32   H. A. Gleason, and A. Cronquist, *Manual of Vascular Plants of Northeastern United States and Adjacent Canada,* D. Van Nostrand, Princeton, N.J., 1968, p. 810.

33   S. E. Weaver and E. L. McWilliams, The biology of Canadian weeds, 44: *Amaranthus retroflexus, A. powellii* and *A. hybridus, Can. J. Plant Sci.,* **60,** 1215 (1980).

# The Nature of Resistance to Triazine Herbicides: Case Histories of Phenology and Population Studies

P. D. PUTWAIN AND K. R. SCOTT

Department of Botany
University of Liverpool
Liverpool, United Kingdom

R. J. HOLLIDAY

Planning Department
Merseyside County Council
Liverpool, United Kingdom

## 1 THE SELECTIVE EFFECTS OF HERBICIDES ON PLANT POPULATIONS

### 1.1 Interspecific Selection by Herbicides

Evolution in weed populations often first comes to the attention of agronomists as increases in population size or density of noncontrolled species or groups of species. Weed infestations that are difficult to control with existing herbicides may simply be a response by various species to the selective effects of herbicides where such species are already resistant or moderately tolerant to herbicides in current use.

Under modern agricultural technology where herbicides are used extensively, interspecific selection by herbicides has become the most significant factor causing changes in the size of weed populations. Fryer and Chancellor[1] and Way and Chancellor[2] have comprehensively reviewed the way in which the history of herbicide development has been a continual adjustment to a changing weed flora. This interspecific selection occurs because many weed species initially possess an extremely low level of inherent tolerance to a herbicide and may take a very long time to evolve resistance (Chapter 17) or to escape the effects of a herbicide by evolving a different life history pattern or morphology. In some species such changes will perhaps never occur. Susceptible species decline in relative abundance and inherently resistant species naturally increase in abundance and often also in geographic range. The studies by Willis[3] in which repeated annual applications of a mixture of maleic hydrazide and 2,4-D to road verges have selected a characteristic grassland flora dominated by *Festuca rubra* and *Poa pratensis* is a classic demonstration of how such interspecific selection may occur. This possible interspecific selection is discussed in relation to other agronomic procedures in Chapter 4.

## 1.2  Intraspecific Selection by Herbicides

Although interspecific selection by herbicides can cause changes in a weed flora, evolutionary changes are not involved, as the characteristics of the populations of weed species do not alter. Evolutionary change may result from a variety of agronomic practices. For example, weed cleaning and crop harvesting procedures appear to have been involved in the selection of early flowering forms of *Avena fatua* in California[4] and *Arabidopsis thaliana* in the United Kingdom.[5] Occasionally, plant morphological changes have evolved in response to changes in agricultural practices. Harper[6] mentions that the introduction of the reaper in the United Kingdom selected dwarf local populations of *Aethusa cynapium* and *Torilis arvensis*. Such evolutionary changes in plant morphology might also involve a change in susceptibility to post-emergence herbicides, although there appears to be no evidence that this has occurred.

## 1.3  Seasonal Patterns of Germination and Phenology

In most plant populations there is a typical, fairly predictable seasonal rhythm of biological events.[7] Seasonal patterns of germination, timing of seedling growth and leaf formation, and flowering and fruiting are all events in the phenology of a species. Annual and perennial weed species are no exception and control measures are usually timed to hit a weed population at a vulnerable time in its phenological cycle, provided that this is convenient in relation to the management of the crop.

In temperate climates, annual species usually germinate during the spring and flower and fruit during the late summer or autumn, depending on the normal life span of the species. This period may be as short as 10 weeks for *Senecio vulgaris* or more typically, 5 to 6 months in other species.

Characteristic seasonal patterns of germination have been described for several species.[8] These germination patterns can be modified by exceptional variation in climate, but they tend to recur from one year to the next.

Other annual species are characterized by autumn germination and spring flowering and fruiting[9,10] and are therefore regarded as winter annuals. In northern temperate climates only a few weed species have the capacity to behave as winter annuals (e.g., *Poa annua, S. vulgaris*). This phenology is much more common in the Mediterranean region, the Near East, California, and other winter rainfall subtropical regions.

Perennial species also have characteristic phenologies which may be of particular significance in determining the response of perennial weeds to herbicide applications. Several authors[11-13] have stressed that evolutionary changes in the phenology of weeds could be an outcome of systematic, repeated herbicide applications, resulting in avoidance of the herbicide phytotoxicity. Certainly, at the interspecific level of selection in California, *Eremocarpus setigerus* has become common on road verges, probably because it has a summer annual phenology. This species, due to late germination of its seeds, "escapes" the winter herbicide sprays which are used to prevent weed growth on verges.[14] Similarly, certain late-emerging perennials such as *Sorghum halepense,* as well as species such as *Convolvulus arvensis*, capable of repeated emergence from an underground "bud bank," became much more common on road verges.[15]

The phenology of these species was "preadapted" to the herbicide regime: the majority of plants within each species already possessed avoidance characteristics. What is the evidence of evolution of phenological changes by selection at the intraspecific level? Cohen[16] suggested that some populations of summer annuals in Israel (e.g., *Solanum hirsutum, Sonchus oleraceus*) became winter annuals. This might have been a result of herbicide applications during the summer months or due to some other aspect of crop management. However, hard evidence of such ecological responses to herbicides has not appeared in the plant protection literature.[17]

Clearly, there are several possible types of response of weed populations to the selective efforts of herbicides. The nature of these responses, and in particular the evolution of herbicide resistance, can be fully understood only if we have information on the inheritance of resistance, that is, the frequency of resistance alleles. We also require detailed actuarial data on survivorship and reproduction of both resistant and susceptible genotypes. These data are required for all seasons of the year and in relation to particular combinations of herbicide/crop/agricultural practices. It will then be possible to analyze critically the whole process of selection for herbicide resistance. Thus there is an urgent requirement for studies of the population dynamics of resistance in weed populations. To the best of our knowledge no studies of this kind have been published. The purpose of this chapter is to outline some of our investigations of the population dynamics of *S. vulgaris* in response to a triazine herbicide, as a useful example. This case study describes a possible approach to the problem of how to gain a general understanding of evolution of herbicide resistance, in a diversity of geographic locations, with a particular emphasis on the significance of plant phenological studies.

## 2  THE INFLUENCE OF SIMAZINE ON THE POPULATION DYNAMICS OF *SENECIO VULGARIS*

Chapters 2 and 3 have shown that the evolution of triazine resistance in weeds is now of widespread occurrence in North America and Europe. Triazine herbicides have been in extensive use in the United Kingdom (e.g., in 1976, 20,000 ha were treated with simazine). Simazine has been used for similar periods of time in the United Kingdom as elsewhere in the world, but natural resistance has not yet become a problem. We have found that the potential for evolution of resistance to simazine occurred in a U.K. population of *S. vulgaris*.[18] This population contained resistance, the presence of which was exposed after one cycle of artificial selection with simazine.

As simazine is a relatively persistent herbicide, it causes a very high rate of mortality of young seedlings of susceptible species at normal agricultural rates. In theory, selection pressures for resistance must be very powerful to bring about rapid enrichment for resistance (Chapter 17). Growers in the United Kingdom usually use other herbicides as well as simazine, and this would negate the selective action of simazine to some extent. However, it is reasonable to assume that the allele(s) which confer(s) resistance must exist in at least several populations of *S. vulgaris* in the United Kingdom. Thus evolution for simazine resistance might have occurred.

As resistance has not become a major problem in the United Kingdom, we must seek explanations for the lack of evolution of resistance. Is selection for simazine resistance, in fact, much less intense than it would appear to be in theory? Are the populations escaping the effects of the herbicide? Detailed studies of the population dynamics of weeds, coupled with a study of the persistence of triazine herbicides in the soil in which the weed populations occur, should provide evidence of the intensity of selection. These studies will also provide evidence of avoidance of selection for resistance, which may occur as an ecological response by weed populations. We might expect selection for changes in patterns of germination, mortality, and timing of reproduction.

We have first[19] examined the population dynamics of *S. vulgaris* in relation to the persistence of simazine phytotoxicity in the soil. The intention was to determine whether phenological escape had occurred. This would provide an explanation for the nonevolution of triazine resistance in orchard populations which have received annual simazine applications for up to 12 consecutive years. This work involved an investigation of the seasonal periodicity of germination, plant mortality, and the timing of reproduction in relation to the time of application of simazine and its persistence in the soil. More recently we have extended the scope of the investigation to determine how the properties of the weed population, including its phenology, change following the introduction of seed of a resistant genotype into the population.

### 2.1  Experimental Sites and Methods

The investigation was made in two parts. In the first, the population dynamics of *S. vulgaris* were studied at two locations. One was a plantation of black currants where simazine had not been applied at all and mechanical weed control was carried out in the autumn. The second location was a black currant plantation where simazine had been applied annually in the spring at rates between 2.24 and 3.36 kg/ha for 10 years. In most years, paraquat was applied as spot treatments in early June prior to harvesting the crop. The persistence of simazine in the soil at the treated site was studied over an 8-month period.

At both locations, the population dynamics of *S. vulgaris* were investigated by monitoring at 14-day intervals. Using either of two methods it was possible to locate the coordinates of individuals to an accuracy of $\pm 1$mm within permanent quadrats. Photographs were taken from a fixed position with the camera located on a metal frame, or an adjustable metal mapping frame was used which incorporated a pointed plumb bob, to locate individuals exactly. The fate of every individual plant that germinated within a permanent quadrat could be followed. If a plant survived long enough, we measured its seed production.

## 2.2 Seasonal Periodicity of Germination

The seasonal pattern of germination of *S. vulgaris* and the number of individuals surviving to reproduction is shown in Fig. 6.1. The basic seasonal pattern of germination was similar in both years.

In the untreated black currants, a flush of seedlings appeared in the spring (April–May), followed by a second period of seedling emergence in July–August, although less germination occurred at this time. The pattern of germination was quite different in the simazine-treated black currants. Here, only a few seedlings appeared in spring and maximum germination occurred in late summer and autumn.

## 2.3 Germination and Development in Relation to Soil Concentrations of Simazine

The mean monthly germination is shown in Fig. 6.2. In addition, the percentage of individuals that survive to reproduce and the concentration of simazine in the soil at the treated site are given.

In the simazine-treated black currants, addition to the population of *S. vulgaris* was from seedlings that appeared when soil concentrations of simazine were below 0.5 kg/ha in the top 7.5 cm of soil. Seedlings that appeared in spring rarely survived for longer than 2 weeks and never lived more than 4 weeks. Clearly, the phytotoxicity of simazine was the major cause of mortality during the spring. In the untreated black currants the mean survival time of groups of seedlings in the same age class which appeared in spring varied between 6 and 10 weeks, although some individuals survived for as long as 26 weeks.

On both the treated and untreated sites, plants that ultimately reproduced and dispersed viable seed originally germinated during very restricted periods

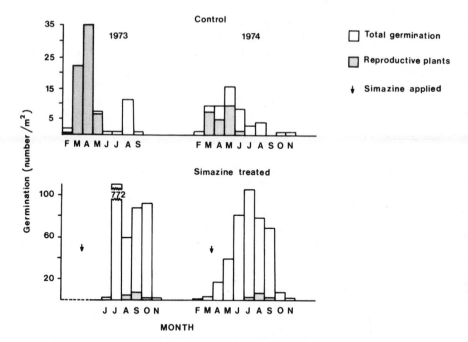

**Figure 6.1** Seasonal pattern of *Senecio vulgaris* germination and number of individuals surviving to reproduction over 2 years, on simazine-treated and control sites (Adapted from Holliday.[19])

of the growing season. In the untreated black currants, seedlings that appeared mainly during April and May grew rapidly and flowered and dispersed most seed in summer. The majority of seedlings survived to become reproductive.

Seedlings that emerged during the summer did not survive to reproduce due to intense competition from other species that had colonized the untreated area. Furthermore, most seedlings became infected by *Puccinia eagenophorae*, which reduced the competitive ability of individuals. Thus *S. vulgaris* acted as a summer annual at this site.

The development of the *S. vulgaris* population in simazine-treated black currants was quite different (Fig. 6.2). Only seedlings that emerged during August and September survived to reproduce and only a few of these ultimately dispersed viable seeds. This was because seedlings overwintered in a dormant vegetative state until the following spring. Clearly, plant mortality was high during the unfavorable cold and wet climate of autumn and winter. In the spring, growth increased rapidly despite the March application of simazine. The deeply rooted plants avoided the phytotoxic activity of simazine, which is confined to the top few centimeters of the soil.

The simazine-treated population of *S. vulgaris* switched completely to a winter annual life cycle, thus avoiding phytotoxic concentrations of simazine in the spring, when newly germinated seedlings are completely vulnerable to

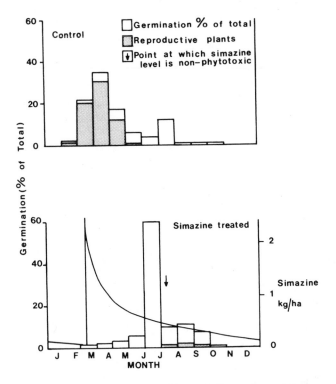

**Figure 6.2** Average monthly germination of *Senecio vulgaris* on simazine-treated and control sites. Mean monthly germination expressed as a percentage of total germination for 1 year and the concentration of simazine in the soil. Replicated soil cores were collected immediately prior to spraying, immediately after spraying and at intervals of 0.5, 1, 2, 3, 4, 6, and 8 months after spraying. The samples were assayed for simazine using two techniques: (a) a bioassay method modified from Holly and Roberts[20] using soybean as the test plant; and (b) chemical analysis using gas-liquid chromatography modified from Bowmer.[21] (Adapted from Holliday.[19])

the herbicide. The maximum age reached by plants of the treated population was 44 weeks and these plants were at least 30 weeks old before the first seeds were dispersed. Maximum dispersal occurred at 36 to 40 weeks of age. In contrast, plants in the untreated black currants dispersed seed between the ages of 4 and 24 weeks, with maximum dispersal between 10 and 12 weeks of age (see Figs. 6.3 and 6.4).

## 2.4 Selection for Late Germination

The continued survival of a population of *S. vulgaris* at the treated site appears to depend on seedlings that are able to escape the most toxic soil concentrations of simazine. There may have been selection for later germination at the

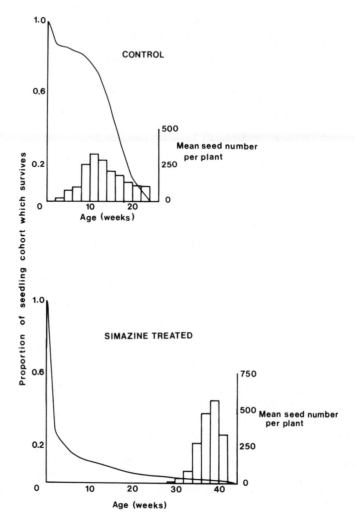

**Figure 6.3** Average survival of cohorts of seedlings of *Senecio vulgaris* that appeared at different times on simazine-treated and control sites. The period of seed dispersal is shown together with the mean seed number per plant. The data are the mean of three replicates for control and simazine-treated sites. Note that some seedling cohorts died without reproducing. (Previously unpublished data.)

treated site, thus fulfilling Harper's[11] prediction of phenological response to regular herbicide applications.

However, selection for late germination would depend on heritable variation in the physiological control of germination and dormancy in *S. vulgaris*. Genetic variation in dormancy is known in other species (e.g., *Papaver* spp.),[22] but in this case maternal influences were also involved in controlling dormancy.

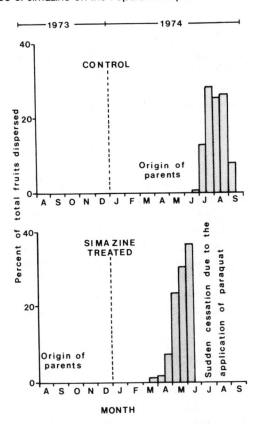

**Figure 6.4** Time of dispersal of seeds of *Senecio vulgaris* in relation to the time when seedlings first emerged. Data are given for a period of 14 months on sima-zine-treated and control sites. (Adopted from Holliday.[19])

Popay and Roberts[23] reported virtually no innate dormancy (i.e., inherited dormancy) in *S. vulgaris* and subsequent work by Holliday[19] confirmed this finding. Seed was hand sown in field plots in July and 99% either germinated or died within 8 months of sowing. If *S. vulgaris* seeds were buried to a depth of 1 cm or more, they did not germinate (i.e., there can be environmentally imposed dormancy). Popay and Roberts[23] have suggested that deep burial en-forces dormancy in *S. vulgaris* due to absence of light. Scott[24] has shown that seed buried at a depth of 15 cm for 2 months possessed 93% germination when retrieved and incubated at 20°C in the light. The remaining 7% of the seed was not viable. Since innate dormancy is absent in *S. vulgaris*, it appears unlikely that the predominantly late summer germination was the outcome of selection by simazine.

The profound change in the phenology of *S. vulgaris* to a winter annual in the simazine-treated site is probably not fixed genetically (i.e., inherited as late summer germination). We suspect that seasonal availability of suitable microsites for germination accounts for the observed patterns of germination in treated and untreated sites. In the simazine-treated black currants, the soil profile is relatively undisturbed, with some surface compaction by tractors but with no tillage. The soil surface was extensively colonized by bryophytes (e.g., *Ceratodon purpureus*), which is typical in orchards and soft fruit plantations where simazine is the main agent of weed control, and thus incorporation of *S. vulgaris* seed beneath the soil surface was substantially inhibited.

In this particular environment, the soil fauna is, without doubt, the principal agent of burial of viable seed. The effect of rain splashing, frost heaving, and so on, on the downward movement of seed will be greatly reduced. Consequently, during the summer, dispersed seeds will tend to remain on the soil surface. They will germinate rapidly if there is a sufficient supply of soil moisture; otherwise, they will be vulnerable to predators or may be dispersed by wind or killed by pathogens. As a result of exposure to hazards on the soil surface, there will be fewer seeds available to germinate in the spring (Fig. 6.1), either as survivors at the soil surface or brought up from below. Seedlings that do emerge in the spring will rapidly succumb to simazine phytotoxicity if they are not resistant. Thus the winter annual characteristics in *S. vulgaris* are a direct consequence of a hazard-free period for seedling establishment in late summer. Some plants survive both the winter and a spring application of simazine, eventually dispersing seed in the late spring of the second year.

Seeds produced in the summer tend to remain in enforced dormancy until the following spring in the untreated site. The presence of vegetation that excludes the light necessary for germination is responsible. Subsequent burial by mechanical cultivation in autumn ensures that there is a large pool of viable seed that will be brought back to the soil surface in the spring by the soil fauna.

## 3  DYNAMICS OF RESISTANCE AND PHENOLOGICAL EVENTS

### 3.1  An Experimental Approach

The phenological change in a population of *S. vulgaris* treated with simazine was not a direct result of selection for resistance by the herbicide. In Chapters 2 and 3 there is ample evidence that herbicide resistance has evolved by selection. If resistance appeared in a previously susceptible orchard population of *S. vulgaris*, would there be further phenological changes different from those already described in this chapter? If they occurred, would such phenological changes be directly attributable to the appearance of a resistance allele?

Answers to these questions have come from a study on the dynamics of a population of *S. vulgaris*. It may evenutally be possible to predict the rate of evolution of resistance from this work, to estimate the selection pressure imposed by the herbicide, and to estimate the relative fitness of resistant and susceptible gentoypes in their natural environments. Fitness is estimated by measuring the survivorship of a group of seedlings with the same age (a "cohort''), and integrating this with measurements of the viable seed produced and dispersed (see Chapters 9 and 15). We initiated a long-term study of the population dynamics of susceptible and resistant genotypes of *S. vulgaris* to provide the necessary information. The first results of our investigation are presented here.

The dynamics of mixed populations of resistant and susceptible genotypes of *S. vulgaris* in populations established in 1979 which received various herbicide treatments were studied. The herbicide treatments were (*a*) early spring simazine application, (*b*) spring application of simazine followed by paraquat spot treatment in July, (*c*) application of simazine in spring and autumn, and (*d*) control—no herbicide applied. Simazine was applied at a rate of 2.24 kg/ha.

The methods used to study the population dynamics of *S. vulgaris* were the same as described in Section 2.1. Seeds (actually achenes) were collected from all surviving plants and their progeny were tested for simazine resistance (2.24 kg/ha) in a greenhouse. As resistance to simazine is inherited maternally in *S. vulgaris*,[25] all progeny must carry the maternal genotype. Thus the relative survival of resistant and susceptible genotypes was assessed in the different herbicide regimes.

Seeds from a population of *S. vulgaris* containing an initial frequency of 2% resistant genotypes were also dispersed in newly established black currant plantations. The resistant genotypes originated from a population of *S. vulgaris* that grew on a commercial fruit farm in Cheshire, U.K. The population was subjected to two generations of artificial selection for simazine resistance, and highly resistant genotypes were obtained.[18] A flush of seedlings appeared during April and May 1980 and monitoring of survivorship and sexual reproduction of *S. vulgaris* in permanent quadrats began in June 1980.

## 3.2 The Influence of Alleles for Resistance on Population Dynamics and Phenology

The seasonal pattern of germination of *S. vulgaris* in simazine-treated and control plots is shown in Fig. 6.5. In the first year, the pattern of germination on both treatments was similar, although more seedlings emerged on the simazine-treated site. Subsequently, a buildup of other weeds on the control site caused low *S. vulgaris* germination due to a lack of light necessary for germination; therefore, in 1980 the number of seedlings in the treated site greatly exceeded the control. Solymosi[26] also noted no significant difference in the seasonal rhythm of germination in resistant and susceptible populations of *Chenopodium album* and *Amaranthus retroflexus* growing in corn.

In the first year, seeds were collected from mid-July onward (Fig. 6.6). These seeds were characterized as resistant and susceptible in greenhouse tests. The selection pressure imposed by the herbicides was considerable in the simazine-treated plots. The proportion of resistant genotypes of surviving plants in the first cohort of seedlings, which appeared before the beginning of June, was 89%, whereas it was only 0.5% in untreated plots.

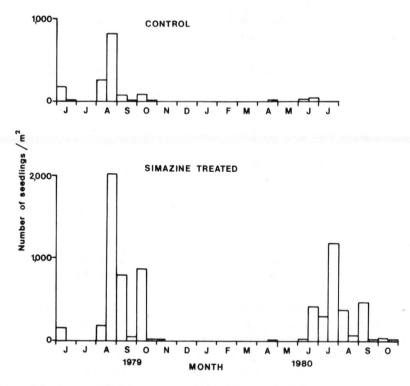

**Figure 6.5** Seasonal pattern of *S. vulgaris* germination on simazine-treated and control sites. On treated sites, after August 1979, 95% of individuals are resistant to simazine, and on control sites, 95% are susceptible. Mean of three replicates. (Previously unpublished data.)

In the second year the proportion of resistant genotypes at the treated site increased to 96.5% of the April seedling population. The equivalent seedling cohort at the control site did not produce seed; thus a comparison of genotype frequency with the treated site is not possible. Selection against susceptible genotypes to simazine-treated plots was clearly very powerful, as we would expect. Nonetheless, escape from simazine phytotoxicity was possible, as a few susceptible individuals remained. Apparently, the distribution of simazine over the permanent quadrats was insufficiently homogeneous to reach all seedlings at a toxic concentration.

It can be seen that in the untreated plots there was only one generation of *S. vulgaris* each year (Fig. 6.6). The species behaved as a summer annual just as it did in the first part of the investigation. However, in the simazine-treated plots, where the *S. vulgaris* population was predominantly composed of resistant genotypes, more than one generation per year was possible. Seedling cohorts that arose during the spring were unaffected by normally phytotoxic concentrations of simazine (Fig. 6.6). The plants dispersed seed during June and July and thus behaved as summer annuals. The seed shed during the summer

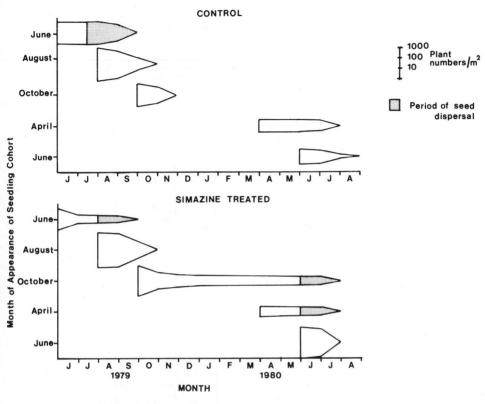

**Figure 6.6** Life span of seedling cohorts of *Senecio vulgaris* that appeared in permanent quadrats at various times during two growing seasons in simazine-treated and control sites. The number of seedlings in each cohort was monitored at 2-week intervals until the last one died. The period of seed dispersal was similarly monitored. (Previously unpublished data.)

germinated in September and October and the seedlings overwintered in the same way as in the simazine-treated commercial black currant plantations in the first part of the investigation. The overwintering plants dispersed seed late in the following spring or early summer. Thus there were two partly overlapping generations per year, one behaving as a summer annual, the other as a winter annual.

The scenario we have described implies that some susceptible individuals survive due to temporal or spatial escape from the herbicide where repeated annual spring application of simazine is the normal practice.

We propose a simple model (Fig. 6.7) in which an equilibrium of resistant and susceptible genotypes is maintained by two niches with temporal separation. Niche 1 occurs in the spring, with high soil concentrations of simazine, where the relative fitness of susceptible genotypes is minuscule. Niche 2 occurs in late summer, autumn, and winter, when simazine phytotoxicity is greatly

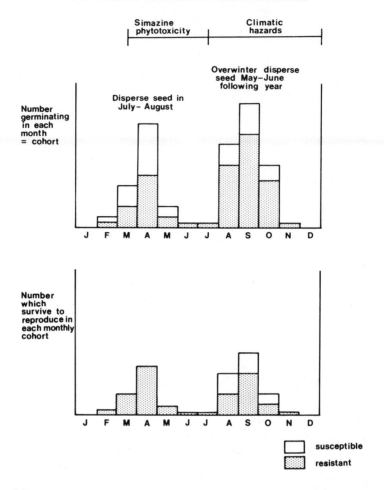

**Figure 6.7**  Two-niche model of the dynamics of simazine-resistant and sima-zine-susceptible genotypes of *Senecio vulgaris* in orchards receiving an annual application of simazine in the spring, without the use of other herbicides.

reduced and climatic hazards are an important source of plant mortality. At this time the population consists of considerably greater numbers of resistant genotypes. Still, the relative fitness of susceptible genotypes is greater[27] and, therefore, they produce relatively greater seed output per unit ground area in late spring. Although the relative frequency of susceptible genotypes may be very low, they are never completely eliminated from the *S. vulgaris* population.

## 4  CONCLUSIONS

The evidence presented in this chapter on the population dynamics of herbicide resistance may be considered to be enigmatic.

Although we have shown experimentally that changes in the phenology of *S. vulgaris* populations may occur, this evidence alone has not provided an adequate explanation of why triazine-resistant populations of this or other annual weed species has not yet occurred in the United Kingdom. One possible answer seems to be that the weed control measures used by almost all fruit growers have utilized a variety of chemicals, in addition to simazine, to maintain a weed-free crop environment. Paraquat, glyphosate, and dichlobenil have been used extensively to eliminate localized infestations of both perennial and annual weeds. Is it unrealistic to suppose that simazine resistance could have occurred in some locations, but that the resistant population (unrecognized as such) was eradicated by another herbicide before it was identified as a case of resistance?

Certain characteristics that are important in determining the life history and phenology of a species may be highly heritable (e.g., time of flowering).[28,29] Evolutionary changes in phenology should be possible. For example, Law et al.[30] showed that populations of *Poa annua* have evolved quite diverse strains which differ substantially in the length of time from germination to flowering and seed dispersal, as well as in the average life span of a plant. Thus the plant phenology of many weed species might be expected to have an evolutionary response to changes in the physical and chemical environment that are imposed by the regular application of herbicides. It is surprising that there are not more reports of such changes.

The phenological changes we have observed in populations of *S. vulgaris* do not provide evidence of evolutionary changes. Nevertheless, our study of the dynamics of *S. vulgaris* populations, although relatively simple, does illustrate one type of approach that is necessary to elucidate the process of evolution of herbicide resistance. It is unfortunate that a relatively rare opportunity to study evolution in action is being passed by almost unnoticed, where resistance to triazine herbicides has recently become a problem, except by a small handful of population biologists.[31,32] Evidence of the factors determining the occurrence and geographical distribution of genes for resistance in natural populations of weed species is very limited.[32] We have no clear understanding of how many foci of resistance there have been in any particular species for a given geographic area. There is no good evidence as to how resistant populations have spread, and there is little evidence as to how selection is really acting.

We shall not be able to accurately predict rates of evolution of herbicide resistance or the behavior of resistant populations until detailed investigations are made on the life histories and dynamics of herbicide-treated populations. It will be necessary to determine the ecological fitness of resistant and susceptible genotypes in the crop environment, the comparative longevity and dynamics of seeds of these genotypes in the soil, and the impact on them of other environmental factors, both natural and human-made.

We have only discussed resistance in *Senecio* in this chapter. *S. vulgaris* has also apparently been increasing in tolerance to triazines in the United Kingdom which is a separate problem and is discussed elsewhere.[33]

## 5. EPILOGUE

Since Chapter 6 was written, triazine-resistant populations of *Senecio vulgaris* have appeared in orchards and a plant nursery in the United Kingdom. Only three resistant populations are known. Cooperative work between Ministry of Agriculture advisors and us confirmed that these first resistant populations are almost unaffected by 5 kg/ha of atrazine and simazine.

At the present time, herbicide resistance in the United Kingdom is known only in *S. vulgaris,* and triazine-resistant populations of this species occur in only two localities (Kent and Staffordshire) more than 120 miles apart. This confirmation of resistance in 1981 follows 14 years of general use of triazine herbicides in these areas.

The main conclusions of Chapter 6 are unaffected by these new developments; we still have to explain the delayed appearance of resistance biotypes in the United Kingdom in comparison with North America and continental Europe.

## REFERENCES

1   J. D. Fryer and R. J. Chancellor, Herbicides and our changing arable weeds, in *The Flora of a Changing Britain,* Symp. Botanical Society of the British Isles, 1969, F. Perring, (Ed.), 1970, p. 105.

2   J. M. Way and R. J. Chancellor, Herbicides and higher plant ecology, in *Herbicides, Physiology, Biochemistry, Ecology,* Vol. 2, L. J. Audus (Ed.), Academic Press, London, 1976.

3   A. J. Willis, Road-verges-experiments on the chemical control of grass and weeds, in *Road Verges, Their Function and Management,* J. M. Way (Ed.), Nature Conservancy, London, 1969.

4   A. G. Imam and R. W. Allard, Population studies in predominantly self pollinating species: VI. Genetic variability between and within natural populations of wild oats from differing habitats in California, *Genetics,* **53**, 633 (1965).

5   M. E. Jones, The population genetics of *Arabidopsis thaliana:* II. Population structure, *Heredity,* **27**, 51 (1971).

6   J. L. Harper, Ecological aspects of weed control, *Outlook Agric.,* **1**, 197 (1957).

7   J. L. Harper, *Population Biology of Plants,* Academic Press, London, 1977, p. 892.

8   H. A. Roberts and P. M. Feast, Seasonal distribution of emergence in some annual weeds, *Exp. Hort.,* **21**, 36 (1970).

9   M. A. Pemadasa and P. H. Lovell, Factors controlling the flowering time of some dune annuals, *J. Ecol.,* **62**, 869 (1974).

10   D. Ratcliffe, Adaption to habitat in a group of annual plants, *J. Ecol.,* **49**, 187 (1961).

11   J. L. Harper, The evolution of weeds in relation to resistance to herbicides, *Proc. 3rd Br. Weed Control Conf.,* 179 (1956).

12   J. L. Hammerton, Past and future changes in weed species and weed floras, *Proc. 9th Br. Weed Control Conf.,* 1136 (1968).

13   L. J. King, Weed ecotypes—a review, *Proc. 20th Northeast Weed Control Conf.,* 604 (1966).

14   H. G. Baker, The evolution of weeds, *Annu. Rev. Ecol. Syst.,* **5**, 1 (1974).

15   H. G. Baker, Migrations of weeds, in *Taxonomy, Phytogeography and Evolution,* D. H. Valentine, (Ed.), Academic Press, London, 1972, p. 327.

16  A. Cohen, The influence of chemical weed control on the weed flora in beet fields in Israel, *2nd Int. Meet. Selective Weed Control Beetcrops,* 235 (1970).

17  R. J. Holliday, P. D. Putwain, and A. Dafni, The evolution of herbicide resistance in weeds and its implications for the farmer, *Proc. 1976 Br. Crop Prot. Conf. Weeds,* 937 (1976).

18  R. J. Holliday and P. D. Putwain, Evolution of resistance to simazine in *Senecio vulgaris* L., *Weed Res.,* **17**, 291 (1977).

19  R. J. Holliday, The potential for evolution of herbicide resistance in annual weeds, Ph.D. thesis, University of Liverpool, 1978.

20  K. Holly and H. A. Roberts, Persistence of phytotoxic residues of triazine herbicides in soil, *Weed Res.,* **3**, 1 (1963).

21  K. H. Bowmer, Measurement of residues of diuron and simazine in an orchard soil, *Aust. J. Exp. Agric. Anim. Husb.,* **12**, 535 (1972).

22  J. L. Harper and I. H. McNaughton, The inheritance of dormancy in inter- and intraspecific hybrids of *Papaver, Heredity,* **15**, 315 (1960).

23  A. I. Popay and E. H. Roberts, Ecology of *Capsella bursa-pastoris* (L.) Medik. and *Senecio vulgaris* L. in relation to germination behavior, *J. Ecol.,* **58**, 123 (1970).

24  K. R. Scott, Botany Department, University of Liverpool, U.K., unpublished results, 1980.

25  K. R. Scott and P. D. Putwain, Maternal inheritance of simazine resistance in a population of *Senecio vulgaris, Weed Res.,* **21**, 137 (1981).

26  P. Solymosi, Research Institute for Plant Protection, Budapest, Hungary, personal communication, 1980.

27  S. G. Conard and S. R. Radosevich, Ecological fitness of *Senecio vulgaris* and *Amaranthus retroflexus* biotypes susceptible or resistant to atrazine, *J. Appl. Ecol.,* **16**, 171 (1979).

28  J. P. Cooper, Studies on growth and development in *Lolium*: IV. Genetic control of heading responses in local populations, *J. Ecol.,* **42**, 521 (1954).

29  J. P. Cooper, Selection and population structure in *Lolium*: II. Genetic control of date of ear emergence, *Heredity,* **14**, 445 (1959).

30  R. Law, A. D. Bradshaw, and P. D. Putwain, Life-history variation in *Poa annua, Evolution,* **31**, 233 (1977).

31  S. I. Warwick and P. B. Marriage, Geographical variation in populations of *Chenopodium album* L. resistant and susceptible to atrazine: 1. Between—and within—population variation in growth and response to atrazine, *Can. J. Bot.* (1982, in press).

32  J. Gasquez and J. P. Compoint, Isoenzymatic variations in populations of *Chenopodium album* L. resistant and susceptible to triazines, *Agro-ecosystems,* **7**, 1 (1981).

33  R. J. Holliday and P. D. Putwain, Evolution of herbicide resistance in *Senecio vulgaris;* variation in susceptibility to simazine between and within populations, *J. Appl. Ecol.,* **17**, 779 (1980).

# Methods of Testing for Herbicide Resistance

## B. TRUELOVE

Agricultural Experiment Station
Department of Botany, Plant Pathology, and Microbiology
Auburn University,
Auburn, Alabama

## J. R. HENSLEY

CIBA-GEIGY Corporation
Greensboro, North Carolina

## 1  INTRODUCTION

The tolerance of some plant species and the susceptibility of others to the same herbicide applied at the same rate has long been recognized. Such differential responses, which may also be seen between biotypes of the same species, can usually be explained in terms of morphological or physiological differences affecting the concentration of herbicide, or phytotoxic herbicide metabolite, reaching the primary site of action (Chapter 8). Recently, however, we have come to recognize a completely new form of herbicide tolerance which has appeared in species that were previously assumed to be completely susceptible. In these plants, resistance is not due to limited absorption and translocation, or to accelerated metabolism of the herbicide, but is the result of a biochemical change at the site of metabolic activity (Chapter 10). So far, this form of resistance has been biochemically characterized in just a few weed biotypes resistant to the s-triazines. With the continued extensive use of herbicides, it seems possible, and perhaps even likely, that resistance to some or all members of certain other herbicide families will arise (Chapter 17).

Developing economically effective strategies to deal with herbicide-resistant weeds will depend upon the ability to recognize the appearance of new, resistant biotypes before they have become established as a major component of the local weed flora.

A variety of laboratory, greenhouse, and field techniques have been devised for recognizing herbicide-tolerant plants. Many of these methods involve challenging plants, or some plant tissue or organ, with a toxic level of the herbicide and recording the effects on plant growth and survival, or some metabolic activity directly related to the primary site of herbicide action. Valuable as such techniques are in detecting herbicide tolerance, they rarely provide definitive information about the causative nature of resistance. Following a preliminary selection of herbicide-tolerant plants through such screening procedures, confirmation of new, resistant biotypes relies on using more sophisticated procedures which look for direct effects of the herbicide on the primary metabolic process known to be affected in susceptible plants.

Because members of the different herbicide families may affect different processes, it follows that selection of the procedures to be used will depend on the mode of action of the herbicide. This chapter is concerned primarily with those methods that might be used to identify s-triazine-resistant biotypes, but the same procedures could be used in looking for resistance to herbicides of other families which also affect the same site in photosystem II. VanAssche[1] and Pfister and Arntzen[2] have presented data suggesting a common photosystem II target area for members of the following herbicide families: N-acylanilides, N-phenylcarbamates, N-phenylureas, pyridazones, s-triazines, s-triazinones, uracils, cyclized ureas, and substituted thiadiazoles. Whether the target area is in all cases a single protein, or several proteins within photosystem II, has not been determined.

Triazine-resistant biotypes were first observed, and have subsequently only been identified, in populations from particular locations in which the number of consecutive applications of an s-triazine herbicide have provided sufficient selective pressure to elevate a resistant biotype from a very low percentage ($<< 0.1\%$) to a major proportion of the population (Chapters 2 and 3). If one is to identify similar low-frequency biotypes, primary emphasis should be placed on studies based on weed populations from areas where the herbicide in question (or members of a herbicide family) has been applied at normally toxic rates over a lengthy period of time and involving several successive generations of plants.

## 2  METHODS OF IDENTIFYING HERBICIDE-RESISTANT BIOTYPES

### 2.1  Field and Greenhouse Testing with Intact Plants

Ryan[3] was the first scientist to publish data on a triazine-resistant biotype. He noted that in a nursery treated over a period of 11 years with triazine herbicides, populations of *Senecio vulgaris* were present that were no longer being controlled by triazines at the recommended rates of application. Seeds of *S. vulgaris* collected from the apparently resistant plants in the nursery, and seeds of *S. vulgaris* collected at a location where triazines had not been in

Table 7.1  Preemergence Control of *Senecio vulgaris* from Two Locations with Atrazine and Simazine

| Herbicide | Rate (kg/ha) | Continuously[a] Treated (% kill[b]) | Control[a] (% kill[b]) |
|-----------|--------------|------------------------------------|------------------------|
| Atrazine  | 0.28  | NA[c] | 46  |
|           | 0.56  | NA    | 69  |
|           | 2.24  | 0     | 100 |
|           | 8.96  | 0     | NA  |
|           | 17.92 | 0     | NA  |
| Simazine  | 0.28  | NA    | 62  |
|           | 0.56  | NA    | 46  |
|           | 1.12  | NA    | 100 |
|           | 2.24  | 0     | 100 |
|           | 8.96  | 0     | NA  |
|           | 17.92 | 12    | NA  |

*Source*: Data of Ryan.[3]

*Note*: Seedlings were grown in the greenhouse in pots containing a peat/sand mixture.

[a]Continuously treated seed came from areas continuously receiving atrazine or simazine since 1958. Control seed came from areas not previously treated with triazine herbicides.

[b]Based on number of surviving seedlings related to number in check pots 6 to 8 weeks after planting.

[c]NA, herbicides not assayed at these rates.

continuous use, were treated preemergence in the greenhouse with simazine and atrazine (Table 7.1). Survival of seedlings from seed collected in the nursery was essentially unaffected by either atrazine or simazine at a rate of 18 kg/ha, whereas seedlings from seed collected at the control location were killed at a herbicide rate of 2.2 kg/ha. Similarly, postemergence applications of atrazine killed susceptible (control) seedlings but had no apparent effect on seedlings grown from the nursery populations.

The simple protocol used by Ryan has, with minor modifications, been used by many scientists interested in identifying herbicide-resistant plants.[4-8] One very serious objection to experiments conducted in this manner is that it is difficult to demonstrate that the applied herbicide is evenly distributed and equally available within the pots of soil, and that it is being absorbed to the same extent by the roots of the suspected resistant and susceptible biotypes. An experimental design modification to overcome this objection is to germinate the seeds on absorbent paper and then transfer the roots of the young seedlings to aerated nutrient solutions containing $^{14}$C-labeled herbicide.[9-12] The relative uptake of herbicide from these solutions by the different biotypes, and the fate of the labeled herbicide within the plants, can be determined in such a system and it can be established whether any resistance recorded is due to physiological differences between plants.

In such experiments it is important to know the uptake and the distribution of the radiolabel within the plants as well as the nature of any metabolites formed, as any of these may account for the relative susceptibility or tolerance of the plants (Chapter 8). For example, it has been shown in soybean cultivars differentially sensitive to metribuzin that although similar amounts of the herbicide are taken up by the plants, in the tolerant cultivar 'Bragg', the herbicide is metabolized more rapidly, and less of the $^{14}C$ label and parent herbicide moves into the leaves, than in the susceptible cultivars 'Semmes' and 'Coker'.[13] This balance-sheet approach has been used as supportive evidence to suggest that the relative tolerance of *Digitaria sanguinalis* to simazine and atrazine is due to differential herbicide detoxification,[7] whereas the resistance to triazines observed in certain biotypes of *S. vulgaris* is due to a difference in the biochemistry of photosynthesis.[10,12,14]

Apart from their not being capable of pinpointing the reason for resistance, another serious disadvantage of the field and greenhouse techniques described is the duration of the experiments. However, because field and greenhouse screening experiments can be conducted concurrently on large numbers of plants using relatively unskilled labor, they will always be of importance in identifying and selecting herbicide-tolerant plants. These field and greenhouse studies are necessary to corroborate the findings of laboratory experiments.

## 2.2  Laboratory Techniques Using Plant Tissues and Organs

### 2.2.1  Chlorophyll fluorescence induction in leaves

Light energy absorbed by chlorophyll is the driving force of photosynthesis. When electron transport on the reducing side of photosystem II is inhibited, chlorophyll-absorbed radiant energy is reemitted as fluorescence. This fluorescence can be detected at the leaf surface. Recording leaf fluorescence levels following the treatment of plants with photosystem II-inhibiting herbicides can be used as a means for recognizing herbicide-tolerant and herbicide-susceptible biotypes.

Workers in France treated young leaves excised in the early morning from triazine-susceptible and triazine-resistant biotypes of several species with aqueous solutions of s-triazines and other herbicides in the dark. The lower surfaces of the leaves were then irradiated with short-wavelength light and the emitted fluorescence was recorded.[4-6,15] The results of one such experiment, in which the fluorescence of atrazine-treated leaves of susceptible and resistant biotypes of *Chenopodium album* was measured, are shown in Fig. 7.1A. Leaves of the resistant biotype showed only a low level of fluorescence, indicating no inhibition of photosystem II, whereas those of the susceptible biotype fluoresced strongly, indicating inhibition of electron transport. Even when atrazine-treated leaves of the susceptible biotype were transferred to water for 15 hr in the dark, they still showed the high level of fluorescence

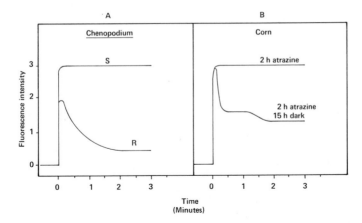

**Figure 7.1** Fluorescence of leaves of the triazine-resistant (R) and triazine-susceptible (S) biotypes of *Chenopodium album* (Fig. A) and of corn (Fig. B). Leaves of both species were soaked for 2 hr in a 30-mg/l solution of atrazine. Fluorescence was measured immediately after removal from the solution. Corn leaf fluorescence was also measured after a further 15 hr dark period in which the leaves were maintained in water. (Redrawn from Ducruet and Gasquez.[15])

associated with inhibition of photosystem II when irradiated with short-wavelength light. In a more recent modification of their technique, they use fiber optics to irradiate from above and a similar fiber to measure the reflected fluorescence.

When atrazine-tolerant corn leaves were examined by this technique,[15] an initially high level of fluorescence was recorded, but after these treated leaves were allowed to stand in water for 15 hr in the dark, only a low level of fluorescence resulted following stimulation with short-wavelength light (Fig. 7.1B). This reduced fluorescence response with time in darkness is thought to reflect the detoxification of atrazine in corn leaves, leading to a reversal of photosystem II inhibition (Chapter 8).

Ahrens et al.[16] have reported a rapid fluorescence assay for detecting triazine resistance using excised leaf disks and a commercially available fluorometer. Sections are excised from young leaves and floated on deionized water in the light for 20 to 60 min. Blotted leaf sections are then placed, adaxial surface upward, on a black cloth and fluorescence is measured. The disks are then placed, adaxial surface down, in solutions containing a low concentration of a surfactant and herbicide. The disks are maintained in the light and removed at periodic intervals for further fluorescence measurements. Fluorescence induction curves[16] of leaf disks from susceptible and resistant biotypes of *Amaranthus hybridus* treated with 50 μM solutions of either atrazine or diuron for up to 5 hr are shown in Fig. 7.2. Leaf disks of resistant and susceptible biotypes floated on water and surfactant only, and leaf disks of the resistant biotype floated on the atrazine solution, showed essentially the same low level of fluorescence, indicating no inhibition of photosystem II. Susceptible

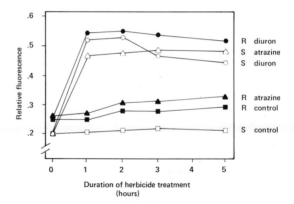

**Figure 7.2**  Fluorescence of leaf disks of resistant (R) and susceptible (S) biotypes of *Amaranthus hybridus* treated for various periods of time with either 50 μ M atrazine or diuron. Leaf fluorescence was measured with a plant productivity fluorometer. (Redrawn from Ahrens et al.[16])

leaf disks floated on the atrazine solution, and disks of both biotypes floated on the diuron solution, showed a high level of fluorescence, indicating inhibition of photosystem II electron transport. The high fluorescence recorded for both biotypes in the presence of diuron is to be expected because it has been shown that triazine-resistant biotypes are usually still susceptible to diuron and other substituted urea herbicides (Chapter 10).

Ahrens et al.[16] used this fluorometer technique to look for atrazine resistance in sorghum varieties, soybean strains, and wheat lines that had been shown in field and greenhouse tests to have wide variations in their tolerance to atrazine. Their conclusions were that in none of these species could the recorded triazine tolerance be ascribed to a failure of atrazine to inhibit photosystem II electron transport.

Ali and Souza Machado[17] have used a similar leaf chlorophyll fluorescence method to monitor the transfer of triazine resistance from *Brassica campestris,* wild turnip rape, to *B. campestris,* cultivated rapeseed, and to *B. napus,* rutabaga. They used a conventional backcross breeding method for gene transfer and monitored triazine resistance in the backcross progeny with the leaf chlorophyll fluorescence method.

Although fluorescence measurements of whole leaves or leaf disks constitute a rapid and convenient procedure for examining plants for herbicide resistance, interpretation of the data may be complex. As Schreiber et al.[18] have pointed out: "Fluorescence induction curves obtained under identical conditions can differ appreciably, even if taken from the same plant or the same leaf, depending for instance on the developmental stage of the leaf, local light intensity, and in particular, the side of the leaf." Ahrens et al.[16] showed that exposure of susceptible leaf sections to atrazine for periods longer than 3 hr decreased the steady-state fluorescence level, probably indicating the onset of secondary damage to the chloroplasts. They consider it essential to conduct a time-course study to determine the time of onset of such secondary effects under any given set of experimental conditions, and then establish an assay time in which only effects due to the primary inhibition of photosynthesis will be measured.

### 2.2.2 Detection of herbicide resistance by the sinking leaf-disk technique

When disks excised from the leaves of a number of species are floated on a phosphate buffer containing surfactant, they continue to float as long as photosynthesis is operating. If they are floated in the dark, or if photosynthesis-inhibiting herbicides are added to the medium, they rapidly lose buoyancy and sink. This phenomenon was discovered by Truelove et al.[19] and used by them as a means of identifying and assaying those herbicides in which the primary mode of action is photosynthesis inhibition (Table 7.2). Flotation appears to depend on maintaining a high ratio of $O_2$ to $CO_2$ in the intercellular air spaces of the leaf disks. Because air-space volume varies in leaves of different plants, not all species have proved usable in this procedure. However, if suitable tissues such as pumpkin[19] or watermelon[20] cotyledons are used, the assay is simple, rapid, and sensitive.

This procedure was used by Gawronski et al.[21] to detect differences in metribuzin tolerance among potato varieties. Data obtained using disks from both field- and greenhouse-grown varieties correlated well with evaluations of the relative tolerance to metribuzin of the varieties in field trials. This technique was not successful, however, when it was used in an attempt to predict cotton varietal sensitivity to a number of known photosynthesis-inhibiting herbicides.[22]

Hensley[23] has recently described an interesting modification of the sinking-disk technique which he has used to distinguish between triazine-resistant and triazine-susceptible biotypes of *Chenopodium album* and *Amaranthus hybridus*. Leaf disks were transferred to tubes containing a solution of the herbicide and placed under vacuum. The leaf disks were rapidly infiltrated by the solution and sank to the bottom of the solution. The vacuum was then released, a bicarbonate solution added, and the tubes were placed in the light. When photosynthesis was not inhibited by the herbicide, photosynthetically generated oxygen within the tissue restored buoyancy and the disks floated to the surface. Atrazine-sensitive leaf disks, and disks of either biotype in the presence of nontriazine, photosynthesis inhibitors, remained at the bottom of the tubes (Table 7.3). Corn leaf disks in this test behaved like triazine-sensitive tissue because the time course of the experiment was too short for the detoxification system of corn to degrade the atrazine.

If a small battery- or hand-operated vacuum pump is used, this technique can be used in the field with leaf disks from plants suspected of being triazine resistant. Because the degradation of triazines to nonphytotoxic metabolites within leaf tissue is a relatively slow process, this procedure will distinguish

Table 7.2  Sinking of Pumpkin Cotyledon Disks
Floated on Herbicide Solutions in the Light[a]

| Chemical Treatment[b] | Cotyledon Disks Sunk[c] (%) |
|---|---|
| *Control* | |
| None (dark) | 87 |
| None (light) | 1 |
| | |
| *Photosynthesis inhibitors* | |
| Atrazine | 95 |
| Prometryn | 96 |
| Fluometuron | 91 |
| Diuron | 95 |
| | |
| *Not photosynthesis inhibitors* | |
| Hydroxyatrazine | 1 |
| DSMA | 3 |
| Potassium azide | 7 |
| 2,4-D | 3 |

*Source:* Data of Truelove et al.[19]

[a]Leaf disks were floated on a medium containing 0.01 M phosphate buffer (pH 6.5) $100 \mu g/ml$ penicillin, $40 \mu g/ml$ chloramphenicol, and 0.01% surfactant. The leaf disks were illuminated with a light intensity of 30 kilolux and maintained at approximately 28°C.

[b]All chemicals were used at a concentration of 0.1 mM.

[c]Data are means of three replicates each of 50 half-disks in one experiment. The data were collected 8 hr after the test was initiated.

between resistant biotypes in which triazines do not inhibit photosystem II, and those species (such as corn) in which tolerance is due to the ability of the plant to detoxify absorbed herbicide.

This procedure was used successfully to confirm the occurrence of triazine resistance in populations of *Kochia scoparia* that had been treated over a period of 13 years with *s*-triazine herbicides at a rate of 13 kg/ha.[24]

## 2.2.3  Measurement of the rate of photosynthesis

Many workers have studied the effects of photosynthesis-inhibiting herbicides on whole plants and leaf sections by direct measurements of the changes they induce in the rate of photosynthesis. Carbon dioxide assimilation by plants can be monitored by following the rate of $CO_2$ uptake with an infrared gas analyzer or by measuring the incorporation of $^{14}CO_2$, and $O_2$ evolution can

Table 7.3  Effects of Atrazine and Fluometuron on the Buoyancy of Leaf Disks of Various Species[a]

| | Leaf Disks Floating after 1-hr Exposure (%)[b] | | | | | | | |
| | Chenopodium album | | Amaranthus hybridus | | | | | |
| Herbicide | Resistant | Susceptible | Resistant | Susceptible | Corn | Cotton | Soybean | Tomato |
|---|---|---|---|---|---|---|---|---|
| Control | 100 | 95 | 100 | 100 | 95 | 100 | 94 | 91 |
| Atrazine, 0.1 mM | 98 | 0 | 95 | 0 | 4 | 0 | 0 | 4 |
| Fluometuron, 0.1 mM | 0 | 0 | 4 | 0 | 0 | 0 | 0 | 0 |

Source: Data of Hensley.[23]

[a]Leaf disks were floated on a medium containing 0.1 M phosphate buffer (pH 6.5), 0.1% surfactant, and 1 drop of antifoam agent. After vacuum infiltration, the leaf disks sank to the bottom of the tubes, 2000 ppm sodium bicarbonate was added, and the disks were illuminated with a fluorescent light source at an intensity of 3 kilolux.

[b]Data are the means of four replications with 20 leaf disks per replication.

be determined manometrically or polarographically. Many standard procedures[25-27] for measuring photosynthesis have been described and will not be repeated here.

For technical reasons, the least used of these procedures is the polarographic measurement of $O_2$ evolution. However, a recent publication[28] describes methods for isolating mesophyll cells from plants and measuring their $O_2$ evolution with an oxygen polarograph equipped with a Clark-type oxygen electrode. In these experiments, photosystem II-mediated $O_2$ evolution was measured in a catalase-containing buffered medium using ferricyanide as the electron acceptor and water as electron donor. Using this technique, $I_{50}$ values for monuron were determined for two types of mesophyll cells obtained from four species. This method may prove to be appropriate for measuring herbicide resistance.

Isolated leaf cells can also be used effectively in $^{14}CO_2$ assimilation studies. Working with mesophyll cells isolated from leaves of susceptible and resistant biotypes of *S. vulgaris,* Radosevich and DeVilliers[14] were able to show that $^{14}CO_2$ assimilation in cells from susceptible plants was reduced by atrazine, but assimilation by cells from resistant plants was unaffected (Table 7.4).

Any procedure measuring the rate of photosynthesis with plant tissues or organs could be used to recognize triazine-resistant plants, but the nature of resistance would not be defined. The major disadvantage of such procedures are: (a) experiments can be complex and lengthy, (b) usually only a few samples can be examined simultaneously, and (c) many of the methods require considerable technical skill and expensive equipment.

### 2.2.4  Measurement of nitrite reductase activity in leaf disks

Klepper[29,30] found lowered activity of the enzyme nitrite reductase and, hence, an increased nitrite concentration in green plant tissue treated with herbicides that inhibit photosynthesis. Nitrite reductase is localized within chloroplasts and has a requirement for the reduced form of ferredoxin. The reduction of

Table 7.4  Effect of Atrazine on Net $CO_2$ Fixation by Isolated Leaf Mesophyll Cells of Susceptible and Resistant Biotypes of *Senecio vulgaris*[a]

|  | Atrazine Concentration ($\mu$M) | | |
| --- | --- | --- | --- |
|  | 0.1 | 1.0 | 10.0 |
| Biotype | ($^{14}CO_2$ fixed as % of controls)[b] | | |
| Susceptible | 83 | 10 | 11 |
| Resistant | 91 | 96 | 108 |

*Source*: Condensed from data of Radosevich and DeVilliers.[14]

[a]Mesophyll leaf cells were isolated from greenhouse grown plants. $^{14}C$-labeled $CO_2$ fixed/mg chlorophyll was measured and related to the untreated control. Cells were allowed to fix $^{14}CO_2$ for 15 min.

[b]$LSD_{0.01} = 36$.

ferredoxin occurs in photosystem I following the transference of electrons between the structural complexes of photosystem II and photosystem I (Chapter 10). Hence inhibition of electron carrier sites in photosystem II decreases the production of the reduced ferredoxin required for reduction of nitrite.

Finke et al.[31] showed that nitrite reductase assays with leaf disk tissue could serve as a means of distinguishing between triazine-resistant and triazine-susceptible biotypes (Table 7.5). Atrazine almost completely inhibited the reduction of nitrite in susceptible *Amaranthus* sp. but had no effect on enzyme activity in the resistant biotype. As anticipated, diuron inhibited the enzyme in both biotypes.

## 2.3 Techniques Using Isolated Chloroplasts

The most discriminating techniques available for studying triazine-resistant weeds involve the use of isolated chloroplasts. Research with chloroplasts[32,33] has demonstrated the resistance to be due to a modification in the photosystem II constituents of the thylakoid membranes, resulting in altered electron transport reactions on the reducing side of photosystem II (see Chapter 10). Moreland[34] has detailed procedures for isolating chloroplasts and using the preparations for studying light-induced electron transport and photophosphorylation. Other applicable procedures are described in the literature cited here.

### 2.3.1 Hill reaction measurements

Electron transport in photosystem II (Hill reaction) can be monitored by measuring the rate of reduction of an electron acceptor, such as DCPIP (2,6-

Table 7.5 Effect of Atrazine and Diuron on the Nitrite Reductase Activity of Triazine-Susceptible and Triazine-Resistant Leaf Tissue of *Amaranthus* sp.[a]

| Chemical | Inhibition of Nitrite Reduction (%) [b] | |
|---|---|---|
| | Susceptible Biotype | Resistant Biotype |
| None | 0 | 0 |
| Atrazine | 92 | 0 |
| Diuron | 118 | 79 |

*Source*: Data of Finke et al.[31]

[a]Leaf tissue of each biotype was vacuum infiltrated with a potassium nitrate medium and 10 ppm of the herbicide. The tissue was illuminated with $230\,\mu$ einsteins/m$^2$/sec of light in the range 400 to 700 nm. Samples of the medium were taken at 30 and 90 min and assayed for nitrite. Controls consisted of the same system without the herbicide added.

[b]Nitrite accumulation in the light compared to the corresponding dark treatment (100% inhibition) was used to calculate the percent inhibition of nitrite reduction due to the herbicide.

dichlorophenolindophenol) or ferricyanide, by chloroplasts in the light.[14,32-38] Because triazine tolerance in corn is due to degradation of the herbicide, rather than an inherent insensitivity of the chloroplasts to triazines, the Hill reaction of isolated corn chloroplasts is inhibited by triazines.[35] The Hill reaction, however, of chloroplasts isolated from most triazine-resistant weed biotypes is not inhibited by triazines at relatively high concentrations[9,14,32,33,36-38] (Fig. 9.4). Measurement of triazine effects on Hill reaction activity of chloroplasts will, therefore, distinguish between resistance due to innate differences in chloroplasts and resistance resulting from cytoplasmic detoxification mechanisms.

The Hill reaction is confined to the internal membrane system of chloroplasts (chloroplast lamellae or thylakoids) and does not require the chloroplast stroma enzyme systems. These experiments can, therefore, be carried out using stroma-free thylakoid membrane preparations.[32,39] The major advantage of using such preparations is that they obviate problems that could arise from failure of herbicides to penetrate freely the two phospholipid membranes that bound the intact chloroplasts.[32]

## 2.3.2  Induction of fluorescence

In Section 2.2, procedures were described for detecting the fluorescence of leaf tissue as a means of monitoring inhibition of photosystem II. Increased fluorescence reflects the reduction of photosystem II electron acceptors in the chloroplast thylakoid membranes. These fluorescence transients can be measured precisely using chloroplast thylakoid preparations. It is believed that the primary target of triazine herbicides is the second electron carrier (component B) on the reducing side of photosystem II (Chapter 10). Procedures such as the measurement of fluorescence, which define the state of reduction of the electron carriers are, therefore, the most sophisticated of the techniques available for studying $s$-triazine inhibition.

In fluorescence induction studies, dark-adapted, chloroplast thylakoid preparations are flash-illuminated with blue light. Fluorescence emission is detected by a photodiode. The transient signals are detected and stored and subsequently transcribed on an x-y recorder.[32] The rate of rise of fluorescence from the low, initial level ($F_0$) through an intermediary level ($F_i$) to the maximal level ($F_M$) indicates the rate of attainment of the state where all the primary electron acceptors of photosystem II are fully reduced. The changes in rates, and the significance of these different transient states, are considered in detail elsewhere (Chapter 10 and refs. 32 and 38). $s$-Triazine herbicides and other compounds blocking electron transport on the reducing side of photosystem II increase the rate at which fluorescence rises from $F_0$ to $F_M$, because the reduced electron carriers cannot perform the normal photochemistry associated with the reduction of $CO_2$.

Chloroplast fluorescence transient studies[32,37] showed that in triazine-susceptible biotypes there was a very rapid rise of fluorescence to the $F_M$ level in the presence of $s$-triazines, but the fluorescence rise seen with resistant chloroplasts was similar in both the presence and absence of $s$-triazines (Fig. 10.6). Diuron induced a very rapid fluorescence rise in both susceptible and resistant biotypes because the electron carrier B of photosystem II, which is changed in the resistant biotype such that triazines are no longer bound, still binds to diuron (see Chapter 10).

## 3 CONCLUDING REMARKS

There is a rapid, or even immediate contact between the applied herbicide and the primary site of action in all the experiments described which use excised plant organs and tissue, isolated cells, or chloroplasts. Hence studies conducted with such systems have to be interpreted with caution. Clearly, finding inhibition of chloroplast activity with triazines is of little significance in relation to the activity of triazines in the field if the herbicide is either not being absorbed by the plant root system or if, after being absorbed, it is either not translocated from the roots or is detoxified before it reaches the site of action. Thus the final word on herbicide resistance must come from field studies, but laboratory studies have a useful place.

## REFERENCES

1  C. J. VanAssche, Characterization of a common molecular target for selected structures of photosynthesis inhibiting herbicides, *Advances in Pesticide Science,* H. Geissbühler (Ed.), Pergamon Press, Oxford, 1979, pp. 494-498.

2  K. Pfister and C. J. Arntzen, The mode of action of photosystem II: specific inhibitors in herbicide-resistant weed biotypes, *Z. Naturforsch.,* **34c**, 996 (1979).

3  G. F. Ryan, Resistance of common groundsel to simazine and atrazine, *Weed Sci.,* **18**, 614 (1970).

4  G. Barralis and J. Gasquez, Comportement écologique et physiologique des dicotyledones résistantes à l'atrazine en France, *Proc. EWRS Symp. The Influence of Different Factors on the Development and Control of Weeds,* 217 (1979).

5  J. Gasquez and G. Barralis, Observation et sélection chez *Chenopodium album* L. d'individus résistants aux triazines, *Chemosphere,* **11**, 911 (1978).

6  J. Gasquez and G. Barralis, Mise en évidence de la résistance aux triazines chez *Solanum nigrum* L. et *Polygonum lapathifolium* L. par observation de la fluorescence de feuilles isolées, *C.R. Acad. Sci. (Paris) Ser. D,* **288**, 1391 (1979).

7  D. E. Robinson and D. W. Greene, Metabolism and differential susceptibility of crabgrass and witchgrass to simazine and atrazine, *Weed Sci.,* **24**, 500 (1976).

8  L. D. West, T. J. Muzik, and R. E. Witters, Differential gas exchange responses of two biotypes of redroot pigweed to atrazine, *Weed Sci.,* **24**, 68 (1976).

9  S. R. Radosevich, Mechanism of atrazine resistance in lambsquarters and pigweed, *Weed Sci.,* **25**, 316 (1977).

10  S. R. Radosevich, Physiological responses to triazine herbicides in susceptible and resistant weed biotypes, *Abstr. Weed Sci. Soc. Am.,* 235 (1979).

11  S. R. Radosevich and A. P. Appleby, Relative susceptibility of two common groundsel (*Senecio*

*vulgaris* L.) biotypes to six *s*-triazines, *Agron. J.,* **65**, 553 (1973).

12   S. R. Radosevich and A. P. Appleby, Studies on the mechanism of resistance to simazine in common groundsel, *Weed Sci.,* **21**, 497 (1973).

13   A. E. Smith and R. E. Wilkinson, Differential absorption, translocation and metabolism of metribuzin by soybean cultivars, *Physiol. Plant.,* **32**, 253 (1974).

14   S. R. Radosevich and O. T. DeVilliers, Studies on the mechanism of *s*-triazine resistance in common groundsel, *Weed Sci.,* **24**, 229 (1976).

15   J. M. Ducruet and J. Gasquez, Observation de la fluorescence sur feuille entière et mise en évidence de la résistance chloroplastique à l'atrazine chez *Chenopodium album* L. et *Poa annua* L., *Chemosphere,* **8**, 691 (1978).

16   W. H. Ahrens, C. J. Arntzen, and E. W. Stoller, Chlorophyll fluorescence assay for the determination of triazine resistance, *Weed Sci.,* **29**, 316 (1981).

17   A. Ali and V. Souza Machado, Rapid detection of 'triazine resistant' weeds using chlorophyll fluorescence, *Weed Res.,* **21**, 191 (1981).

18   U. Schreiber, R. Fink, and W. Vidaver, Fluorescence induction in whole leaves: differentiation between the two leaf sides and adaptation to different light regimes, *Planta,* **133**, 121 (1977).

19   B. Truelove, D. E. Davis, and L. R. Jones, A new method for detecting photosynthesis inhibitors, *Weed Sci.,* **22**, 15 (1974).

20   J. F. DaSilva, R. D. Fadayomi, and G. F. Warren, Cotyledon disc bioassay for certain herbicides, *Weed Sci.,* **24**, 250 (1976).

21   S. W. Gawronski, R. H. Callihan, and J. J. Pavek, Sinking leaf-disc test for potato variety herbicide tolerance, *Weed Sci.,* **25**, 122 (1977).

22   J. L. Davis, J. R. Abernathy, and J. R. Gipson, Sinking leaf-disc test for cotton varietal herbicide tolerance, *Proc. South. Weed Sci. Soc.,* **32**, 331 (1978).

23   J. R. Hensley, A method for identification of triazine resistant and susceptible biotypes of several weeds, *Weed Sci.,* **29**, 70 (1981).

24   C. R. Salhoff, A. R. Martin, and J. S. Beitler, Two techniques for detecting atrazine resistance in (*Kochia scoparia* L. Schrader), *Abstr. Weed Sci. Soc. Am.,* 95 (1981).

25   D. E. Davis and B. Truelove, The measurement of photosynthesis and respiration using whole plants or plant organs, *Research Methods in Weed Science,* B. Truelove (Ed.), Southern Weed Science Society, Auburn, Ala., 1977, pp. 119–129.

26   Z. Sesták, J. Catský, and P. G. Jarvis (Eds.), *Plant Photosynthetic Production: Manual of Methods,* Dr. W. Junk, The Hague, 1971 p. 818.

27   W. W. Umbreit, R. R. Burris, and J. F. Stauffer, *Manometric and Biochemical Techniques,* 5th ed., Burgess, Minneapolis, Minn., 1972, p. 387.

28   N. Malakondaiah and S. C. Fang, Influence of monuron on photosystem II and light-dependent $^{14}CO_2$ fixation in isolated cells of $C_3$ and $C_4$ plants, *Pestic. Biochem. Physiol.,* **9**, 33 (1978).

29   L. Klepper, A mode of action of herbicides: inhibition of the normal process of nitrite reduction, *Nebr. Agric. Exp. Stn. Bull.,* **259**, (1974), p. 42.

30   L. Klepper, Inhibition of nitrite reduction by photosynthetic inhibitors, *Weed Sci.,* **23**, 188 (1975).

31   R. L. Finke, R. L. Warner, and T. J. Muzik, Effect of herbicide on *in vivo* nitrate and nitrite reduction, *Weed Sci.,* **25**, 18 (1977).

32   V. Souza Machado, C. J. Arntzen, J. D. Bandeen, and G. R. Stephenson, Comparative triazine effects upon photosystem II photochemistry in chloroplasts of two common lambsquarters (*Chenopodium album*) biotypes, *Weed Sci.,* **26**, 318 (1978).

33   S. R. Radosevich, K. E. Steinback, and C. J. Arntzen, Effect of photosystem II inhibitors on thylakoid membranes of two common groundsel (*Senecio vulgaris*) biotypes, *Weed Sci.,* **27**, 216 (1979).

34 D. E. Moreland, Measurements of reactions mediated by isolated chloroplasts, in *Research Methods in Weed Science*, B. Truelove (Ed.), Southern Weed Science Society, Auburn, Ala, 1977, pp. 141–148.

35 D. E. Moreland and K. L. Hill, Interference of herbicides with the Hill reaction of isolated chloroplasts, *Weeds,* **10**, 229 (1962).

36 K. Pfister, S. R. Radosevich, and C. J. Arntzen, Modification of herbicide binding to photosystem II in two biotypes of *Senecio vulgaris* L., *Plant Physiol.,* **64**, 995 (1979).

37 C. J. Arntzen, C. L. Ditto, and P. E. Brewer, Chloroplast membrane alterations in triazine-resistant *Amaranthus retroflexus* biotypes, *Proc. Natl. Acad. Sci. USA,* **76**, 278 (1979).

38 P. E. Brewer, C. J. Arntzen, and F. W. Slife, Effects of atrazine, cyanazine, and procyazine on the photochemical reactions of isolated chloroplasts, *Weed Sci.,* **27**, 300 (1979).

39 J. J. Burke, C. L. Ditto, and C. J. Arntzen, Involvement of the light harvesting complex in cation regulation of excitation energy distribution in chloroplasts, *Arch. Biochem. Biophys.,* **187**, 252 (1978).

# CHAPTER EIGHT

# The Roles of Uptake, Translocation, and Metabolism in the Differential Intraspecific Responses to Herbicides

K. I. N. JENSEN

Agriculture Canada
Kentville, Nova Scotia, Canada

## 1 INTRODUCTION

Resistance to triazine herbicides has been reported in a number of monocotyledonous and dicotyledonous weed biotypes that were generally considered to be controlled by this group of herbicides (see Chapters 2 and 3). The resistance of these biotypes is unique because it is based on differences at the active site, namely a change within the chloroplast thylakoid system which prevents the binding of triazine herbicides that is associated with inhibition of photosynthesis (Chapter 10).

Differential response to herbicides has generally been assumed to result from differences between and within species in those morphological and physiological factors that affect the levels of herbicidal compound accumulating at the active site. Of the latter, uptake, translocation, and particularly metabolism have repeatedly been shown to account for differential response. In the case of triazines, resistance of such species as corn,[1,2] sorghum,[3-5] and sugarcane[6] is based on rapid herbicide metabolism.

Differential response to numerous herbicides, ranging from tolerance to complete resistance, has been reported for many weed species and crop cultivars (Appendix Tables A1 and A2). This chapter deals with the importance of uptake, translocation, and metabolism in determining intraspecific responses to herbicides. It is hoped that it may serve as a background against which to compare and contrast the unique type of resistance found among the more recently identified triazine-resistant weed biotypes. Because the main emphasis of this chapter is on triazine resistance at levels other than plastid binding, considerable detail has been included on the behavior of triazines in susceptible and resistant species. Herbicides belonging to other chemical groups are also discussed.

## 2  s-TRIAZINE AND as-TRIAZINONE HERBICIDES

### 2.1  Uptake and Translocation

When considering uptake and translocation, it has been customary to consider two separate routes of translocation in plants.[7,8] The first is the apoplast, or that "inert continuum comprising the xylem, cell walls and cuticle of a plant that surrounds the protoplast." The second is the symplast, or that "living continuum of interconnected protoplasm within the plasmalemma."[7]

The s-triazines[9,10] and the as-triazinones[11,12] have been considered apoplastic in that their translocation and patterns of distribution in plants reflect movement with water in the xylem and cell walls. However, as triazines and other apolar herbicides readily penetrate the symplast of roots, leaves, and probably vascular tissue, it is unlikely that translocation occurs exclusively within the apoplast. The criterion for translocation within the apoplast is not exclusion from the symplast, but rather an inability of the symplast to retain and accumulate these pesticides within the plasmalemma. Edgington and Peterson[7] claim that the triazines and triazinones can readily shuttle between adjacent xylem and phloem, but as the rate of transpiration exceeds the rate of phloem transport, there is net movement of herbicide in the apoplast. This type of translocation has been termed "pseudoapoplastic."[7] When transpiration was inhibited, basipetal movement of triazines was observed.[9] Despite penetration into the symplast, movement and distribution of these herbicides are governed by the movement of water in the xylem and cell walls.

The triazines and the triazinone metribuzin are readily taken up by the roots and translocated via the xylem to the shoots.[8] Within the root, these herbicides probably penetrate the symplast of the endodermis and diffuse to the xylem. Triazine herbicides, applied to the roots in nutrient solutions, can be detected in the shoots of young plants within 20 to 30 min by their inhibition of photosynthesis.[13,14] Factors affecting water movement or transpiration would, in turn, affect herbicide movement. Following a short, initial period of rapid intake associated with the filling of adsorptive sites in the roots, uptake and translocation of triazines is directly proportional to the volume of water transpired.[14-16] This, in turn, is influenced by such factors as temperature,[16] humidity,[14,16] light intensity,[14] and stomatal aperture.[13-15] Their uptake is passive and is not influenced by metabolic inhibitors, insofar as these do not affect transpiration.[15] The concentration of atrazine and simazine in the xylem exudate does not exceed that of the treatment solution, although the higher concentrations in the root tissue may be due to adsorption.[15]

Bukovac[17] has reviewed the properties and structure of the shoot surface in relation to herbicide penetration following foliar treatment. The lipoidal cuticle, particularly the outer layer of waxes, is believed to be the major barrier to most foliar applied pesticides.[7,17] Factors that enhance triazine penetration into the leaves include gentle abrasion of the leaf surface, the use of surfactants, increased humidity and rewetting of the spray deposit.[17] It has been reported that substitutions which increase the water-soluble properties of the triazine herbicides enhance their foliar uptake.[18] Triazines are believed to penetrate the cuticle via an "aqueous route" consisting possibly of a hydrated cuticular pore system or ectodesmata, which makes certain areas of the cuticle preferentially permeable.

Autoradiographic studies following root uptake of triazines and triazinones in susceptible dicotyledonous species have demonstrated that detectable $^{14}$C label is first observed in the veins, but with time it becomes interveinal and ultimately accumulates at the leaf margin, where transpiration is the greatest.[9,11,16] In susceptible grasses, the initial accumulation occurs at the leaf apex and continues basipetally.[10,16] However, in triazine- and triazinone-resistant species which rapidly metabolize the parent herbicide, such as corn[9] and soybean,[11] respectively, distribution of $^{14}$C label following herbicide uptake is diffuse, suggesting that metabolites may not be as readily mobile within the plants.

Although the differential response to triazines and triazinones generally involves differential herbicide metabolism, cases implicating differential uptake and translocation have also been reported. Care must be taken in interpreting results of $^{14}$C assays of plant parts or autoradiographs following uptake of $^{14}$C herbicides unless the $^{14}$C label is identified. Differences in membrane permeability or solubility between the parent herbicide and its metabolites would affect translocation and distribution patterns. In this regard, conjugation of triazines in corn leaves may account for the uniform distribution of $^{14}$C label in the leaves, as opposed to more localized accumulations in susceptible species following root uptake. Herbicide metabolism in the roots appears to result in accumulation of $^{14}$C label in the roots and a corresponding decrease in transport to the leaves.[11,19]

Studies with a triazine-susceptible and a triazine-tolerant line of corn have also shown that the susceptible line absorbed and translocated more $^{14}$C label from a nutrient solution than did the resistant one.[20] However, further work[21] with these essentially isogenic lines showed that the rate of atrazine detoxification in the tolerant line greatly exceeded that in the susceptible one. It was metabolism and not uptake that accounted for differences in tolerance.

Another example of differential translocation affecting selectivity involves metribuzin and differentially tolerant soybeans.[11] Uptake of $^{14}$C-metribuzin from a $10^{-6}$ M $^{14}$C-metribuzin solution had no effect on a tolerant cultivar but severely injured two sensitive ones. There was no difference among cultivars in entry into the roots. Autoradiographs of plants harvested after 24 hr indicated significantly greater $^{14}$C label in the leaves of the sensitive cultivars than in leaves of the tolerant one (Fig. 8.1). In the tolerant cultivar, the $^{14}$C label was restricted primarily to the vascular tissue, whereas in the sensitive ones it was distributed throughout the leaves. Assays indicated $^{14}$C label was three to

**Figure 8.1** Autoradiographs (top) of shoots of three soybean cultivars (bottom) that had absorbed $^{14}$C-metribuzin for 24 hr from a nutrient solution. 'Bragg' soybean is tolerant to metribuzin; 'Coker-102' and 'Semmes' are sensitive. (Reproduced by permission from *Physiologia Plantarum* from an article by Smith and Wilkinson.[11])

four times higher in the leaves of the susceptible cultivars compared to the tolerant one (Fig. 8.2). This was explained by the larger percentage of $^{14}$C label found in metribuzin metabolites in the stems (95%) and roots (76%) of the tolerant cultivar when compared to the sensitive ones, where the metabolites in stems and roots ranged from 70 to 76% and 26 to 36%, respectively (Fig. 8.2). All three cultivars had the ability to metabolize metribuzin, but it was metabolized faster in the root and stem (and leaf) tissue of the tolerant cultivar. This resulted in less translocation of $^{14}$C label to the leaves and affected distribution within the leaves. In the tolerant cultivar, the $^{14}$C label was primarily in the form of a water-soluble glucosidic conjugate (see Section 2.3), which may have limited mobility within the plant. In another study, Stephenson et al.[12] found no differences in uptake or translocation of metribuzin in differentially tolerant tomato cultivars.

Shone and Wood[22] reported that differential translocation and accumulation may account for differential response between triazine-tolerant black currant, and susceptible red currant. In black currant, $^{14}$C-simazine was not readily translocated out of the roots, and in the foliage, $^{14}$C label accumulated primarily in the vascular tissue. Of this $^{14}$C label, 40% was in the form of water-

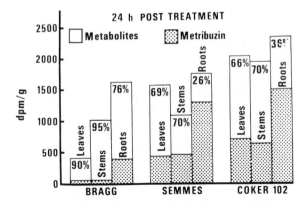

**Figure 8.2** Extractable radioactivity (dpm/g dry weight) incorporated into metribuzin and its metabolites by leaves, stems, and roots of three soybean cultivars with differential tolerance to metribuzin, 24 hr post-treatment. 'Bragg' soybean is tolerant; 'Coker-102' and 'Semmes' are sensitive. (Reproduced by permission from *Physiologia plantarum* from an article by Smith and Wilkinson.[11])

soluble metabolites, which suggests that retention and detoxification of simazine in the vascular tissue may account for the tolerance of black currant.

Werner and Putnam[23] screened the world's cucumber collection and found differential atrazine tolerance. Growth analysis studies with an accession that tolerated 0.6 kg/ha atrazine and a susceptible cultivar showed that the former had a significantly higher leaf area ratio but a lower root biomass than the latter. However, this did not appear to result in significant differences in total [14]C-atrazine uptake from a nutrient solution. After 24 hr, [14]C-levels were higher in the leaves of the susceptible cultivar. In the tolerant accession, [14]C label was localized within the vascular tissue, whereas in the susceptible cultivar the [14]C label was distributed throughout the leaf. Symptom development of treated plants also reflected differential distribution.

The accumulation of lipid-soluble 2-chloro- and 2-methoxy-*s*-triazines in the lysigenous glands on the epidermis of roots, stems, and leaves of cotton has been thought to contribute to the partial tolerance of certain cotton cultivars to these herbicides.[9,16,24] Distribution of [14]C label in leaves of glandless cotton cultivars is not localized, but resembles that of sensitive broad-leaved species following root uptake of triazines.[25] Foy[24] reported that studies on the comparative susceptibilities of normal glanded and genetically glandless cultivars indicated a relationship between the presence of glands and increased tolerance to triazines. In screening 130 glanded and glandless commercial and experimental lines, Abernathy et al.[26] found a direct relationship between tolerance to the herbicide fluridone and gossypol content. Most glanded cultivars showed less than 20% injury when treated with 0.8 kg/ha fluridone, whereas injury to glandless cultivars generally exceeded 60%.

Although accumulation in lysigenous glands may contribute to partial tolerance of cotton cultivars to triazines, other factors may be involved. Shimabukuro and Swanson[25] reported that the presence of glands on roots and stems of cotton did not decrease transport of [14]C-atrazine to the leaves, nor did it influence tolerance between two isogenic lines which differed only in the presence or absence of glands. There was no significant difference in the metabolism of [14]C-atrazine, and both lines proved equally tolerant following 8 days in 0.1 to $10\mu$ M atrazine nutrient solution. It should be pointed out that this study involved a cultivar not used in the earlier studies and the relative importance of triazine metabolism and accumulation may depend on the cultivar. It may also depend on method of treatment, since root uptake from a nutrient solution would be much greater and faster than from soils following field treatment. Rapid uptake may mask a differential tolerance that is expressed under field conditions. Under field conditions, cotton cultivars have shown a wide variation in response to atrazine and propazine.[27] A combination of factors, including metabolism, translocation, and accumulation in lysigenous glands, probably contribute to triazine tolerance.

Studies with triazine-resistant biotypes of *Senecio vulgaris,*[28] *Chenopodium album,*[29,30] and *Amaranthus* sp.[30,31] have also found that there were no differences in uptake and translocation of herbicides among susceptible and resistant biotypes. After uptake from a nutrient solution, no differences were observed in the amount of [14]C-simazine absorbed or translocated to the foliage of *Senecio* biotypes.[28,32] Early herbicide uptake into the leaves of the resistant biotype appeared to be localized in isolated accumulations, but with time the [14]C label accumulated along the leaf margins of both biotypes, reflecting the typical pattern of apoplastic translocation found in other broadleaved species.[32] In a similar study with resistant and susceptible *C. album* biotypes, Jensen et al.[29] found that there was, over a 48-hr period, no significant difference between biotypes in root uptake of [14]C-atrazine. Similarly, there was little difference between the two *C. album* biotypes in the distribution of [14]C label in roots, stems, and leaves. There was no further accumulation in roots and stems after 12 hr, but accumulation of the [14]C label continued over the 48-hr test period in the leaves of both biotypes. The greatest portion of extractable [14]C label following root uptake in resistant and susceptible *C. album* was in the form of unaltered triazine.[29,30]

Differences in foliar uptake of triazine herbicides may significantly affect differential tolerance between species. For example, foliar uptake of cyanazine from a 5-$\mu$l droplet was approximately 5 and 18 times greater in *Setaria viridis* and *Panicum dichotomiflorum,* respectively, than in the more tolerant corn.[19] However, in a similar study with *C. album* biotypes, no differences were found in the rate of foliar uptake.[29] In both biotypes, 80% of the applied [14]C-atrazine was absorbed within 6 hr and greater than 90% had been absorbed within 12 hr. Foliar uptake is affected by such leaf properties as cuticular characteristics, size and distribution of stomates, leaf morphology, and metabolism.[7,17] The total effect of these factors appeared to be similar in both biotypes, and differences in response of these biotypes to foliar applications of atrazine could not be related to differences in foliar uptake.

## 2.2 Metabolism of 2-Chloro-s-triazines

With the exception of the recently found resistant weed biotypes (see Chapter 10), triazine and triazinone herbicides inhibit photosynthesis in both tolerant and susceptible species provided that the dosage is sufficiently high.[2,3,13,21] Phytotoxic concentrations in the chloroplast cause a series of changes that result in the deterioration and ultimate destruction of the chloroplast membrane system,[33] suggesting that more than simple carbohydrate starvation is involved in the death of the plant tissue. Most resistant plants have the ability to metabolize these herbicides rapidly, and the gradual recovery of photosynthesis in herbicide-treated resistant plants has been correlated with a concomitant reduction in levels of parent herbicide by means of metabolism. This recovery of photosynthesis indicates that binding at the active sites is reversible and that phytotoxic levels of herbicide in the leaf can be reduced by detoxification.[2,3,13,21] In susceptible species, metabolism is slow and photosynthesis remains suppressed.

The degradation of s-triazine herbicides in higher plants has been extensively reviewed.[34,35] Pathways of atrazine metabolism have been studied in great detail and this has been summarized by Lamoureux et al.[5] In higher plants, degradation of s-triazine herbicides can begin by one of three initial reactions: (a) nonenzymatic hydroxylation at the 2-position, (b) N-dealkylation or other modifications of the N-alkyl side chains, and (c) conjugation with glutathione (Fig. 8.3).

Early studies indicated that 2-chloro-s-triazines could be converted to 2-hydroxy derivatives.[36] These derivatives did not inhibit photosynthesis and were not phytotoxic.[18,37] Hydroxylation was shown to be catalyzed by benzoxazinone (2,4-dihydroxy-7-methoxy-1,4-benzoxazine-3-one) or its 2-glucoside.[38] Early attempts to relate benzoxazinone content with tolerance were inconclusive. For example, approximately equal levels of benzoxazinone or its derivatives were found in tolerant and susceptible lines of corn,[38] and none was found in triazine-tolerant sorghum.[39] Futhermore, hydroxylation was not evident in such partially tolerant species as peas[40] or cotton.[25] However, when treated with atrazine, some resistant species, such as corn or *Coix lachryma-jobi,* form significant amounts of 2-hydroxyatrazine (I)*[13,39] especially in the roots. Therefore, hydroxylation may contribute to triazine resistance, especially when the triazines are taken into the plants via the roots.[2,19,39]

N-dealkylated metabolites have been identified in most s-triazine metabolism studies and N-dealkylation probably occurs in all plants.[35] Although preliminary N-dealkylation results in only partial loss of phytotoxicity,[37,41] it contributes significantly to the partial tolerance of peas[40] and cotton.[25] For

---

*Roman numerals in this section refer to metabolites in Fig. 8.3.

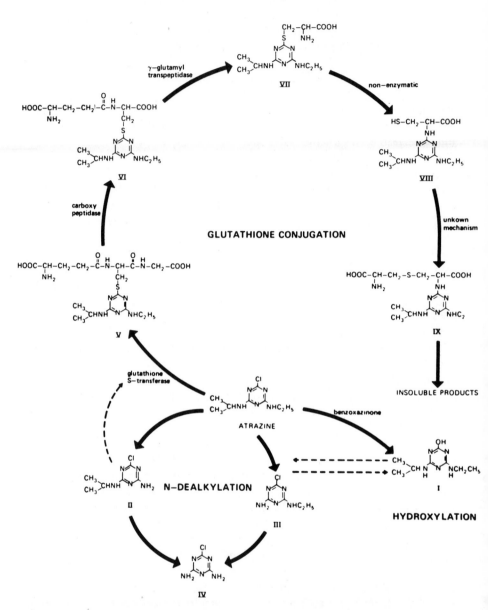

**Figure 8.3** Summary of the three pathways of atrazine degradation which may be found in higher plants: hydroxylation, N-dealkylation, and conjugation to glutathione and the subsequent catabolism of this conjugate. The occurrence and relative importance of any pathway varies markedly between species. The solid lines represent major pathways. The hatched lines indicate that an intermediate of one pathway may enter another. The metabolites are: 2-hydroxy-4-ethylamino-6-isopropylamino-s-triazine (I); 2-chloro-4-amino-6-isopropylamino-s-triazine (II); 2-chloro-4-amino-6-ethylamino-s-triazine (III); 2-chloro-4,6-diamino-s-

140

example, 2-chloro-4-amino-6-isopropyl-s-triazine (II) was the predominant [14]C-atrazine metabolite extracted from mature pea plants after 48 hr of root uptake from a nutrient solution. This, too, was generally the most common monodealkylated product formed in leaf sections of a number of grasses, indicating preferential dealkylation of the N-ethyl side chain.[13] However, in resistant species such as corn[13,21] or sorghum,[4,5,13] N-dealkylation is of minor importance. N-dealkylation occurs in both roots and shoots,[29,30] and it is enzymatically catalyzed. N-dealkylation has been shown to occur not only with 2-chloro-s-triazines, but also with 2-methylthio-, 2-methoxy-, and 2-hydroxy-s-triazines.[1,34,42] Complete N-dealkylation resulting in the formation of 2-chloro-4,6-diamino-s-triazine (IV)[3] or its hydroxylated derivative 2-hydroxy-4,6-diamino-s-triazine[35] has also been observed in plants. Loss of both N-alkyl side chains results in complete loss of phytotoxicity.[41] Other side-chain modifications may occur, depending on structure of the triazine.[34,35]

Early studies by Shimabukuro[37] showed that unidentified water-soluble metabolites were present in both roots and shoots of corn, sorghum, peas, soybeans, and wheat. These were the major metabolites in the tolerant species. Shimabukuro and Swanson[3] later demonstrated that recovery of photosynthesis in atrazine-treated sorghum leaf disks was directly related to a concomitant conversion of atrazine to a water-soluble metabolite(s). Lamoureux et al.[4] subsequently identified two water-soluble peptide conjugates of atrazine in sorghum as S-(4-ethylamino-6-isopropylamino-s-triazinyl-2)glutathione (V) and γ-L-glutamyl-S-(4-ethylamino-6-isopropylamino-s-triazinyl-2)cysteine (VI). The γ-glutamyl cysteine conjugate was rapidly formed when the purified glutathione conjugate was injected directly into the stems of sorghum plants, indicating a product–precursor relationship between the two which is probably catalyzed by a nonspecific carboxypeptidase.[5] Conjugation to glutathione is a common mechanism of detoxifying halogenated compounds in animals, but this was the first case where a similar mechanism was identified in plants.[35]

The initial detoxification reaction was enzymatically catalyzed by a glutathione S-transferase which displaces the 2-chloro group with reduced glutathione. High enzyme activity was found in partially purified leaf extracts of such tolerant grasses as corn, sorghum, Sorghum halapense, sudangrass, and sugarcane; and relatively little activity was found in the roots of the same species.[43] No detectable enzyme activity was found in the leaves of such sensitive species as peas, oats, wheat, and Amaranthus retroflexus, but some water-soluble metabolites were extracted from some of these species.[6,13,37] This glutathione S-transferase had a specific requirement for a 2-chloro group, as well as the presence of both N-alkyl side chains.[43] There was negligible activ-

triazine (IV); S-(4-ethylamino-6-isopropylamino-s-triazinyl-2)glutathione (V); γ-glutamyl-S-(4-ethylamino-6-isopropylamino-s-triazinyl-2)cysteine (VI); S-(4-ethyl-amino-6-isopropylamino-s-triazinyl-2)cysteine (VII); N-(4-ethylamino-6-isopropyl-amino-s-triazinyl-2)cysteine (VIII); N-(4-ethylamino-6-isopropyl-amino-s-triazinyl-2) lanthionine (IX). (Redrawn, in part, from Lamoureux et al.[5])

ity when 2-hydroxy or monodealkylated derivatives were used as substrate. The enzyme was inhibited by 2-methylthio- and 2-methoxy-s-triazines.

The initial conjugation to glutathione effectively detoxifies s-triazines. Therefore, differences in the subsequent metabolism of the glutathione conjugate among plant species may not affect tolerance, but it would effect the terminal, or final insoluble residues of the herbicide.[35] The ultimate fate of the herbicide appears to be associated with a decrease with time in the levels, but not numbers, of water-soluble metabolites. This is linked to a concomitant increase in the amount of methanol-insoluble residues.[4,5,37,41] Figure 8.3 provides further details on specific intermediates and pathways in s-triazine herbicide metabolism.

The role of ring cleavage in s-triazine degradation or the mechanism by which the ring is oxidized is as yet unclear. Conflicting reports of $^{14}CO_2$ evolution following uptake of $^{14}C$-ring labeled s-triazines may result from differences in species, herbicides, or experimental conditions. Triazine ring cleavage in plants has been reviewed by Esser et al.[34]

Initial studies into the mechanism of resistance in the recently discovered triazine-resistant weed biotypes naturally investigated metabolism.[28-31] In a *Senecio vulgaris* biotype that had previously survived 9 kg/ha atrazine, Radosevich and Appleby[28] found that after 24 hr of root uptake and 96 hr of incubation, unchanged simazine still accounted for more than 80% of the extractable $^{14}C$ label in the roots and shoots. There was essentially no difference in the rate of simazine degradation between the resistant and susceptible *S. vulgaris* biotype. Further studies with resistant and susceptible biotypes of *Amaranthus* sp.[30,31] and *Chenopodium album*[30] collected in Washington State revealed no significant differences between biotypes in the rate of conversion of atrazine to water-soluble metabolites. In this case, the susceptible *A. retroflexus* biotype was killed by 0.6 kg/ha atrazine, whereas the resistant biotype survived 18 kg/ha.[31]

Jensen et al.[29] found that over a 30-hr incubation period, detoxification of atrazine occurred somewhat faster in leaf disks of a resistant *Chenopodium album* biotype collected in Ontario than in those from a susceptible one. This may have been due to greater deterioration of the susceptible leaf disks in the light, as atrazine inhibition of photosynthesis in sensitive plants induces disruption of cellular membrane systems.[33] In resistant biotypes, photosynthesis is not affected by s-triazines (Chapters 9 and 10). The studies by Jensen et al.[29] showed that the rate of detoxification of atrazine was similar in stems and leaves of both biotypes, although degradation was faster in roots of the resistant biotype. However, this was of little significance in terms of resistance of *C. album* since the increased rate of root metabolism had no noticeable effect on translocation of $^{14}C$-atrazine to the shoots. N-dealkylated, hydroxylated, and conjugated metabolites were found in the roots, stems, and leaves of both biotypes. After 12 hr of root uptake, N-dealkylated metabolites accounted for 7% of the total extractable activity in plant parts of both biotypes, whereas hydroxylated and conjugated products represented less than 3% (Table 8.1).

In contrast to these studies with weed biotypes, Shimabukuro et al.[21] ex-

**Table 8.1 Metabolites of Atrazine Extracted from Leaves of Resistant and Susceptible Corn Lines and *Chenopodium album* Biotypes Following Uptake of $^{14}$C-atrazine**

| Species | Response | Total Extractable $^{14}$C Radioactivity (%) | | | | |
| | | Unaltered Atrazine | N-Dealkylated Atrazine | 2-Hydroxy Derivatives | Peptide Conjugates | Unidentified Metabolites |
|---|---|---|---|---|---|---|
| Corn[a] | Susceptible | 83.8 | Trace | 5.0 | 7.4 | 3.8 |
| | Resistant | 7.2 | Trace | 10.0 | 63.2 | 19.6 |
| C. *album*[b] | Susceptible | 83.0 | 6.8 | 1.9 | 3.1 | 5.2 |
| | Resistant | 84.1 | 7.3 | 1.6 | 2.6 | 4.4 |

[a] Metabolites were extracted from excised corn leaves that had taken up $^{14}$C-atrazine for 24 hr through their cut ends. (Data from Shimabukuro et al.[21])

[b] Metabolites extracted from leaves of intact plants after 12 hr of root uptake from a $^{14}$C-atrazine solution. (Data from Jensen et al.[29])

amined the degradation of $^{14}$C-atrazine in two inbred lines of corn that had previously been shown to vary greatly in their response to triazines (Table 8.1). Atrazine was shown to inhibit photosynthesis in leaf disks of the susceptible line at concentrations that had no effect on the resistant line. This was directly related to the levels of glutathione S-transferase, which was more than 50 times greater in the leaves of the resistant line than in the susceptible one. After 24 hr of uptake of $^{14}$C-atrazine into excised corn leaves, the glutathione conjugate represented 63.2% and 7.4% of the total extractable radioactivity in the resistant and susceptible lines, respectively. In this case, differences between the two corn lines in their response to atrazine was clearly related to differential rates of glutathione conjugation. Differential tolerance of zoysiagrass cultivars has also been related to metabolism,[44] as has the phytotoxic reaction of *Cirsium arvense* biotypes to the fungicide anilazine [2,4-dichloro-6-(2-chloroanilino)-1,3,5-triazine].[45]

The conclusion of all the studies investigating triazine metabolism in most of the resistant and susceptible annual weed biotypes was that, unlike corn and other resistant or tolerant grasses previously studied, herbicide metabolism plays no role in the differential response of triazine-resistant weed biotypes to these herbicides. It was apparent from these early studies that a new and unique mechanism, possibly at the active site, was responsible for the extremely high level of resistance exhibited by these biotypes. This is discussed at length in Chapters 9 and 10.

## 2.3  Metabolism of Metribuzin

In studying the behavior of metribuzin in a tolerant and a sensitive species, Hargroder and Rogers[46] found that adsorption, translocation, and metabolism all contributed to differential tolerance. Although the metabolites were not identified, resistant soybean converted twice as much metribuzin to water-soluble metabolites as did sensitive *Sesbania exaltata*.

Differential response to metribuzin has also been reported among soybean cultivars.[47-49] Smith and Wilkinson[11] reported that although both sensitive and resistant cultivars deaminated metribuzin to its diketo derivatives, the resistant one formed significantly more of a water-soluble metabolite. The data strongly suggested that this metabolite was the N-glucoside conjugate of metribuzin. In this regard, Frear[50] purified and characterized a UDP-glucose : arylamine N-glucosyl transferase from soybean with a broad specificity toward arylamine acceptors which catalyzed the formation of 2-glucoside conjugates. In studies with cell cultures of resistant and susceptible soybean cultivars, sensitive cells could be "protected" from the phytotoxic effect of metribuzin when incubated with extracts of the resistant cultivar.[51] Furthermore, extracts of the sensitive cultivar reduced the tolerance of the resistant cell cultures. This suggested that the enzyme detoxification system that determines tolerance was found in both sensitive and resistant cultivars. It appears, however, that an inhibitor of this enzyme system may be present in both cultivars, but only the resistant cultivar is capable of rapidly reducing the level of this enzyme inhibitor, allowing conjugation of unchanged metribuzin to the 2-glycoside. In another study employing a different resistant soybean cultivar, the deaminated diketo derivate of metribuzin was found to be the predominant metabolite.[52]

Differential response to metribuzin has also been reported for tomato and potato cultivars. Stephenson et al.[12] reported that the greater tolerance of a tomato cultivar in its early seedling stage was related to the rapid formation of a conjugate from which metribuzin was liberated by mild hydrolysis. A similar metabolite was found to be formed five times faster in leaves of a metribuzin-tolerant potato cultivar than in a sensitive one.[53] Increasing the light flux favored the formation of this metabolite, suggesting that it was a conjugate of some product of photosynthesis, possibly glucose.

While there is abundant evidence that the relative tolerance of several crop varieties and cultivars to metribuzin is influenced by metabolism, the increased tolerance to metribuzin exhibited by the triazine-resistant weed biotypes is a function of plastid differences and probably not of increased herbicide metabolism (see Chapter 10).

## 3 OTHER HERBICIDES

### 3.1 Amitrole

Unlike most other herbicides, amitrole is readily translocated in both the xylem and phloem.[54] The characteristic lag period preceding the translocation of amitrole from the treated leaf has been associated with the formation of a mobile, phytotoxic metabolite.[55,56] However, Smith et al.[57] found no evidence of such a metabolite in *Cirsium arvense* ecotypes. Amitrole and/or its metabolites accumulate in areas of rapid cell division. Amitrole is thought to inhibit carotenoid synthesis. The typical achlorophyllous symptoms in newly developed tissue results from the photooxidation of chlorophyll and flavin systems in the absence of the protective carotenoid pigments.[54]

Differences in translocation and metabolism of amitrole accounts for differential response among species. Slower penetration into the leaf, coupled with faster degradation, accounted for the greater tolerance of *Convolvulus arvensis* to amitrole than of *Cirsium arvense*.[56] Amitrole could be detected 20 days after foliar treatment in *Cirsium arvense*, but in *Convolvulus arvensis* amitrole disappeared within 4 days. Two major inactive metabolites were extracted from both species. One was chromatographically similar to the alanine conjugate of amitrole, and the other, although not identified, appeared to be related to the levels of free reducing sugars in the leaves. Glucose conjugates of amitrole have been reported, but Carter[54] maintained that this type of conjugation product is an artifact of extraction. A third metabolite was extracted from sensitive *Cirsium arvense*, which was more phytotoxic than amitrole.[56] The role of this metabolite is not clear, even in *Cirsium arvense*, as it has not been isolated from all species exhibiting sensitivity to amitrole, including the sensitive *Cirsium arvense* biotypes.[57] However, the differential selectivity of amitrole among species appears to be dependent on some change in the parent herbicide, whether this results in the formation of toxic or nontoxic metabolites, or both.

Differential intraspecific response of *Cirsium arvense* to amitrole has been reported for many North American ecotypes.[57,58] Unlike the annual weed biotypes previously discussed, which are distinguishable primarily by their response to triazines, the *C. arvense* biotypes constitute true ecotypes that are distinguishable by differences in morphology and phenology. Ecotypes responded differently to cultivation and to amitrole applications.[58] After three consecutive annual applications of 4.5 kg/ha amitrole, control of the ecotypes ranged from 29 to 86% of their untreated controls.

Smith et al.[57] studied the behavior of amitrole in several of these ecotypes. Amitrole was rapidly absorbed by excised *C. arvense* leaves and only 4 to 8% could be washed from the leaves after 6 hr, indicating little difference between ecotypes in foliar uptake. Two major nontoxic metabolites (designated unknown I and II) were extracted from the excised leaves of all three ecotypes. Other metabolites were detected, but their levels remained low over the 6-day incubation period. Unknown II was tentatively identified as the alanine-amitrole conjugate and appeared to be a precursor for unknown I. The formation of the latter was enhanced by light, possibly through some product of photosynthesis. Herrett and Linck[56] have suggested that the formation of this metabolite may be necessary for the translocation of amitrole in the phloem. In this regard, the most sensitive ecotype converted almost twice as much amitrole to this metabolite than the more tolerant ecotypes.[57] Although differences in metabolism existed between ecotypes, differences were relatively small, suggesting that other factors may be involved.

## 3.2 Barban

Foliar-applied barban disturbs the meristems of sensitive species and axillary buds fail to develop. A number of carbamate herbicides are known to interfere with normal mitotic activity.[59] Foy[60] concluded that the differential response among several oat and barley cultivars was not related to differential translocation, as movement of $^{14}$C-barban was apoplastic, with little moving from the treated leaves. Still and Herrett[59] reported that differential metabolism accounted for the differences among species in their response to barban. For example, Riden and Hopkins[61,62] reported a rapid decrease in levels of foliarly applied barban in both monocotyledonous and dicotyledonous crops that was paralleled by a corresponding increase in a short-lived, unidentified, nontoxic water-soluble metabolite(s) that contained a 3-chloroaniline moeity. Although these water-soluble metabolites have not yet been identified, Still and Herrett[59] argue that they cannot result from hydroxylation or conjugation as with the detoxification of related herbicides, such as propham and chloropropham.

Intraspecific variation in tolerance to carbanilate herbicides has been reported among barley cultivars[63] and *Avena fatua* selections.[63-65] Rydrych and Seeley[65] reported that differences in tolerance among 63 *A. fatua* selections to propham could be related to seed characteristics, but in a similar study with 214 *A. fatua* selections from the Red River Valley, seed characteristics did not correlate with herbicide tolerance to barban and several dithiocarbamate herbicides.[65] In the latter study, growth suppression following a foliar application of 0.7 kg/ha barban ranged from 10 to 90% of the untreated control. Further studies with the two *A. fatua* selections giving extremes in response and two barley cultivars with differential tolerance were reported by Jacobsohn and Andersen.[63] The differential response was observed only if barban was applied to the foliage. Susceptibility of the tolerant plants increased significantly when barban was applied to the roots. In both sensitive and tolerant *A. fatua* and barley plants, free barban could not be detected after 3 days of root uptake from a 1-ppm $^{14}$C-barban solution. Degradation product(s) with similar properties to Riden and Hopkins water-soluble, 3-chloroaniline containing compound(s) were identified as the major form of $^{14}$C activity when $^{14}$C-barban was applied to both roots and leaves. However, following foliar uptake, less free $^{14}$C-barban was extracted from the leaves of the tolerant plants 12 and 24 hr after application. However, after longer periods there was little difference in the level of the water-soluble metabolite(s). A small, but in the case of barley, significant difference was found in the rate at which tolerant plants metabolized barban beyond metabolites containing the 3-chloroaniline moeity. The authors speculated that the buildup of these metabolites might in turn reduce the rate of barban metabolism that could account for the differences observed after 12 and 24 hr of treatment. Still the relatively small differences in rates of barban metabolism cannot account entirely for the differences in response of sensitive and tolerant biotypes to barban. Other factors residing in the leaves must also influence response to barban.[62]

### 3.3 Bentazon

Bentazon initially inhibits photosynthesis in both susceptible and tolerant plant species.[66] Mine et al.[66] investigating the 100-fold difference in tolerance between susceptible *Cyperus serotinus* and resistant rice found that, despite an initial inhibition of $CO_2$ assimilation following bentazon treatment, photosynthesis recovered in rice but remained suppressed in *C. serotinus*. Recovery was related to a significantly greater rate of bentazon conjugation in the shoots of rice to $\beta$-glucopyranose forming 6-(3-isopropyl-2,1,3-benzo-thio-diazin-4-one-2,2-dioxide)-*O*-$\beta$-glucopyranoside. This conjugated product was not found in three other susceptible species. Only a minor, unconjugated metabolite was found in the susceptible species. Conjugation of bentazon with endogenous metabolites plays a determining role in bentazon tolerance between species.

Differences in tolerance to bentazon have also been reported among soybean cultivars.[67,68] Wax et al.[68] screened more than 330 named North American soybean cultivars and found that all except one was tolerant to 3.4 kg/ha bentazon. In the same trial, all of 10 Japanese lines tested were killed by this rate. Further studies with a resistant North American and a susceptible Japanese line revealed a 100-fold difference in tolerance. Eight days after foliar application, 86% of the [14]C label extracted from the resistant line was in the form of a major metabolite that was not found in the Japanese line. This metabolite, although not identified, was labile to glycolytic enzymes, strongly suggesting a glucose conjugate. Small amounts of a minor metabolite that was not hydrolyzed by these enzymes were found in both lines. In addition, slightly more [14]C label was transported out of the treated leaf of the sensitive cultivar than the leaf of the tolerant one. The mechanism of tolerance in soybean cultivars closely parallels that reported for rice and *Cyperus serotinus.* *Cirsium arvense* ecotypes were also shown to differ significantly in their ability to metabolize bentazon.[69]

Radin and Carlson have regenerated bentazon-resistant mutants of tobacco by isolating and culturing green-cell clones from bentazon-treated irradiated haploid plants (Chapter 14).

### 3.4 Butylate and Other Thiocarbamates

Low levels of butylate and other thiocarbamates inhibit growth and induce morphological abnormalities.[70] Banting[71] observed thickened chromosomes, malformed nuclei, and doubled chromosome number in meristem tissue of sensitive *Avena fatua,* but not in tolerant wheat, exposed to 4 ppm of diallate and triallate. Lower levels of herbicide having no effect on mitosis greatly reduced shoot elongation, and Banting concluded that the primary effect of these herbicides was on cell elongation rather than on mitotic activity. Thiocarbamate herbicides may also act at the hormonal level as gibberellic acid and the auxinlike 2,4-D can sometimes nullify their effect.[70]

Thiocarbamate herbicides can be altered by two types of metabolic pathways. One involves an initial hydrolysis at the ester linkage.[70] Another pathway involves the conversion of the parent herbicide to a highly toxic sulfoxide that may interfere with lipid metabolism. In tolerant species, such as corn, this sulfoxide is rapidly inactivated by conjugation with glutathione, a process catalyzed by a glutathione S-transferase.[72] Fang[70] has concluded that differences in the rates of detoxification account for differences in tolerance between species.

Intraspecific variation in response to diallate, triallate, and butylate have been reported (see Appendix Tables A1 and A2). Wright and Rieck[73] reported that T-cytoplasm corn hybrids were generally less tolerant to butylate than N-cytoplasm hybrids, although environmental factors inducing stress decreased overall tolerance. Comparative studies with a butylate sensitive and a tolerant hybrid showed that after 10 hr of root uptake, the concentration of

$^{14}$C-butylate was 66% higher in the susceptible than in the tolerant hybrid.[74] The greatest proportion of $^{14}$C-activity in both hybrids was found in the roots. As $^{14}$CO$_2$ evolution was not monitored, it could not be ascertained whether this difference resulted from differences in metabolism or uptake. The distribution of $^{14}$C label among the roots, stems, and leaves of both hybrids was similar, although the levels of $^{14}$C-activity was higher in the sensitive hybrid. Metabolic release of $^{14}$CO$_2$ from $^{14}$C-butylate in cell-free root and shoot extracts of the tolerant hybrid was three times higher than in the sensitive hybrid. Butylate detoxification in both hybrids was greater in the roots than in the shoots. It appears that differential metabolism is an important factor in the response of corn hybrids to thiocarbamate herbicides.

## 3.5 Chloramben

Chloramben, like other benzoic acid herbicides (e.g., dicamba and 2,3,6-TBA), is translocated in both xylem and phloem and is an effective growth regulator. Chloramben inhibits growth and elongation of roots and shoots of many plants, and at low concentrations, chloramben promotes auxinlike growth.[75] However, unlike other benzoic acid herbicides, chloramben has an amino group which appears to affect its translocation as well as its behavior in plants because it offers a site for N-glucosyl conjugation.

Following imbibition of $^{14}$C-chloramben into resistant soybean seed, it was found that $^{14}$C activity was restricted primarily to the cotyledons and roots.[76] Following root uptake, the $^{14}$C activity was found to be concentrated in the roots. Chloramben is translocated unchanged in the xylem.[77] However, in the leaves it did not accumulate along the margins but was distributed throughout the leaf tissue, suggesting that it entered and was immobilized in the cytoplasm. There was little or no translocation of $^{14}$C label from the leaves to the meristems.

Differential response to chloramben has been attributed to its differential translocation.[75,77-79] Following root uptake of $^{14}$C-chloramben from a nutrient solution, 1.4% and 74.3% of the absorbed $^{14}$C activity was translocated to the shoot in tolerant wheat and susceptible *Echinochloa crus-galli*, respectively. This could account for the 50-fold difference in tolerance between the two species.[77] In a similar study,[79] susceptible *Amaranthus retroflexus* translocated five times more $^{14}$C activity to the shoots than did resistant soybeans. Retention of chloramben in the roots has been associated with the conversion of this herbicide to the N-glucosyl conjugate.[77,79] N-glucosyl chloramben is nonphytotoxic and appears to undergo no further metabolism.

Miller et al.[78] investigated the physiological basis for tolerance and susceptibility of four lines of cucumber to chloramben methyl ester after preliminary studies had revealed a three- to seven-fold difference in tolerance among 200 cucumber lines. In this case the role of uptake and translocation in 4-week-old plants was not clear cut. In two of the four lines, tolerance was related to translocation of chloramben from the roots. When roots of a susceptible line were grafted to scions of the tolerant line, there was increased uptake of chloramben into the shoots. Similarly, the tolerant rootstock decreased uptake into susceptible scions. In the other two lines, however, there was no relationship between translocation and tolerance. However, between these lines [14]C-activity in the tolerant line accumulated within the vascular tissue of the leaves following root uptake. In addition, five metabolites, including three conjugates, were isolated from the roots and shoots of cucumber. However, there were no major differences in chloramben metabolism among the four lines which could account for differences in uptake or translocation.

### 3.6  Chlorfenprop-methyl and Diclofop-methyl

Chlorfenprop-methyl gives selective postemergence control of *Avena fatua* and other annual grasses in certain cereals and other crops. It has a unique selectivity among *Avena* species. Of the wild oats, *A. fatua* is sensitive and *A. sterilis* and *A. ludoviciana* are tolerant.[80] Commercial oat cultivars have also shown differential tolerance.[80,81]

The herbicidal effect of chlorfenprop-methyl and its initial deesterified free acid metabolite chlorfenprop [collectively chlorfenprop(-methyl)] has been linked to its inhibitory effect on auxin-mediated systems.[80,81] Chlorfenprop (-methyl) does not act as an anti-auxin nor does it have auxin activity. In coleoptile tissue or extracts, chlorfenprop(-methyl) interfered with auxin-induced elongation,[80,81] auxin transport, proton excretion, and binding of naphthyl-1-acetic acid to supposedly auxin-specific sites.[80] Similar inhibitory effects were measured in tolerant and susceptible species and oat cultivars, but higher concentrations of chlorfenprop(-methyl) were required in tolerant tissue.[80] Both chlorfenprop-methyl and the free acid have herbicidal properties with similar selectivities when applied to whole plants, but the free acid is less effective in *in vitro* and *in vivo* systems. This is due perhaps to slower penetration of the hydrophilic metabolite to the active sites in the lipoidal membrane system.[80]

Fedtke and Schmidt[80] found no difference between tolerant and sensitive oat cultivars in penetration of [14]C-chlorfenprop-methyl following foliar application. Following uptake into the leaves, chlorfenprop-methyl was rapidly and almost completely hydrolyzed to the free acid in all cultivars, suggesting that this may be the herbicidal compound. With time, an increasing proportion of [14]C label was found in water-soluble metabolites, whereas chlorfenprop(-methyl) remained in a small and stable pool. It appears that accumulation of chlorfenprop(-methyl) at the active site(s) is similar in both tolerant and sensitive cultivars. These sites are stereo specific since only the L(-)enantiomer is active. For these reasons, Fedtke and Schmidt[80] hypothesized that the basis for the selective action of chlorfenprop(-methyl) is a difference at the site of action between sensitive and tolerant species and cultivars.

The mode of action of the closely related diclofop-methyl shares many similarities.[82] Shimabukuro et al.[83] have proposed another mechanism of tolerance between wheat and *A. fatua* based on metabolism. Both species form a free acid as above, but in wheat the phytotoxic free acid undergoes aryl hydroxylation, resulting in a phenol which is probably nontoxic. This in turn undergoes phenol conjugation, forming an O-glycoside. In A. fatua the free acid directly undergoes ester conjugation, and this O-glycoside may later be hydrolyzed, regenerating the phytotoxic free acid. This interesting possibility has not been investigated in tolerant and susceptible oat cultivars treated with chlorfenprop-methyl, as the water-soluble metabolites have not yet been identified.

### 3.7  2,4-D and Related Phenoxy Herbicides

Two primary sites of action have been proposed for phenoxy herbicides. One appears to involve a direct effect on cell wall elongation followed by a stimulatory effect on RNA metabolism. After a recent review on phenoxyalkanoic herbicides, Loos[84] concluded that the general selectivity of this group between dicotyledonous and monocotyledonous plants was determined primarily by differences in plant structure and herbicide translocation. However, many dicotyledonous species also exhibit differential tolerance, and among these species, tolerance is governed by differences in herbicide penetration, translocation, and metabolism. Inactivation of 2,4-D by adsorption to cellular components has also been suggested,[85] and adsorption to such sites has been suggested in cases where the foregoing factors could not account for differential tolerance.

In general, translocation of 2,4-D and related herbicides can be classified as symplastic with transport of the herbicide to metabolic "sinks."[84,86] Lack of phloem transport is an important mechanism in determining differential sensitivity to 2,4-D among plant species. Another factor governing sensitivity is metabolic detoxification by one or more of these processes: (a) side-chain decarboxylation, (b) side-chain lengthening, (c) conjugation to such naturally occurring compounds as glucose and aspartate, and (d) ring hydroxylation.[84]

Luckwill and Lloyd-Jones[87] studied the metabolism of 2,4-D and other phenoxy herbicides in tolerant and sensitive apple and strawberry cultivars. Following uptake of [14]C-carboxyl-labeled 2,4-D, it was shown that [14]$CO_2$ evolution in the dark was related to tolerance. For example, after 92 hr 59% of the absorbed [14]C-2,4-D had been decarboxylated in a tolerant apple compared to only 2% in a sensitive one. Decarboxylation and hence tolerance can be greatly reduced by substituting fluorine at the 4-position.[88] When treated with 2-chloro-4-fluorophenoxyacetic acid, a tolerant and a sensitive apple responded similarly.

Tolerance was found to be heritable. Decarboxylation was found to be greater in 10 apple cultivars having Cox in their parentage, as compared with 13 more sensitive cultivars that were not related to Cox. Extraction of [14]C-carboxyl-labeled metabolites from the leaves also revealed the formation of two water-soluble 2,4-D conjugates (possibly glucose conjugates) that had no auxin activity and an unidentified methylene chloride-soluble metabolite that did.[89] There was little difference between tolerant and susceptible cultivars in the levels of these other metabolites.

Similar patterns of metabolism were found among strawberry cultivars.[87] Although decarboxylation was shown to be the basis of differential tolerance within apples and strawberries, Luckwill and Lloyd-Jones[87] also presented data from 17 other species in which there was no clear association between sensitivity and decarboxylation of 2,4-D. Weintraub et al.[90] also failed to find any correlation between tolerance and decarboxylation among 12 corn cultivars with differing 2,4-D tolerance. In this case, 2,4-D was more readily translocated in the 2,4-D sensitive corn cultivars.

Tolerance of apple, strawberry, and currant cultivars to 2,4-D appears to be more complex than decarboxylation alone. For example, while these plants are sensitive to 2,4,5-T, they decarboxylate it as readily as 2,4-D.[87] The product of decarboxylation of 2,4-D is probably 2,4-dichlorophenol, which was phytotoxic to 2,4-D-susceptible plants. Leaves of 2,4-D-tolerant plants were unaffected by this compound. The dichlorophenol is rapidly conjugated, possibly to glucose, in 2,4-D-tolerant apple and strawberry cultivars. The toxicity of 2,4,5-T in these cultivars is not expressed with the typical phenoxy-type symptoms, which suggests that 2,4,5-T itself is not phytotoxic. Tolerance to 2,4-D among these cultivars, therefore, appears to be dependent on the rapid decarboxylation of 2,4-D, which prevents the shift to an abnormal growth response followed by inactivation, probably by conjugation, of the resulting phytotoxic phenol.[89]

Whitworth and Muzik[91] reported that 51 clones of *Convolvulus arvensis* differed widely in their response to 2,4-D. Differential tolerance could not be related to differences in leaf pubescence, stomatal number, plant vigor, or adsorption and translocation of the herbicide.[92] The differential response remained in tissue culture of the sensitive and tolerant biotypes.[91,92] The response of the tissue to 2,4-D could be modified by addition of glutamine and glutamate to the medium. These compounds did not influence uptake or metabolism of [14]C-2,4-D in the callus tissue. The authors concluded that 2,4-D must be deactivated within cells of the tolerant tissue by binding to nonactive sites such as proteins. Brian[85] studied the adsorption of numerous aromatic acids, including a number with herbicidal activity, to monolayers of tissue squashes. He found that there may be a broad correlation between nonspecific adsorption and tolerance to 2,4-D and related herbicides. Adsorption to nonactive sites has also been proposed as the mechanism of tolerance in a 2,4-D-tolerant biotype of *Daucus carota*,[93] but experimental evidence of this was not presented.

Differential response to phenoxy herbicides has been reported for other weed species (Appendix Tables A1 and A2). In regard to differential response, two generalizations can be made. First, there appears to be no correlation between morphological characteristics and tolerance. One exception is the cuticle thickness among *Cirsium arvense* ecotypes with modified responses to 2,4-D,[94] but even this is an undependable characteristic, as cuticle properties are strongly influenced by environmental conditions. This lack of distinguishable morphological characteristics suggests that differential tolerance may be based on internal physiological factors. Second, differential tolerance appears to be heritable. For example, Bell et al.[95] found that the variability in response of *Sonchus arvensis* to phenoxyalkanoic herbicides might have been the result of hybridizing with the more tolerant *Sonchus oleraceus.*

### 3.8  Difenzoquat and Other Bipyridylium Herbicides

Resistance to the bipyridylium herbicides paraquat, diquat, and morfamquat is discussed in Chapter 11. Their mode of action and the mechanism of resistance of certain perennial ryegrass lines are also discussed. Sensitive and resistant plants do not differ in the penetration, translocation, or metabolism of these herbicides. They inhibit electron transport in photosynthesis by intercepting electrons, thereby forming bipyridylium free radicals. These are instantly reoxidized by molecular oxygen, with the production of the superoxide radical ion, which, in turn, undergoes dismutation, forming hydrogen peroxide. These free radicals and oxidants peroxidate constituents of the various cellular membrane systems, causing a rapid collapse of the cells. Leaves of resistant lines of perennial ryegrass were shown to contain significantly higher levels of the enzymes superoxide dismutase, catalase, and peroxidase, which reduce the levels of hydrogen peroxide in cells (Chapter 11).

Differential response of sensitive *Avena fatua* and resistant barley to difenzoquat was not based on differential uptake, translocation, or metabolism.[96] Although wheat is generally tolerant, difenzoquat was shown to be phytotoxic to some wheat cultivars in a screening test involving 56 cultivars.[97] Again, no significant differences could be established in terms of herbicide retention, adsorption, translocation, or metabolism, which could account for variation in response among spring wheat cultivars.[98] Perhaps a mechanism as described above for paraquat resistance which does involve differential accumulation of herbicide at the active site could account for these differences in response.

### 3.9  Diuron

Two factors have been implicated in the selectivity of phenylurea herbicides: (*a*) differential uptake and/or translocation and (*b*) differential herbicide metabolism. Distribution patterns following either root or foliar uptake reflect apoplastic translocation as previously described for the *s*-triazine herbicides, with little or no evidence of symplastic or basipetal movement.[99] Root accumulation or lack of translocation out of the roots has been related to the tolerance of several crop species to phenylurea herbicides.[100]

Phenylurea herbicides inhibit photosynthesis, and $CO_2$ uptake is blocked in both susceptible and tolerant species.[101] The recovery in photosynthetic activity following limited root uptake of monuron in *Plantago lanceolata*[101] was later shown to be dependent on the capacity of this species to convert monuron and diuron to their respective monomethyl derivatives and certain water-soluble metabolites.[102] The latter were glucosidic conjugates of hydroxymethyl intermediates of the demethylation process. Information on the uptake, translocation, and metabolism of substituted phenylurea herbicides is extensive and it has been reviewed in detail.[99]

Differential response to diuron of a sensitive sugarcane cultivar and a tolerant one was found after field applications of 10.1 kg/ha, as well as uptake from a nutrient solution.[103] When [14]C-diuron was applied in a nutrient solution, there was no difference between the cultivars in root uptake over a 10-day period. However, the distribution of [14]C label within the plants differed following root uptake. The concentration of [14]C label in the roots of the tolerant cultivar was twice that of the susceptible one.

The tolerant and susceptible sugarcane cultivars metabolized diuron to monomethyl- and demethylated diuron. Loss of the single methyl group resulted in about 50% loss of phytotoxicity. After 3 weeks of root uptake, levels of unchanged diuron were lower in the roots, stems, and particularly leaves of the tolerant cultivar. A significant proportion of the [14]C label was in the form of an unidentified metabolite, presumably a glucosidic conjugate. Differences in diuron phytoxicity between these two cultivars could, in part, be attributed to differences in the rates of diuron degradation.[103]

Intraspecific differences in tolerance have been reported for other phenylurea herbicides (see Appendix Tables A1 and A2). Greater root accumulation resulting in decreased translocation to the leaves has also been implicated in the greater tolerance of some cotton cultivars to diuron and fluometuron.[104] This also appears to be an important factor in the tolerance of carrots to linuron.[100] In another comparative study,[105] absorption of [14]C-fluometuron from a nutrient solution was significantly greater in a sensitive cotton cultivar than in a more tolerant one.

Available data show that the triazine-resistant weed biotypes have generally been susceptible to diuron, confirming that the nature of the binding sites for PS II inhibition by these two major classes of herbicides is not identical. On the other hand, urea herbicides, such as diuron and chloroxuron, have shown reduced effectiveness in chloroplasts from triazine-resistant biotypes compared to susceptible chloroplasts (see Chapter 10). There is increasing evidence that cross-resistance to both s-triazines and urea herbicides can occur (see Chapters 2 and 18). As in the case of other herbicides, the degree of resistance or reduced sensitivity in these biotypes is due to changes in the nature of the thylakoid membrane binding site rather than differences in herbicide movement or metabolism.

### 3.10 Nitrofen

Nitrofen is a contact herbicide that does not readily penetrate leaves of either susceptible or tolerant species.[106,107] There was a trend toward acropetal translocation of nitrofen following absorption into the leaf with slight movement of the [14]C label out of the [14]C-nitrofen-treated leaf in tolerant rape and susceptible *Amaranthus retroflexus* and *Setaria viridis*. In these three species, four major metabolites or groups of metabolites were found 16 days after foliar application of [14]C-nitrofen, but they were not positively identified.[107] It was hypothesized that at least two of these metabolites were either lipid-nitrofen conjugates or nitrofen polymers, and the other, smaller metabolites may have resulted from cleavage of the ether linkage. There was little difference among the three species in the amount of these metabolites. Therefore, it would appear that neither uptake nor metabolism determines the wide range in response to nitrofen among these species.

Hopen[108] reported differential responses ranging from no effect to severe injury among 36 cabbage cultivars treated with 5.6 kg/ha nitrofen. There was a general trend toward increased tolerance with later-maturing cultivars, and despite significant post-treatment injury, even severely injured cultivars recovered. He further demonstrated that the tolerant variety could be affected at increasing rates and that the tolerance of the susceptible one increased with age.[109] Selectivity could be modified by changing formulations, light conditions prior to treatment, and by disruption or removal of the cuticle. All treatments that increased wetting, or modified or reduced the thickness of the cuticle in the tolerant cultivar, increased its susceptibility.

Differences in response to nitrofen were closely related to differences in the cuticles of the two cultivars. The cuticle of the tolerant one was characterized by a white glaucous bloom, which had more wax than the cuticle of the susceptible cultivar, which had a bluish bloom. Certain individual plants of the tolerant cultivar had cuticles with a bluish bloom and these were found to be susceptible. Furthermore, tolerance of the susceptible cultivar increased with age due to additional cuticular waxes. Studies with [14]C-nitrofen showed no difference in translocation within the two cultivars, but penetration into the leaves were almost two times greater in the susceptible cultivar (Fig. 8.4). Cuticular waxes limit nitrofen uptake, and therefore cuticle thickness determined the differential response of the two cultivars.[109]

Wilkinson[110] reported differences not only in the total wax content of the cuticle of *Tamarix pentandra* ecotypes, but also in the alkane and fatty acid content of their cuticles. He suggested that these differences may play a role in differential response of ecotypes to postemergence herbicides.

### 3.11 Pyrazon

Pyrazon inhibits the Hill reaction in isolated chloroplasts extracted from both tolerant and susceptible species,[111] and like other inhibitors of photosynthesis eventually induces a disruption of the chloroplast membranes.[33] It is readily translocated in the xylem to the shoots following root uptake, and no basipetal movement was observed when applied to the leaves.[112,113] Susceptible tomato[113] and *Chenopodium album*[112] accumulated more [3]H label following root uptake of [3]H-pyrazon than were accumulated by both tolerant red and sugar beets.

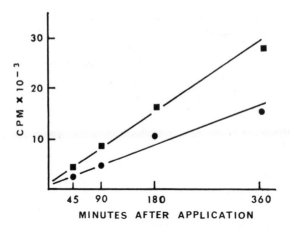

**Figure 8.4** Uptake of $^{14}$C-nitrofen into leaves of a susceptible (■) and a tolerant (●) cabbage cultivar. (Redrawn from Pereira et al.[109])

Stephenson and Ries[113] found that a major metabolite in red beet accounted for 50 to 60% of the extractable $^3$H activity after 72 hr. This metabolite was not found in tomatoes. This metabolite was subsequently identified as N-glucosyl pyrazon,[114,115] and like other glucose conjugates, rates of conjugation in tolerant species were related to the carbohydrate status of the plant. There was no further degradation of the N-glucosyl pyrazon after 50 days,[115] although other minor metabolites have also been found.[113]

Stephenson et al.[116] examined pyrazon metabolism in nine crop species and found N-glucosyl pyrazon only in red beet, although other minor metabolites were found in all species. Only red beet was tolerant, and after 24 hr of incubation, 82 to 94% of the $^3$H activity extracted from $^3$H-pyrazon-infiltrated leaf disks of the eight susceptible species was found as unchanged pyrazon. After 10 hr of incubation with $^3$H-pyrazon, leaf disks from nine inbred lines of red beet showed that the percent conversion to N-glucosyl pyrazon ranged from 44 to 76%.[116] These differences in detoxification among inbred lines could account for differences in growth reduction when the beets were grown in sand culture in the presence of pyrazon. These differences may not influence field tolerance of these selected lines under normal condition, but it does demonstrate significant variability in N-glucosyl pyrazon formation.

## 4  SUMMARY AND CONCLUSION

Within any species there appears to be considerable variation in response among individuals or populations to herbicides. This has been amply documented for weed species and crop cultivars alike (see Chapters 2, 3, 12, and 13). Differences in response to herbicides between different species has generally been shown to depend on factors that influence levels of herbicide reaching the phytotoxic site of action—most important, differential uptake, translocation, and metabolism of the herbicide. In most cases, these factors have been implicated in the differential response within species as well. Rapid herbicide metabolism, and its possible effect on translocation, was generally the main factor involved when wide differences in tolerance existed between weed populations or crop cultivars. The recent appearance of resistance of many weed biotypes to triazine herbicides, on the other hand, involves differences that prevent binding at the active site. In the case of these weeds, there were no significant differences between resistant and susceptible biotypes in the uptake, translocation, or metabolism of the herbicide.

## REFERENCES

1  R. H. Shimabukuro, Atrazine metabolism in resistant corn and sorghum, *Plant Physiol.,* **43**, 1925 (1968).

2  R. H. Shimabukuro, H. R. Swanson, and W. C. Walsh, Glutathione conjugation-atrazine detoxification mechanism in corn, *Plant Physiol.,* **46**, 103 (1970).

3  R. H. Shimabukuro and H. R. Swanson, Atrazine metabolism, selectivity and mode of action, *J. Agric. Food Chem.,* **17**, 199 (1969).

4  G. L. Lamoureux, R. H. Shimabukuro, H. R. Swanson, and D. S. Frear, Metabolism of 2-chloro-4-ethylamino-6-isopropylamino-*s*-triazine in excised sorghum leaf sections, *J. Agric. Food Chem.,* **18**, 81 (1970).

5  G. L. Lamoureux, L. E. Stafford, R. H. Shimabukuro, and R. G. Zaylskie, Atrazine metabolism in sorghum: catabolism of the glutathione conjugate of atrazine, *J. Agric. Food Chem.,* **21**, 1020 (1973).

6  G. L. Lamoureux, L. E. Stafford, and R. H. Shimabukuro, Conjugation of 2-chloro-4, 6-bis(alkylamino)-*s*-triazines in higher plants, *J. Agric. Food Chem.,* **20**, 1004 (1972).

7  L. V. Edgington and C. A. Peterson, Systemic fungicides: theory, uptake and translocation, in *Antifungal Compounds,* Vol. 2: *Interactions in Biological and Ecological Systems,* M. R. Siegle and H. D. Sisler (Eds.), Marcel Dekker, New York, 1977, p. 51.

8  J. R. Hay, Herbicide transport in plants, in *Herbicides: Physiology, Biochemistry, Ecology,* Vol. 1, L. J. Audus (Ed.), Academic Press, London, 1976, p. 365.

9  D. L. Davis, N. E. Funderberk, and S. G. Sansing, The absorption and translocation of $^{14}$C-labeled simazine by corn, cotton and cucumber, *Weeds,* **7**, 300 (1959).

10  R. H. Shimabukuro and A. J. Linck, Root absorption and translocation of atrazine in oats, *Weeds,* **15**, 175 (1967).

11  A. E. Smith and R. E. Wilkinson, Differential absorption, translocation and metabolism of metribuzin by soybean cultivars, *Physiol. Plant.,* **32**, 253 (1974).

12  G. R. Stephenson, J. E. McLeod, and S. C. Phatak, Differential tolerance of tomato cultivars to metribuzin, *Weed Sci.,* **24**, 161 (1976).

13  K. I. N. Jensen, G. R. Stephenson, and L. A. Hunt, Detoxification of atrazine in three Gramineae subfamilies, *Weed Sci.,* **25**, 212 (1977).

14  J. L. P. van Öorschot, Effect of transpiration rate of bean plants on inhibition of photosynthesis by some root-applied herbicides, *Weed Res.,* **10**, 230 (1970).

15  M. G. T. Shone and A. V. Wood, Factors affecting absorption and translocation of simazine by barley, *J. Exp. Bot.,* **23**, 141 (1972).

16 T. J. Sheets, Uptake and distribution of simazine by oat and cotton seedlings, *Weeds,* **9**, 1 (1961).

17 M. J. Bukovac, Herbicide entry into plants, in *Herbicides: Physiology, Biochemistry, Ecology,* Vol. 1, L. J. Audus (Ed.), Academic Press, London, 1976, p. 335.

18 H. Gysin and E. Knüsli, Chemistry and herbicidal properties of triazine derivatives, *Adv. Pest Control Res.,* **3**, 289 (1960).

19 A. D. Kern, W. F. Meggitt, and D. Penner, Uptake, movement and metabolism of cyanazine in fall panicum, green foxtail and corn, *Weed Sci.,* **23**, 277 (1975).

20 E. F. Eastin, Absorption and translocation of atrazine by resistant and susceptible lines of *Zea mays, Abstr. Meet. Weed Sci. Soc. Am.,* No. 184 (1969).

21 R. H. Shimabukuro, D. S. Frear, H. R. Swanson, and W. C. Walsh, Glutathione conjugation: an enzymatic basis for atrazine resistance in corn, *Plant Physiol.,* **47**, 10 (1971).

22 M. G. T. Shone and A. V. Wood, Factors responsible for the tolerance of black currants to simazine, *Weed Res.,* **12**, 337 (1972).

23 G. M. Werner and A. R. Putnam, Differential atrazine tolerance within cucumber (*Cucumis sativus*), *Weed Sci.,* **28**, 142 (1980).

24 C. L. Foy, Accumulation of *s*-triazine derivates in lysigenous glands of *Gossypium hirsutum* L., *Plant Physiol.* (Suppl. to vol. 37), xxv (1962).

25 R. H. Shimabukuro and H. R. Swanson, Atrazine metabolism in cotton as a basis for intermediate tolerance, *Weed Sci.,* **18**, 231 (1970).

26 J. R. Abernathy, J. W. Keeling, and L. L. Ray, Response of 130 cotton varieties to fluridone, *Proc. South. Weed Sci. Soc.,* **31**, 77 (1978).

27 J. R. Abernathy, J. W. Keeling, and L. L. Ray, Cotton cultivar response to propazine and atrazine, *Agron. J.,* **71**, 929 (1979).

28 S. R. Radosevich and A. P. Appleby, Studies on the mechanism of resistance to simazine in common groundsel, *Weed Sci.,* **21**, 497 (1973).

29 K. I. N. Jensen, J. D. Bandeen, and V. Souza Machado, Studies on the differential tolerance of two lambsquarters selections to triazine herbicides, *Can. J. Plant Sci.,* **57**, 1169 (1977).

30 S. R. Radosevich, Mechanism of atrazine resistance in lambsquarters and pigweed, *Weed Sci.,* **25**, 316 (1977).

31 L. Thompson, R. W. Schumacher, and C. E. Rieck, An atrazine resistant strain of redroot pigweed, *Abstr. Meet. Weed Sci. Soc. Am.,* No. 196 (1974).

32 S. R. Radosevich and O. T. Devillers, Studies on the mechanism of *s*-triazine resistance in common groundsel, *Weed Sci.,* **24**, 229 (1976).

33 F. E. Ashton, E. M. Gifford, and T. Bisalputra, Structural changes in *Phaseolus vulgaris* induced by atrazine: II. Effect on fine structure of chloroplasts, *Bot. Gaz.,* **124**, 336 (1963).

34 H. O. Esser, G. Dupuis, E. Ebert, G. Marco, and E. Vogel, *s*-Triazines, in *Herbicides: Chemistry, Degradation and Mode of Action,* Vol. 1, P. C. Kearney and D. D. Kaufman (Eds.), Marcel Dekker, New York, 1975, p. 129.

35 R. H. Shimabukuro, G. L. Lamoureux, D. S. Frear, and J. E. Bakke, Metabolism of *s*-triazine and its significance in biological systems, in *Pesticide Terminal Residues,* (suppl. to *Pure Appl. Chem.*), 323 (1971).

36 P. Castelfranco, C. L. Foy, and D. B. Deutsch, Nonenzymatic detoxification of 2-chloro-4, 6-bis(ethylamino)-*s*-triazine (simazine) by extracts of *Zea mays, Weeds,* **9**, 580 (1961).

37 R. H. Shimabukuro, Atrazine metabolism and herbicidal selectivity, *Plant Physiol.,* **42**, 1269 (1967).

38 R. D. Palmer and C. O. Grogan, Tolerance of corn lines to atrazine in relation to content of benzoxazinone 2-glucoside, *Weeds,* **13**, 219 (1965).

39 R. H. Hamilton, Tolerance of several grass species to 2-chloro-*s*-triazine herbicides in relation to degradation and content of benzoxazinone derivatives, *J. Agric. Food Chem.,* **12**, 14 (1964).

40 R. H. Shimabukuro, R. E. Kadunce, and D. S. Frear, Dealkylation of atrazine in mature pea plants, *J. Agric. Food Chem.*, **14**, 392 (1966).

41 R. H. Shimabukuro, W. C. Walsh, G. L. Lamoureux, and L. E. Stafford, Atrazine metabolism in sorghum: chloroform-soluble intermediates in the N-dealkylation and glutathione conjugation pathways, *J. Agric. Food Chem.*, **21**, 1031 (1973).

42 P. W. Müller and P. H. Payot, Fate of $^{14}$C-labeled triazine herbicides in plants, *Proc. IAEA Symp. Isotopes Weed Res.*, Vienna, 61 (1966).

43 D. S. Frear and H. R. Swanson, Biosynthesis of S-(4-ethylamino-6-isopropylamino-2-s-triazino)glutathione: partial purification and properties of a glutathione S-transferase from corn, *Phytochemistry*, **9**, 2123 (1970).

44 W. F. Smith and E. D. Ilnicki, Intraspecific selectivity of atrazine to *Zoysia japonica* I. Toxicity studies, *Proc. Northeast. Weed Sci. Soc.*, 145 (1973).

45 P. B. Marriage, Herbicidal activity and metabolism of dyrene in Canada thistle, *Weed Sci.*, **21**, 389 (1973).

46 T. G. Hargroder and R. L. Rogers, Behavior and fate of metribuzin in soybean and hemp sesbania, *Weed Sci.*, **22**, 238 (1974).

47 W. S. Hardcastle, Soybean (*Glycine max*) cultivar response to metribuzin in solution culture, *Weed Sci.*, **27**, 278 (1979).

48 W. S. Hardcastle, Differences in the tolerance of metribuzin by varieties of soybeans, *Weed Res.*, **14**, 181 (1974).

49 L. M. Wax, E. W. Stoller, and R. L. Bernard, Differential response of soybean cultivars to metribuzin, *Agron. J.*, **68**, 484 (1976).

50 D. S. Frear, Herbicide metabolism in plants: I. Purification and properties of UDP-glucose: arylamine N-glycosyl-transferase from soybean, *Phytochemistry*, **7**, 381 (1968).

51 T. H. Oswald, A. E. Smith, and D. V. Phillips, Phytotoxicity and detoxification of metribuzin in dark-grown suspension cultures of soybean, *Pestic. Biochem. Physiol.*, **8**, 73 (1978).

52 B. L. Mangeot, F. E. Slife, and C. E. Rieck, Differential metabolism of metribuzin by two soybean (*Glycine max*) cultivars, *Weed Sci.*, **27**, 267 (1979).

53 T. G. Hinks, The nature of the tolerance of potatoes to metribuzin herbicide, M.Sc. thesis, University of Guelph, Ontario, 1977.

54 M. C. Carter, Amitrole, in *Herbicides: Chemistry, Degradation and Mode of Action,* Vol. 1, P. C. Kearney and D. D. Kaufman (Eds.), Marcel Dekker, New York, 1975, p. 377.

55 R. A. Herrett and W. P. Bagley, Metabolism and translocation of 3-amino-1,2,4-triazole by Canada thistle, *J. Agric. Food Chem.*, **12**, 17 (1964).

56 R. A. Herrett and A. J. Linck, The metabolism of 3-amino-1,2,4-triazole by Canada thistle and field bindweed and the possible relation to its herbicidal action, *Physiol. Plant.*, **14**, 767 (1961).

57 L. W. Smith, D. E. Bayer, and C. L. Foy, Metabolism of amitrole in excised leaves of Canada thistle ecotypes and bean, *Weed Sci.*, **16**, 523 (1968).

58 J. M. Hodgson, The response of Canada thistle ecotypes to 2,4-D, amitrole and intensive cultivation, *Weed Sci.*, **18**, 253 (1970).

59 G. G. Still and R. A. Herrett, Methylcarbamates, carbanilates and acylanilides, in *Herbicides: Chemistry, Degradation and Mode of Action,* Vol. 2, P. C. Kearney and D. D. Kaufman (Eds.), Marcel Dekker, New York, 1976, p. 609.

60 C. L. Foy, Uptake of radioactive 4-chloro-butynyl N-(3-chlorophenyl)carbamate (barban) and translocation of $^{14}$C in *Hordeum vulgare* and *Avena* spp., *West. Weed Control Conf.*, 96 (1961).

61 J. R. Riden and T. R. Hopkins, Decline and residue studies on 4-chloro-2-butanyl N-(3-chlorophenyl)carbamate, *J. Agric. Food Chem.*, **9**, 47 (1961).

62   J. R. Riden and T. R. Hopkins, Formation of a water-soluble, 3-chloroaniline-containing substance in barban treated plants, *J. Agric. Food Chem.,* **10**, 455 (1962).

63   R. Jacobsohn and R. N. Andersen, Intraspecific differential response of wild oat and barley to barban, *Weed Sci.,* **20**, 74 (1972).

64   R. Jacobsohn and R. N. Andersen, Differential response of wild oat lines to diallate, triallate and barban, *Weed Sci.,* **16**, 491 (1968).

65   D. J. Rydrych and C. I. Seeley, Effect of IPC on selections of wild oats, *Weeds,* **12**, 265 (1964).

66   A. Mine, M. Miyakado, and S. Matsunaka, The mechanism of bentazon selectivity, *Pestic. Biochem. Physiol.,* **5**, 566 (1975).

67   R. M. Hayes and L. M. Wax, Differential intraspecific responses of soybean cultivars to bentazon, *Weed Sci.,* **23**, 516 (1975).

68   L. M. Wax, R. L. Bernard, and R. M. Hayes, Response of soybean cultivars to bentazon, bromoxynil, chloroxuron and 2,4-DB, *Weed Sci.,* **22**, 35 (1974).

69   D. Penner, Bentazon selectivity between soybean and Canada thistle, *Weed Res.,* **15**, 259 (1975).

70   S. C. Fang, Thiocarbamates, in *Herbicides: Chemistry, Degradation and Mode of Action,* Vol. 2, P. C. Kearney and D. D. Kaufman (Eds.), Marcel Dekker, New York, 1976, p. 323.

71   J. D. Banting, Effect of diallate and triallate on wild oat and wheat cells, *Weed Sci.,* **18**, 80 (1970).

72   M. M. Lay and J. E. Casida, Dichloroacetamide antidotes enhance thiocarbamate sulfoxide detoxification by elevating corn root glutathione content and glutathione S-transferase activity, *Pestic. Biochem. Physiol.,* **6**, 442 (1976).

73   T. H. Wright and C. E. Rieck, Factors affecting butylate injury to corn, *Weed Sci.,* **22**, 83 (1974).

74   T. H. Wright and C. E. Rieck, Differential butylate injury to corn hybrids, *Weed Sci.,* **21**, 194 (1973).

75   D. S. Frear, The benzoic acid herbicides, in *Herbicides: Chemistry, Degradation and Mode of Action,* Vol. 2, P. C. Kearney and D. D. Kaufman (Eds.), Marcel Dekker, New York, 1976, p. 541.

76   D. G. Swan and F. W. Slife, The absorption, translocation and fate of amiben in soybeans, *Weeds,* **13**, 133 (1965).

77   E. W. Stoller, Mechanism for the differential translocation of amiben in plants, *Plant Physiol.,* **46**, 732 (1970).

78   J. C. Miller, D. Penner, and L. R. Baker, Basis for variability in cucumber for tolerance to chloramben methyl ester, *Weed Sci.,* **21**, 207 (1973).

79   S. R. Colby, Mechanism of selectivity of amiben, *Weeds,* **14**, 197 (1966).

80   C. Fedtke and R. R. Schmidt, Chlorfenprop-methyl: its hydrolysis *in vivo* and *in vitro* and a new principle for selective herbicidal action, *Weed Res.,* **17**, 233 (1977).

81   G. K. Andreev and N. Amrhein, Mechanism of action of the herbicide 2-chloro-3(4-chlorophenyl)propionate and its methyl ester: interaction with cell responses mediated by auxin, *Physiol. Plant.,* **37**, 175 (1976).

82   M. A. Shimabukuro, R. H. Shimabukuro, W. S. Nord, and R. A. Hoerauf, Physiological effects of methyl 2-[4(2,4-dichlorophenoxy)phenoxy]propionate on oat, wild oat and wheat, *Pestic. Biochem. Physiol.,* **8**, 199 (1978).

83   R. H. Shimabukuro, W. C. Walsh, and R. A. Hoerauf, Metabolism and selectivity of diclofop-methyl in wild oat and wheat, *J. Agric. Food Chem.,* **27**, 615 (1979).

84   M. A. Loos, Phenoxyalkanoic acids, in *Herbicides: Chemistry, Degradation and Mode of Action,* Vol. 1, P. C. Kearney and D. D. Kaufman (Eds.), Marcel Dekker, New York, 1975, p. 1.

**85** R. C. Brian, Action of plant growth regulators: IV. Adsorption of unsubstituted and 2,6-dichloro-aromatic acids to oat monolayers, *Plant Physiol., 42,* 1209 (1967).

**86** M. M. Robertson and R. C. Kirkwood, The mode of action of foliage-applied translocated herbicides with particular reference to the phenoxy-acid compounds: II. The mechanism and factors influencing translocation, metabolism and biochemical inhibition, *Weed Res., 10,* 94 (1970).

**87** L. C. Luckwill and C. P. Lloyd-Jones, Metabolism of plant growth regulators: II. Decarboxylation of 2,4-dichlorophenoxyacetic acid in leaves of apple and strawberry, *Ann. Appl. Biol., 48,* 626 (1960).

**88** L. J. Edgerton and M. B. Hoffman, Fluorine substitution affects decarboxylation of 2,4-dichlorophenoxyacetic acid in apples, *Science, 134,* 341 (1961).

**89** L. C. Luckwill and C. P. Lloyd-Jones, Metabolism of plant growth regulators: I. 2,4-dichlorophenoxyacetic acid in leaves of red and black currant, *Ann. Appl. Biol., 48,* 613 (1960).

**90** R. L. Weintraub, J. H. Reinhart, and R. A. Scherff, Role of entry, translocation and metabolism in specificity of 2,4-D and related compounds, *Radioactive Isotopes in Agriculture,* Rep. No. T10-7512, U.S. Atomic Energy Commission, 203 (1956).

**91** J. W. Whitworth and T. J. Muzik, Differential response of selected clones of bindweed to 2,4-D, *Weeds, 15,* 275 (1967).

**92** R. G. Harvey and T. J. Muzik, Effects of 2,4-D and amino acids on field bindweed *in vitro, Weed Sci., 21,* 135 (1973).

**93** C. W. Whitehead and C. M. Switzer, The differential response of strains of wild carrot to 2,4-D and related herbicides, *Can. J. Plant Sci., 43,* 255 (1963).

**94** J. M. Hodgson, Lipid deposition on leaves of Canada thistle ecotypes, *Weed Sci., 21,* 169 (1973).

**95** A. R. Bell, J. D. Nalewaja, S. Alam, A. B. Schooler, and J. S. Hsieh, Herbicidal response and morphology of interspecific sowthistle crosses, *Weed Sci., 21,* 189 (1973).

**96** M. P. Sharma, W. H. Vanden Born, H. A. Friesen, and D. K. McBeath, Penetration, translocation and metabolism of $^{14}$C-difenzoquat in wild oat and barley, *Weed Sci., 24,* 379 (1976).

**97** J. E. Hill, J. D. Prato, and K. Bedane, A comparison of several wheat cultivars in response to difenzoquat, *Abstr. Meet. Weed Sci. Soc. Am.,* No. 21 (1977).

**98** S. E. Blank, Studies on the intraspecific response of spring wheat cultivars to difenzoquat, *Diss. Abstr. Int. B, 37,* 2607 (1976).

**99** H. Geissbuhler, H. Martin, and G. Voss, The substituted ureas, in *Herbicides: Chemistry, Degradation and Mode of Action,* Vol. 1, P. C. Kearney and D. D. Kaufman (Eds.), Marcel Dekker, New York, 1975, p. 209.

**100** A. Walker and R. M. Featherstone, Absorption and translocation of atrazine and linuron in plants with implications concerning linuron selectivity, *J. Exp. Bot., 24,* 450 (1973).

**101** J. L. P. van Öorschot, Selectivity and physiological inactivation of some herbicides inhibiting photosynthesis, *Weed Res., 5,* 81 (1965).

**102** C. R. Swanson and H. R. Swanson, Metabolic fate of monuron and diuron in isolated leaf disks, *Weed Sci., 16,* 137 (1968).

**103** R. V. Osgood, R. R. Romanowski, and H. W. Hilton, Differential tolerance of Hawaiian sugarcane cultivars to diuron, *Weed Sci., 20,* 537 (1972).

**104** A. A. Saeed, S. O. S. Ahmed, and H. Idris, Differential response of cotton varieties to some urea herbicides, *Sudan Agric. J., 8,* 33 (1973).

**105** D. A. McCall, P. W. Santelman, and L. M. Verhalen, Genetic variation in cotton susceptibility to fluometuron, *Proc. South. Weed Sci. Soc., 29,* 85 (1976).

106    D. Hawton and E. H. Stobbe, Selectivity of nitrofen among rape, redroot pigweed and green foxtail, *Weed Sci.,* **19,** 42 (1971).

107    D. Hawton and E. H. Stobbe, The fate of nitrofen in rape, redroot pigweed and green foxtail, *Weed Sci.,* **19,** 555 (1971).

108    H. J. Hopen, Selectivity of nitrofen between cabbage cultivars, *HortScience,* **4,** 119 (1969).

109    J. F. Pereira, W. E. Splittstoesser, and H. J. Hopen, Mechanism of intraspecific selectivity of cabbage to nitrofen, *Weed Sci.,* **19,** 647 (1971).

110    R. E. Wilkinson, Ecotypic variation of *Tamarix pentandra* epicuticular wax and possible relationship with herbicide sensitivity, *Weed Sci.,* **28,** 110 (1980).

111    R. Frank and C. M. Switzer, Effects of pyrazon on growth, photosynthesis and respiration, *Weed Sci.,* **17,** 34 (1969).

112    R. Frank and C. M. Switzer, Absorption and translocation of pyrazon by plants, *Weed Sci.,* **17,** 365 (1969).

113    G. R. Stephenson and S. K. Ries, The movement and metabolism of pyrazon in tolerant and susceptible species, *Weed Res.,* **7,** 51 (1967).

114    G. R. Stephenson and S. K. Ries, Metabolism of pyrazon in sugar beets and soil, *Weed Sci.,* **17,** 327 (1969).

115    S. K. Ries, M. J. Zabik, G. R. Stephenson, and T. M. Chen, $N$-glucosyl metabolism of pyrazon in red beets, *Weed Sci.,* **16,** 40 (1969).

116    G. R. Stephenson, L. R. Baker, and S. K. Ries, Metabolism of pyrazon in susceptible species and inbred lines of tolerant red beet (*Beta vulgaris* L.), *J. Am. Soc. Hort. Sci.,* **96,** 145 (1971).

# CHAPTER NINE

# Physiological Responses and Fitness of Susceptible and Resistant Weed Biotypes to Triazine Herbicides

S. R. RADOSEVICH and J. S. HOLT

Department of Botany
University of California
Davis, California

## 1  INTRODUCTION

The basis of tolerance to triazine herbicides in various crop species has been under investigation for many years and has been adequately explained by differential detoxification. Such studies are described in depth in Chapter 8. The first report of the seeming evolution of a triazine-resistant weed, *Senecio vulgaris,* in 1970[1] instigated a series of experiments to ascertain whether the mode of resistance was the same as that within the crops. Plants of both biotypes were grown in nutrient culture and subjected to various *s*-triazine herbicides,[2] confirming Ryan's initial observations.[1] Plants of the susceptible biotype became chlorotic and died after the addition of a *s*-triazine herbicide, whereas plants of the resistant biotype did not (Table 9.1, Fig. 9.1). These experiments excluded seed depth, germination characteristics, and some morphological differences as factors affecting herbicide tolerance. Effective control was obtained when either biotype was subjected to nontriazine herbicides,[1,2] suggesting that the mechanism of resistance was physiological and limited to the triazine structure.

Since the original discovery in 1970, biotypes of many other weed species, in addition to *S. vulgaris,* have been discovered to be resistant to *s*-triazine herbicides from formerly susceptible populations (see Chapters 2 and 3). Extensive efforts have been made since 1970 to determine the physiological mechanism of triazine resistance in the resistant biotypes of these weed species. Most early studies involved comparisons of differential triazine uptake, translocation,

**Table 9.1** Dosages of Six s-Triazine Herbicides Required to Cause Complete Necrosis ($LD_{100}$) in Biotypes of *Senecio vulgaris* Grown in Greenhouse Nutrient Cultures[a]

| Herbicide | $LD_{100}$ (mg/1) | | |
| --- | --- | --- | --- |
| | Susceptible Biotype | Resistant Biotype[b] | Resistance Factor |
| Simazine | 0.1–0.5 | 4[b] | > 8 |
| Atrazine | 0.1–0.5 | 30 | > 60 |
| Secbumeton | 0.5–1.0 | 30 | > 30 |
| Prometon | 0.1–1.0 | 30 | > 30 |
| Prometryn | 1.0–4.0 | 30 | > 8 |

*Source:* Data were modified from Radosevich and Appleby.[2]

[a]$LD_{100}$ estimated visually in four replications in each of two experiments. Exposure times and experimental conditions were identical for both biotypes.

[b]The maximum concentration of simazine used was 4 (mg/1) because of its low water solubility.

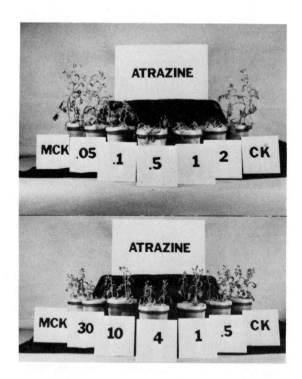

**Figure 9.1** Effect of atrazine on a susceptible (top) and a resistant (bottom) biotype of S. *vulgaris*. Rates are in mg/l in nutrient solution. Ck refers to plants grown in untreated nutrient solution. MCk (methanol check) refers to plants subjected to the amount of methanol required in the highest atrazine rates. (Reproduced with permission from Radosevich and Appleby.[2])

and metabolism between resistant and susceptible biotypes of *S. vulgaris,*
*Chenopodium album,* and *Amaranthus* sp.* More recent studies have involved
differences between susceptible and resistant biotypes at the site of action of
the *s*-triazine herbicides.

## 2  UPTAKE, TRANSLOCATION, AND METABOLISM

The differential herbicide uptake, translocation, and degradation of the triazine herbicides to nontoxic metabolites by susceptible and tolerant plant species
has been extensively reviewed in Chapter 8. These same mechanisms have been
thoroughly studied in weed biotypes, and will be discussed below.

### 2.1  Uptake and Translocation by Resistant Weeds

To determine whether the weed biotypes absorbed and translocated triazines
differentially, they were exposed to a nutrient solution containing [14]C-sima-
zine. With *S. vulgaris,* no differences in either the rate or the amount of [14]C-
simazine uptake[3] was observed between the two biotypes (Fig. 9.2). Subsequent
experiments, using autoradiographic techniques, revealed no differences between the ability of the *Senecio* biotypes to absorb and translocate [14]C-sima-
zine.[4] Similar results from more extensive studies have been reported for susceptible and resistant biotypes of *C. album*[5,6] and *Amaranthus* sp.[5]

### 2.2  Metabolism of *s*-Triazines by Resistant Weeds

Herbicide metabolism studies in crops with different tolerances to triazines
have revealed several metabolic pathways for *s*-triazine (e.g., atrazine) degradation in plants (Fig. 8.3). These include dealkylation of *N*-alkyl side
chains (forming chloroform soluble metabolites), hydroxylation at the 2-posi-
tion, and glutathione conjugation (both forming water-soluble metabolites).
When metabolism studies were conducted with the resistant and susceptible
biotypes of *S. vulgaris, Amaranthus* sp., and *C. album* (Table 9.2), the occurrence of water-soluble metabolites varied among the species but without
statistically significant differences between the two biotypes of any species.
The greatest concentration of radioactivity (from applied [14]C-simazine or [14]C-
atrazine) was usually located in the chloroform partition of foliage extracts
from each biotype. Much less [14]C activity was observed in root extracts from
both biotypes. Although some [14]C residues were isolated in the water portions
of the plant extracts from each species, time-course studies revealed no differential increase in water-soluble simazine or atrazine metabolites by either
resistant or susceptible biotypes of any of the three species (Table 9.2). Radioactivity in the chloroform fractions of the plant extracts was determined by

*Because of the present ambiguous status of the Amaranthacae (see Chapter 5), what was pub-
lished as *A. retroflexus* will be denoted as *Amaranthus* sp.

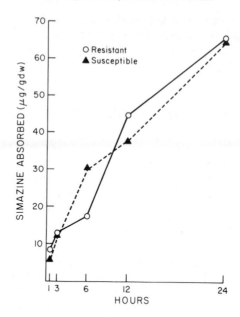

**Figure 9.2** [14]C-simazine absorption by the two biotypes of *S. vulgaris* as a function of time. Points represent averages of results from two experiments, each with two replications. (Plotted from data in Table 1 of Radosevich and Appleby.[3])

thin-layer chromatography to be intact [14]C-simazine or [14]C-atrazine.[3,4] The results suggest that the weeds do not dealkylate the side chains to a significant extent, as do the crops, within the time periods studied.

## 2.3  Differences Between Triazine-Resistant Crops and Weeds

Even though the weeds were resistant solely to the triazine herbicides, the mechanism of resistance was not based on differential uptake, translocation, or degradation of the herbicide. These observations were in contrast to those of earlier workers who studied the cause for differential triazine tolerance between and within crop species as well as other weeds (Chapter 8). It is now generally agreed that herbicide uptake, translocation, and metabolism to nontoxic metabolites play an insignificant role in the differential response between these recently discovered triazine-resistant and triazine-susceptible weed biotypes to triazine herbicides.

The crops being bred for resistance to *s*-triazine herbicides using genetic material from other *crops* (Chapter 12) may be expected to work at the level of detoxification. Those being bred using genetic material introduced from resistant weed biotypes, (Chapter 13), are expected to lack effective mechanisms for herbicide exclusion or degradation, as is apparent with the weeds.

Table 9.2  Distribution of Radioactivity from Applied s-Triazine $^{14}$C in Foliage Extracts of *Senecio vulgaris*, *Chenopodium album*, and *Amaranthus* sp. Biotypes[a]

| Species | Biotype | Exposure Time (hr) | Chloroform[b] (Triazine) (%) | Water (Metabolites) (%) |
|---------|---------|--------------------|------------------------------|-------------------------|
| | | | $^{14}$C Recovered from Foliage Extraction | |
| *S. vulgaris* | Susceptible | 24 | 79 | 9 |
| | Resistant | 24 | 72 | 12 |
| | Susceptible | 48 | 81 | 14 |
| | Resistant | 48 | 82 | 13 |
| *C. album* | Susceptible | 12 | 95 | 5 |
| | Resistant | 12 | 93 | 7 |
| | Susceptible | 36 | 88 | 12 |
| | Resistant | 36 | 91 | 9 |
| | Susceptible | 60 | 85 | 12 |
| | Resistant | 60 | 81 | 19 |
| *Amaranthus* sp. | Susceptible | 12 | 82 | 18 |
| | Resistant | 12 | 84 | 16 |
| | Susceptible | 36 | 64 | 36 |
| | Resistant | 36 | 55 | 45 |
| | Susceptible | 60 | 56 | 44 |
| | Resistant | 60 | 50 | 50 |

*Source:* Data condensed from Radosevich and Appleby[3] and Radosevich.[5]

[a]Plants of both biotypes of each species were grown in aerated nutrient solution and later exposed to either $^{14}$C-simazine (*Senecio*) or $^{14}$C-atrazine (*Chenopodium* and *Amaranthus*). Differences between biotypes of each species were not statistically significant at the 5% level of probability.

[b]The $R_f$ of the single spot appearing on chromatograms of concentrated chloroform spots was identical to that of the s-triazine used with each biotype.


## 3  EFFECTS OF TRIAZINES ON PHOTOSYNTHESIS IN RESISTANT WEED BIOTYPES

The triazine herbicides are potent inhibitors of photosynthesis. To study the effects of triazines on photosynthesis in susceptible and resistant weed biotypes, gas exchange was measured in the presence and absence of herbicide. Research in this area has increased in complexity from experiments on whole plants to studies with isolated chloroplasts, and has led to a better understanding of the mechanism of resistance (Chapter 10). A review of the differential effect of triazines on photosynthesis in weed biotypes follows.

## 3.1 Gas Exchange in Response to Triazines

Photosynthetic gas exchange was measured in the leaves of susceptible and resistant biotypes of *S. vulgaris* and *Amaranthus* sp.[3,7] With *S. vulgaris,* plants were exposed to simazine in aqueous culture during measurement, and removed from the herbicide one day after treatment. Leaves of *Amaranthus* were first treated with atrazine and then net photosynthesis was measured. Rapid inhibition of net $CO_2$ fixation occurred when the susceptible biotype of either species was subjected to simazine[3] (shown for *S. vulgaris* in Fig. 9.3) or to atrazine.[7] Net photosynthesis of the resistant weed biotypes was not altered by the presence of the herbicides[3,7] (Fig. 9.3). Photosynthesis of the susceptible biotype proceeded toward normal rates when simazine-treated plants were washed and returned to an aqueous medium without the herbicide (Fig. 9.3). This indicates that the inhibitory effect of simazine is reversible in susceptible species or biotypes.

## 3.2 Photosynthesis in Isolated Chloroplasts

When plastids were isolated from leaves of triazine tolerant and susceptible crop species, photosynthesis was inhibited by the triazines (Chapter 8). Similar studies have been carried out on weed biotypes. Chloroplasts were isolated from leaves of *S. vulgaris* to determine if resistant biotypes had the same sensitivity to triazines as those from susceptible biotypes. Atrazine did not inhibit the photochemical activity of isolated chloroplasts of the triazine-resistant *S. vulgaris* biotype, whereas the photochemical activity of chloroplasts from the susceptible biotype was severely inhibited.[4] To preclude the possibility that some stromal component confers the resistance to the tolerant biotypes, further experiments were performed with isolated thylakoids. These more extensive studies[8-10] indicated that the differences in triazine inhibitor sensitivity between biotypes was related to inherent differences between susceptible and resistant thylakoid membranes. These herbicides inhibited photosynthetic electron transport on the reducing side of photosystem II in thylakoid membranes from susceptible biotypes (Chapter 10). Thylakoids isolated from resistant biotypes were 60 to 3200 times more tolerant to *s*-triazine herbicides than were thylakoids from susceptible biotypes, based on relative rates of photoreduction of DCPIP[9,10] (Fig. 9.4).

Several studies have attempted to relate chemical structure to the activity of various Hill reaction inhibitors.[11-13] Thylakoid suspensions of both biotypes of *C. album*[9] and *S. vulgaris*[10] were most affected by the methylthio-*s*-triazines and least affected by the methoxy-*s*-triazines. Chloro-*s*-triazines were intermediate in effect (Fig. 9.4). These observations agree with those of other workers concerning the activity of numerous substituted *s*-triazines.[11] The Hill reaction inhibition of susceptible *S. vulgaris* thylakoids may be favored by asymmetric alkyl amino substitution at the 4- and 6-positions of the triazine ring (Fig. 9.4). These data support the observations of Moreland[13] and indicate

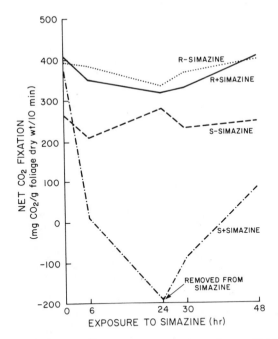

**Figure 9.3**  Net $CO_2$ fixation in resistant (R) and susceptible (S) biotypes of *S. vulgaris*, each exposed to 0.5 mg/l of simazine or untreated (Ck). Averages of two replicated experiments. The susceptible biotype was removed from simazine after 24 hr. LSD (0.05) values were 115, 71, 100, and 67 mg $CO_2$/g dry foliage/10 min for 6-, 24-, 30-, and 48-hr exposure times, respectively.[3] (Reproduced with permission from Radosevich and Appleby.[3])

that orientation of the triazine molecule at the active site may be controlled by alkyl amino substitutes at ring positions 4 and 6. This relationship was not apparent with the resistant biotype because of its greater herbicide tolerance. Resistant-type thylakoid membranes of *S. vulgaris* were also more tolerant of bromacil and slightly more tolerant of diuron than were triazine-susceptible thylakoids. Of the herbicides studied, diuron was the most effective inhibitor of the resistant biotype, whereas bromacil was intermediate in activity. All photosystem II inhibitors (e.g., ureas, uracils, and triazines) have the $-C-NH$ group as a basic structural element, and the lipophilic substituent at the nitrogen is usually an aromatic ring.[12] These data suggest that the inhibitor affinity or orientation for the reactive site may have been altered in resistant thylakoids but not in susceptible thylakoids.[9,10]

The mechanism for herbicide resistance appears to be due to specific structural or conformational changes in one or more of the thylakoid membrane constituents. This alteration apparently reduces the affinity and/or orientation of the triazine herbicide molecule at the reactive site itself. This subject is thoroughly reviewed in Chapter 10.

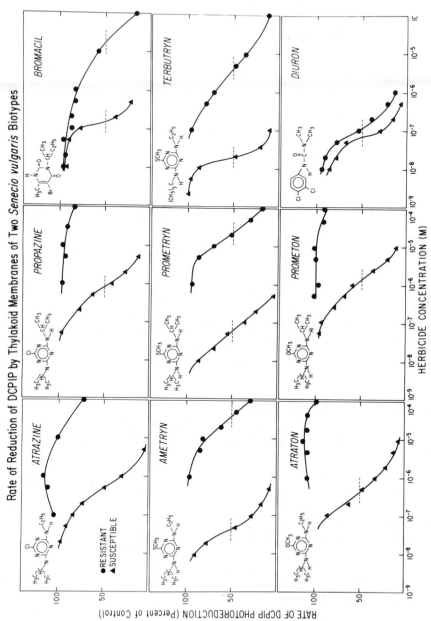

**Figure 9.4** Rate of reduction of dichlorophenolindophenol (DCPIP) by isolated chloroplasts of atrazine-resistant (●) and atrazine-susceptible (▲) biotypes of

## 4  COMPETITION AND ECOLOGICAL FITNESS

It is a common occurrence throughout biology that whenever a particular trait is selected for, the selected individuals are often ecologically less "fit." This means that whenever the selective agent is missing, the wild type will outcompete the selected strains and thus maintain those strains at low frequency within the population.

This lack of fitness has been described in plants in a system analogous to herbicides (i.e., with heavy metal resistance; see Chapter 15). It had been predicted that herbicide-resistant weeds, once selected, would also be less fit than their susceptible counterparts. This was taken into account in models on the rate at which resistance might be expected to be enriched in a population from repeated herbicide use (see Chapter 17).

### 4.1  Measurements of Fitness

Even under noncompetitive conditions, the dry matter production per plant of resistant *S. vulgaris* and *Amaranthus* sp. was 25 and 40% less, respectively, than that of susceptible plants.[14] This may be a function of the modified photosynthesis described in Chapter 10, and more fully in the following sections.

Fitness is usually measured by planting different ratios of plants with constant total plant number and area and determining biomass production. Plants of *S. vulgaris* and *Amaranthus* sp. were thus cultivated in a greenhouse and the results for both species were similar (Fig. 9.5). The data indicate that the resistant biotype is more affected by competition from the susceptible biotype of the same species than by competition with itself.[14] This suggests that plants of the susceptible biotype of each species are better competitors than resistant plants in addition to being more vigorous in noncompetitive conditions.

The long-term outcome of competition depends on the ultimate reproductive output of the plants involved. The proportion of the total seed output (dry weight) produced by each weed biotype under varying levels of interbiotype competition was measured (Fig. 9.6). For both species the proportion of total seed produced by susceptible plants was always greater than the proportion of susceptible plants present at any level of competition. These data also support

---

*S. vulgaris* with various concentrations of chloro-*s*-triazines (atrazine, propazine), methylthio-*s*-triazines (ametryn, prometryn, terbutryn), methoxy-*s*-triazines (atraton, prometon), bromacil, and diuron. Chloroplasts were isolated from leaves of 6-week-old plants of both S and R biotypes. The chloroplasts were osmotically disrupted, centrifuged, and the stroma-free thylakoid membranes were resuspended in buffer with herbicides. Photosystem II mediated DCPIP reduction was measured at 580 nm immediately after addition of the herbicides. Data points represent means of replicated observations. Standard errors of the mean are no larger than the plotted points. (Reproduced with permission from Radosevich et al.[10])

Physiological Responses and Fitness

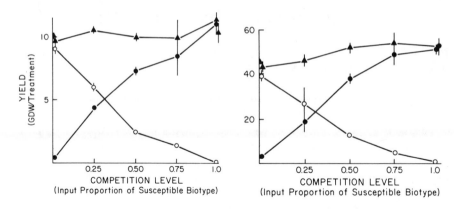

**Figure 9.5** The dry weight of biotypes of *S. vulgaris* (left) and *Amaranthus* sp. (right) that are susceptible (●) or resistant (○) to atrazine. Seedlings of susceptible and resistant biotypes were planted in varying proportions on a 3 X 3 cm grid in metal flats. Each flat contained 48 plants (equivalent to 1111 plants/m²). The experiment for each species was arranged in a randomized complete block design with three replications in the greenhouse. Plants of *S. vulgaris* were harvested after 68 days and *Amaranthus* after 86 days. The yields of the mixture of susceptible and resistant biotypes (▲) are also shown. Vertical lines represent 95% confidence limits. (Reproduced with permission from Conard and Radosevich.[14])

the conclusion that susceptible biotypes are more fit than the resistant biotypes of the same species.

Population changeover from 98% resistant to 98% susceptible biotype would require 9 to 10 generations for *S. vulgaris* and 10 to 11 generations for *Amaranthus* sp., based on predictions from Fig. 9.6. Thus, under field conditions in which competition for space is a factor, the frequency of resistant biotypes should be readily suppressed by the susceptible biotypes. A fitness differential of this magnitude is sufficient to explain why there would be no conversion to dominance by the resistant biotypes under normal conditions (i.e., without atrazine application).

## 4.2  Measurements of Resource Allocation

The relative success of a species (or biotypes of the same species) in mixed populations may be influenced by the allocation of fixed carbon to various portions of the plant.[15] Resource allocation patterns of atrazine susceptible and resistant biotypes under varying levels of interbiotype competition were measured.[14] Most allocation patterns of the resistant *S. vulgaris* biotype did not change with increasing levels of susceptible biotype competition. An increase in resistant biotype shoot allocation was observed at the 0.5 proportion of the susceptible biotype but not at higher or lower proportions. The repro-

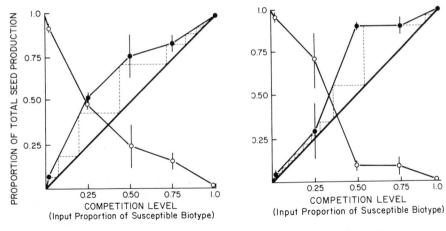

**Figure 9.6** The proportion of total seed production (dry weight) of *S. vulgaris* biotypes (left) and *Amaranthus* sp. (right) that are susceptible (●) and resistant (O) to atrazine. The biotypes were grown together at various proportions at constant density, as described in Fig. 9.5. After harvest, reproductive portions were separated into seeds (achenes) and chaff. Three samples of 50 seeds of each biotype were weighed for each treatment. Average seed weights were used to determine seed production. The predicted number of generations for the population to change from 98% resistant to 98% susceptible biotype is indicated by the dashed line. Theoretical yield of the biotypes, if they had identical competitive abilities, is indicated by the solid diagonal line. Vertical lines represent 95% confidence limits. (Reproduced with permission from Conard and Radosevich.[14])

ductive allocation of the resistant *Amaranthus* biotype remained constant while shoot biomass allocation tended to increase at high levels of interbiotype competition. However, at the 0.98 proportion of susceptible biotype there was little resistant *Amaranthus* seed production. Even high levels of resistant biotype competition did not affect resource allocation of the susceptible biotype of *Amaranthus*.[14]

Resource allocation patterns for susceptible and resistant biotypes of *S. vulgaris* and *Amaranthus* grown in pure stands in productivity experiments at low density (123 plants/m²) and in competition experiments at high density (1111 plants/m²) are shown in Fig. 9.7. Both biotypes of each species have a relative increase in shoot material at high density. Again, the fitness advantage of the susceptible biotype cannot be ascribed to a more advantageous redistribution of resources within the plant which provide a competitive advantage.

## 5   SOURCES OF COMPETITIVE ADVANTAGE BY SUSCEPTIBLE BIOTYPES

Two possibilities have been promulgated to date which can explain some aspects of the competitive advantage of the susceptible over the resistant bio-

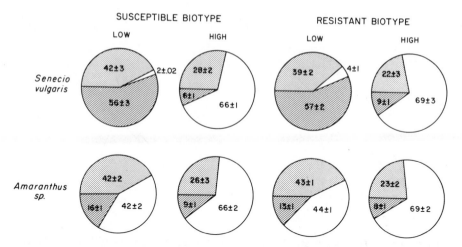

**Figure 9.7** Resource allocation pattern (dry weight) of biotypes of *S. vulgaris* and *Amaranthus* sp. that are susceptible or resistant to atrazine when grown separately at high density (1111 plants/m²) and low density (123 plants/m²). ▦ reproductive biomass, ☐ shoot biomass, ▨ root biomass. (Reproduced with permission from Conard and Radosevich.[14])

types. It is clear that at least in some species, the growth rate of the herbicide-resistant biotype is less than that of the susceptible biotype. This is not the "classical" case, as is apparent with heavy metal resistance (compare Fig. 9.5 and 9.6 with Fig. 15.4). The lesser growth may be ascribed to a lower photosynthetic potential (see Chapter 10). This possibility, as well as preliminary data showing that the susceptible biotype may exert an allelopathic advantage over the resistant biotype, are discussed below. Such an allelopathic interaction between susceptible and resistant weeds has been proposed only with *Amaranthus* spp. to date. It is possible that both types of effects are compounded in mixed field conditions.

## 5.1 Possible Allelopathic Interactions in *Amaranthus*

Allelopathy is defined as the direct or indirect harmful effect of one plant on another through the production of chemical compounds that escape into the environment. This definition separates the effect of physical competition from the effects of chemical growth inhibition. The observation of the strong competitive interaction caused one group[16] to consider allelopathy as a possible reason for the advantage of the susceptible biotypes of *Amaranthus* over the resistant ones.

Historically, it has been difficult to separate plant interactions into various types and to obtain a good understanding of whether growth inhibition is from competition or is truly an allelopathic interaction. To separate competition for

light, water, and minerals from allelopathy, the resistant and susceptible biotypes were grown in separate, well-spaced pots with a circulating mineral nutrient supply. In the single case described, resistant *A. hybridus* seed from
Maryland was compared with susceptible *Amaranthus* sp. seed obtained from
Valley Seed Service, California, which was assumed to be *A. retroflexus.* Both
*Amaranthus* spp. were assumed to be *A. retroflexus* at the time, but this was
later determined to be incorrect for the resistant biotype.

Seedlings were planted in a series of pots either in pure series (all S and all R)
or in mixed series (S-R-S) and (R-S-R). The fresh weights of susceptible seedlings grown in pure series were significantly greater than those of resistant
seedlings grown in pure series (Fig. 9.8). Seedling dry weights gave similar
results to the fresh weights. When the seedlings in the mixed series were compared to their respective controls, an inhibitory effect by the susceptible seedlings on the resistant seedlings was suggested. The fresh weights of resistant
seedlings grown in conjunction with two pots containing susceptible seedlings,
(S-R-S), were only 33% of those for resistant seedlings grown in pure series,
whereas there was no reduction for the susceptible seedlings grown in this series.
When the series contained two pots of resistant and one pot of susceptible,
(R-S-R), the fresh weights of susceptible seedlings were 78% of their control,
and the resistant seedlings' fresh weights were reduced to 60% of their control
seedlings. These authors[16] believe that the resistant seedlings may have had a
detrimental effect on the susceptible seedlings, but this effect only occurred
when there was a greater proportion of resistant than susceptible seedlings.
Furthermore, the degree of growth inhibition, in this case, was not as great as
the effect of the susceptible seedlings on the resistant ones.

Although care was taken to separate nutritional effects from other causes of
interference, the significance of this study is uncertain since biotypes of the
same species were not used. In addition, no attempt was made to isolate or
identify the postulated plant exudates that may have been responsible for
these observations. Unfortunately, this line of research has not been continued.

## 5.2 Measurements of Photosynthetic Gas Exchange in *Senecio* Biotypes

In order to better understand the physiological basis for the differences in
competitive ability between susceptible and resistant biotypes, another approach has been taken by Holt et al.[17] Characteristics of gas exchange were
measured on biotypes of *Senecio vulgaris* in the absence of all herbicide. These
studies involved measurements on whole leaves and on isolated thylakoids,
and will be reviewed below.

### 5.2.1 Whole-leaf $CO_2$ assimilation

The photosynthetic responses to light of whole attached leaves of the susceptible
and resistant biotypes of *S. vulgaris* are shown in Fig. 9.9. The susceptible biotype had higher rates of net photosynthesis than the resistant biotype at all

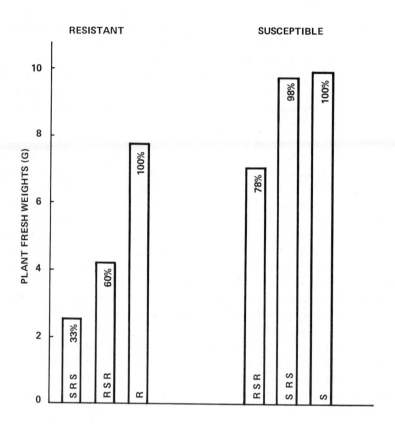

**Figure 9.8** Allelopathic interactions between triazine-resistant and triazine-susceptible *Amaranthus* spp. The susceptible seed (assumed to be *A. retroflexus*) was obtained from Valley Seed Company in California and the resistant seed (confirmed to be *A. hybridus*) was collected in Maryland. A stairstep system with three pots containing sand constituted each series and four series were run at one time. This system was fashioned after a method previously published by D. T. Bell and D. E. Koeppe, *Agron. J.*, **64**, 321 (1972), whereby nutrient solution was dripped from one pot to the next and periodically pumped back to the upper pot. Two *Amaranthus* seedlings (50 mm high) were transplanted into each pot in the stairstep system and plant heights (data not shown) were measured at 3, 4, and 5 weeks after the transplanting. Plant fresh and dry weights were obtained 5 weeks after the seedlings were transplanted. Each series of the stairstep contained one of the following arrangements of seedlings: (1) all resistant, (2) all susceptible, (3) susceptible-resistant-susceptible, or (4) resistant-susceptible-resistant. The data are the means of four replications and all tests were conducted in an air-conditioned greenhouse (85°F and 72°F maximum and minimum temperatures, respectively; 14-hr photoperiod). (This previously unpublished figure is reproduced with the courtesy and permission of J. R. Hensley and C. J. Counselman.[16])

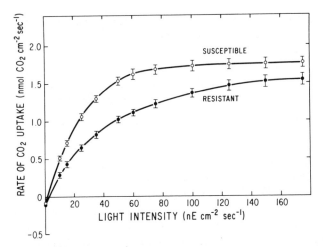

**Figure 9.9** The light response of $CO_2$ uptake by triazine-susceptible and triazine-resistant biotypes of *S. vulgaris* in the absence of herbicide. Plants used for gas exchange measurements were 30 to 60 days old from the time of seed sowing. An open-system gas exchange apparatus was used for measurements of light-dependence curves. A young, fully expanded, attached leaf was inserted into the leaf chamber. Response of net photosynthesis to light flux was measured by exposing the leaf initially to maximum irradiance from a 1500-W mercury vapor lamp (approximately 200 nE/cm²/sec). Wire-mesh screens were used to lower incident flux density. Vapor pressure deficits (5–10 mbar), leaf chamber temperatures (25°C), and ambient $CO_2$ concentrations (330–360μl/l) were held constant during measurements. Each point is an average of measurements from five individual plants. Each vertical bar depicts one standard error above, and one below, the mean. (Reproduced with permission from Holt et al.[17])

light intensities. Additionally, the resistant biotype did not become light saturated until a high light intensity was reached, at which point $CO_2$ was the limiting factor. The half-saturation light intensity for the susceptible biotype was 20 nE/cm²/sec; resistant plants had a more gradual light-dependent phase and reached half-saturation at 32 nE/cm²/sec. At similar conditions of illumination, water supply, temperature, $CO_2$ concentration, and leaf development, the photosynthetic capacity of resistant plants was markedly lower than that of susceptible plants.

Differences in net carbon fixation between the two biotypes were most pronounced at light intensities below saturation. Further measurements of gas exchange at very low light (less than 12 nE/cm²/sec), combined with measurements of leaf absorptance, demonstrated that the quantum yields of the two biotypes were different (0.070 and 0.056 moles $CO_2$/moles absorbed photons for the susceptible and resistant biotypes, respectively). Reporting photosynthesis on an absorbed light rather than an incident light basis eliminates the effect of leaf characteristics determining light absorptance, which may contribute to differences in photosynthetic rates between plants. These results suggest an intrinsic alteration in photosynthetic light harvesting ability in the resistant biotype.

Because the above measurements can be affected by differences in leaf thickness, chlorophyll content and structure, further measurements of photosynthesis were made with isolated chloroplast membranes.

## 5.2.2  Oxygen evolution experiments

The chloroplast alteration conferring resistance is thought to be in the region of photosystem II (Chapter 10). Therefore, photosynthetic oxygen evolution of photosystem II (PS II) was studied to characterize chloroplast-level differences between biotypes, and to determine the consequences of resistance on electron transport.[17] Oxygen evolution by isolated thylakoids as a function of light intensity is shown in Figure 9.10. The resistant biotype had lower rates of oxygen production than the susceptible biotype at all light intensities. On a percentage basis, the difference between biotypes was greater than two-fold at all light intensities; however, as in whole leaf measurements, the difference was greatest at low light and decreased with increasing light. The difference between biotypes at identical light levels and greater difference at low light indicate that fewer light harvesting reaction centers are operating in the resistant biotype than in the susceptible biotype.

Another technique involving a flashing light regime was used to study the precise nature of the oxygen evolving system. When subjected to a series of brief, saturating flashes after dark adaptation, broken chloroplasts show a highly reproducible pattern of oxygen emission. Oxygen yields oscillate with a periodicity of four, with a maximum after the third flash, and eventually damp to steady state after about 25 flashes.

This pattern was interpreted by Kok et al.[18] in their "S-state" model of oxygen evolution. In each reaction center complex, four photochemical reactions are required to split water and produce one oxygen molecule. In each step, one positive charge is added to the oxygen-evolving enzyme, "S", and one electron is transferred from $S_n(n = 1 \rightarrow 4)$ to $P_{680}$ in photosystem II. Each enzyme complex stores oxidation equivalents until four are accumulated ($S_4$); then oxygen is produced and the enzyme returns to state $S_0$. These reactions are accompanied by the transfer of electrons along PS II.

Two imperfections in the system, the "transition parameters", help account for the eventual damping to steady state yield. During a short flash some centers do not react, probability of misses $\alpha$; some centers react twice, probability of double hits $\gamma$; and some react normally, advancing one step, probability of single conversions $\beta$. Finally, only states $S_0$ and $S_1$ are stable in the dark; after around 10 minutes of darkness, states $S_3$ and $S_2$ deactivate to $S_1$, which consequently accounts for the maximum yield on the third flash after dark adaptation.

Figure 9.11a shows oxygen yields for *S. vulgaris* biotypes under a regime of 3-$\mu$ sec long flashes, spaced 1-sec apart, after 10 minutes of dark adaptation.

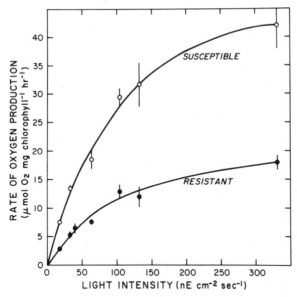

**Figure 9.10** The rate of oxygen evolution as a function of light intensity in *S. vulgaris* L. chloroplasts, measured in continuous light. Reduced light intensities were attained with neutral density filters. Measurements were made with a Clark-type oxygen electrode. The reaction mixture contained 10 mM $NaH_2PO_4$ buffer (pH 6.8), 10 mM NaCl, 5 mM $MgCl_2$, 100 mM sorbitol, 0.5 mM $K_3Fe(CN)_6$, 10 mM methylamine, and 40 $\mu$g Chl/ml for both biotypes. Initially the samples were anaerobic. Each point represents an average of four measurements. Each vertical bar depicts one standard error above, and one below, the mean. (Reproduced with permission from Holt et al.[17])

Susceptible chloroplasts displayed a normal pattern of oxygen yield, while the resistant biotype showed a rapid damping of the sequence after only 9 flashes. In Figure 9.11b, oxygen yield was plotted as a function of flash number. The two different ordinates, 0 to 10 for susceptible and 0 to 2 for resistant, indicate that in resistant plants the total oxygen yield per mg chlorophyll and the steady state yield were less than 20% that of susceptible chloroplasts. Therefore, just as in continuous light, in flashing light fewer reaction centers are evolving oxygen in resistant chloroplasts.

The parameters of charge accumulation in oxygen evolution in *S. vulgaris* chloroplasts are also dramatically different between biotypes (Table 9.3). Susceptible chloroplasts exhibited the same parameters as described for normal systems,[18] which are 10% misses, 5% double hits, and 85% single hits. Resistant chloroplasts are very different, with an $\alpha$ of 24%, $\gamma$ of 10%, and only 67% of the flashes yield single photoconversions ($\beta$). Therefore, under a flashing light regime, the reduced number of reaction centers which continue to operate in resistant plants do so abnormally. The larger $\alpha$ and $\gamma$ in resistant plants may

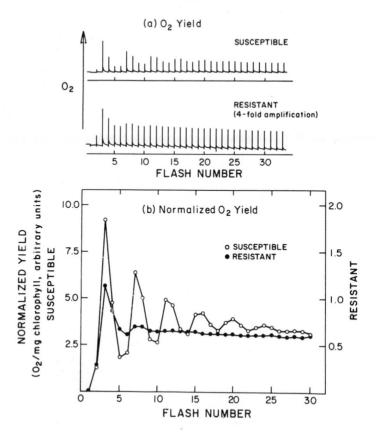

**Figure 9.11**  Oxygen evolution in flashing light after 10 min of dark equilibration at 25°C, measured in broken chloroplasts of *S. vulgaris* biotypes. (a) Representative recorder traces for oxygen evolution yield sequences in the presence of saturating 3-μsec light flashes at 1-sec intervals. (b) Oxygen yield sequences normalized to an average steady-state yield of oxygen for each biotype as a function of flash number, derived from (a). The reaction mixture contained 100 mM $NaH_2PO_4$ buffer (pH 7.0), 100 mM NaCl, and either 0.285 mg Chl/ml for the resistant biotype or 0.103 mg Chl/ml for the susceptible biotype. Values in (b) were corrected for differences in chlorophyll concentration and recorder sensitivity settings between biotypes in (a). The experiments were replicated six times; however, error bars for standard errors are smaller than the plotted points. (Reproduced with permission from Holt et al.[17])

help account for the lowered yield and rapid damping to steady state after only 9 flashes. $S_0^{(o)}$ and $S_1^{(o)}$ represent the fractions of reaction centers in those oxidation states after 10 minutes dark. Both biotypes have a dark distribution close to the 25% $S_0^{(o)}$/75% $S_1^{(o)}$ ratio of normal systems. After dark adaptation the reaction centers of both types are at the same oxidation level, yet in the light they behave very differently.

Table 9.3  Kinetics of Charge Accumulation in Photosynthetic $O_2$ Evolution of Biotypes of *S. vulgaris* L[a]

| Parameter | Biotype | |
| --- | --- | --- |
| | Susceptible | Resistant |
| $\alpha$ (Misses) | .100 ± .004 | .235 ± .011 |
| $\gamma$ (Double hits) | .035 ± .005 | .099 ± .005 |
| $\beta$ (Single hits) | .865 ± .005 | .666 ± .012 |
| $S_0^{(o)b}$ | .24 ± .016 | .27 ± .005 |
| $S_1^{(o)}$ | .76 ± .016 | .73 ± .005 |

*Source:* Reproduced with permission from Holt, Stemler, and Radosevich.[17]

[a]Values for transition probabilities were calculated from non-normalized $O_2$ yields in response to 3-$\mu$ sec flashes of saturating light at 1-sec intervals, after 10 min of dark equilibration at 25C. The experiments were replicated six times. Standard deviations are presented.

[b]$S_0^{(o)}$ and $S_1^{(o)}$ represent fractions of the reaction centers in the $S_0$ and $S_1$ states, respectively, after 10 min of dark equilibration at 25C.

Lowered yields and aberrant patterns of oxygen production indicate that the resistant biotype, with fewer reaction centers operating at any given time, was less photoefficient compared to the susceptible biotype. Two explanations were offered for this difference.[17] First, the proposed chloroplast membrane alteration in the binding protein on the reducing side of PS II has also affected electron transport. Chlorophyll fluorescence data (Chapter 10) indicated the presence of higher concentrations of reduced Q ($Q^-$) in dark adapted resistant chloroplasts, and slower rates of $Q^-$ reoxidation by electron transfer to B, compared to susceptible chloroplasts. This in turn could limit the number of charge separations from $P_{680}$ to Q occurring during a light flash, and result in a "miss" in the reaction center, thereby affecting oxygen evolution as well.

A second explanation is possible. The thylakoid membrane alteration which confers resistance also may have resulted in a modification on the oxidizing side of PS II. This could result in the abnormal patterns of oxygen evolution seen in resistant chloroplasts, which affect the entire electron transport chain. A proposed modification of the oxygen evolving mechanism must remain tentative, however, until further studies are done.

### 5.2.3  Implications of the gas exchange data

These results indicate that in several facets of light response, resistant plants of *S. vulgaris* are different from susceptible ones. Oxygen yield is reduced and patterns of oxygen emission are abnormal in resistant plants. This suggests an intrinsic inefficiency in photosynthetic light harvesting and electron transferring capability, which is closely linked to the mechanism of resistance. This inefficiency, in turn, may be responsible for lowered quantum yields of carbon fixation, high intensity required for light saturation, and lowered maximum net carbon fixation in resistant plants. Other possible causes of variations in rates of photosynthesis between plants, such as stomatal number and arrangement, and leaf and mesophyll conductances, are thought not to be a factor.[17]

Photosynthetic inefficiency in the resistant biotype of *S. vulgaris* could certainly result in its less vigorous, less competitive nature compared to the susceptible. As a recently developed biotype, resistant *S. vulgaris* may not have had sufficient time, in an evolutionary sense, to respond to selection pressures which might result in a more efficient organism capable of competing with the susceptible biotype. Poor photosynthetic performance may continue to limit competition with the susceptible biotype. At the cost of lowered vigor, photosynthetic performance, and competitive fitness, triazine resistance is apparently only of benefit to the plant in field situations where triazine herbicides are repeatedly used. This information is particularly relevant to studies on transferring the resistance trait to crop species, for the possibility of transferring the trait of decreased fitness cannot be overlooked.

## REFERENCES

1  G. F. Ryan, Resistance of common groundsel to simazine and atrazine, *Weed Sci.*, **18**, 614 (1970).

2  S. R. Radosevich, and A. P. Appleby, Relative susceptibility of two common groundsel biotypes to six *s*-triazines, *Agron. J.*, **65**, 553 (1973).

3  S. R. Radosevich, and A. P. Appleby, Studies on the mechanism of resistance to simazine in common groundsel, *Weed Sci.*, **21**, 497 (1973).

4  S. R. Radosevich, and O. T. DeVilliers, Studies on the mechanism of *s*-triazine resistance in common groundsel, *Weed Sci.*, **24**, 229 (1976).

5  S. R. Radosevich, Mechanism of atrazine resistance in lambsquarters and pigweed, *Weed Sci.*, **25**, 316 (1977).

6  K. I. N. Jensen, J. D. Bandeen, and V. Souza Machado, Studies on the differential tolerance of two lambsquarters selections to *s*-triazine herbicides, *Can. J. Plant Sci.*, **57**, 1169 (1977).

7  L. D. West, T. J. Muzik, and R. E. Witter, Differential gas exchange responses of two biotypes of redroot pigweed to atrazine, *Weed Sci.*, **24**, 68 (1976).

8  C. J. Arntzen, C. L. Ditto, and P. E. Brewer, Chloroplast membrane alterations by triazines in *Amaranthus retroflexus*, *Proc. Natl. Acad. Sci. USA*, **76**, 278 (1979).

9  V. Souza Machado, C. J. Arntzen, J. D. Bandeen, and G. R. Stephenson, Comparative triazine effects upon system II photochemistry in chloroplasts of two common lambsquarters (*Chenopodium album*) biotypes, *Weed Sci.*, **26**, 318 (1978).

10  S. R. Radosevich, K. E. Steinback, and C. J. Arntzen, Effect of photosystem II inhibitors on thylakoid membranes of two common groundsel (*Senecio vulgaris* L.) biotypes, *Weed Sci.*, **27**, 216 (1979).

11  C. J. VanAssche and E. Ebert, Photosynthesis, *Residue Rev.*, **65**, 2 (1976).

12  C. Hansch, Theoretical considerations of the structure–activity relationship in photosynthesis inhibitors, *Prog. Photosynth. Res.*, **3**, 1685 (1969).

13  D. E. Moreland, Inhibitors of chloroplast electron transport: structure–activity relations, *Prog. Photosynth. Res.*, **3**, 1693 (1969).

14  S. G. Conard, and S. R. Radosevich, Ecological fitness of *Senecio vulgaris* and *Amaranthus retroflexus* biotypes susceptible and resistant to atrazine, *J. Appl. Ecol.*, **16**, 171 (1979).

**15** G. L. Stebbins, *Variation and Evolution in Plants*, Columbia University Press, New York, 1950.

**16** J. R. Hensley, and C. J. Counselman, Allelopathic interaction between triazine resistant and susceptible strains of redroot pigweed (*Amaranthus retroflexus* L.), *Weed Sci. Soc. Am. Abstr.*, (1979), p. 110.

**17** J. S. Holt, A. J. Stemler, and S. R. Radosevich, Differential light responses of photosynthesis by triazine-resistant and triazine-susceptible *Senecio vulgaris* biotypes, *Plant Physiol.*, **67**, 744 (1981).

**18** B. Kok, B. Forbush, and M. McGloin, Cooperation of charges in photosynthetic oxygen evolution. I. A linear four-step mechanism, *Photochem. Photobiol.*, **11**, 457 (1970).

# CHAPTER TEN

# The Mechanism of Chloroplast Triazine Resistance: Alterations in the Site of Herbicide Action

CHARLES J. ARNTZEN, KLAUS PFISTER, and KATHERINE E. STEINBACK

MSU/DOE Plant Research Laboratory
Michigan State University
East Lansing, Michigan

## 1 INTRODUCTION

Earlier chapters have outlined the evolution of triazine-resistant weed biotypes throughout the world. Initial attempts at understanding the mechanism(s) responsible for herbicide resistance in the newly discovered weed biotypes lead a number of laboratories to investigate alterations in uptake, translocation, or metabolism of the applied triazines. In the cases examined to date, only small differences between susceptible and resistant biotypes have been established and these differences have been regarded as insufficient in explaining the mechanism for extreme herbicide resistance as has been described in detail in Chapter 9. Other means of attaining resistance to triazines are discussed in Chapter 8.

The primary mode of action of the $s$-triazines is to inhibit photosynthesis. This has directed attention to analysis of chloroplast reactions in the resistant weeds. These studies, summarized below and in Chapter 9, demonstrate that there are alterations in the chloroplast membranes in resistant weed biotypes; $s$-triazine binding is severely reduced, explaining the resistance.

Our purpose is to document the data that allowed an evaluation of triazine binding effects upon chloroplasts in normal and resistant weeds. As an understanding of the validity of these data requires an appreciation of the biochemistry of photosynthetic processes, we will (*a*) provide a background for the

---

Klaus Pfister is now at the Basic Botany Laboratories, Ciba-Geigy Ltd., Ch-4003, Basle, Switzerland.

understanding of biochemistry and biophysics of photosynthesis, (*b*) describe the experimental procedures used to detect how a herbicide affects photosynthetic reactions, and (*c*) describe changes in the chloroplast membranes which have been observed in triazine-resistant weeds.

## 2  SITES WHERE HERBICIDES MIGHT AFFECT PHOTOSYNTHESIS

Many commercial herbicides act by inhibiting photosynthesis. As a general framework for describing the mode of action of triazines and related inhibitors of photosystem II (PS II), the following information is included to provide a general description of our current understanding of the biochemical steps involved in the overall photosynthetic process.

The pigments and enzymes that catalyze photosynthesis are compartmentalized within the cell in chloroplasts (see Fig. 10.1*a*). Isolated, intact chloroplasts

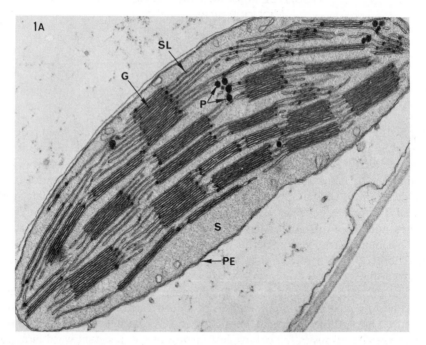

**Figure 10.1**  Structure of chloroplasts. (*a*) Transmission electron micrograph of an intact higher plant chloroplast. The internal thylakoid membrane network consists of highly ordered appressed lamellae which form grana stacks (G); these are interconnected by unappressed stroma lamellae (SL). Surrounding these membranes is the soluble matrix or stroma (S) of the chloroplast and lipid body inclusions termed plastoglobuli (P). Both thylakoid membranes and chloroplast stroma are spacially separated from the rest of the plant cell by a double limiting plastid envelope membrane (PE).

are capable of light-mediated $CO_2$ fixation as well as other metabolic processes, including protein synthesis. The chloroplast can, therefore, be considered as a partially autonomous cellular organelle, which is the sole location for all molecular events involved in radiant energy capture.

Photosynthesis can be subdivided into two general processes: light-dependent production of ATP and NADPH, and "dark reactions" which utilize the NADPH and ATP for $CO_2$ fixation into sugars. Separation of the two processes for biochemical characterization can be easily accomplished. Intact chloroplasts can be broken to release the soluble, stroma enzymes that catalyze $CO_2$ fixation, whereas the stroma-free membranes can be easily centrifuged out of solution for analysis. These thylakoid membranes contain chlorophyll and proteins needed for electron transport and photophosphorylation (i.e., the light reactions) (see Fib. 10.1$b$).

Many commercial herbicides that act by inhibition of photosynthesis affect the light reactions.[1,2] Others can affect processes as diverse as pigment biosynthesis. The next sections of this discussion are devoted to a characterization of the enzyme components of the chloroplast membranes, with special attention devoted to those constituents that appear to be the primary site of action of herbicides such as triazines.

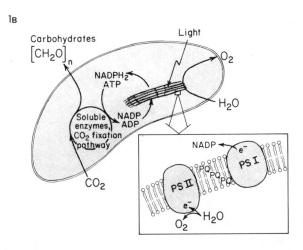

(b) Structural compartmentalization of chloroplast photosynthetic functions. The chlorophylls and enzymes of photosynthetic electron transport are localized in the thylakoid membranes; these mediate the light reactions, which convert light energy into utilizable chemical energy (ATP, $NADPH_2$) via two enzyme–pigment complexes (PS II and I) which are embedded in a lipid bilayer and are interconnected in series by a lipid-soluble plastoquinone pool (PQ) (see the insert). The ATP and $NADPH_2$ are utilized in the soluble stroma of the chloroplast in the formation of metabolic intermediates by the dark-reaction enzymes of the $CO_2$ fixation pathway.

### 3  LIGHT REACTIONS OF PHOTOSYNTHESIS

The thylakoids are made up of both proteins, which carry out specific functions of the light reactions of photosynthesis, and lipids, with less-defined functions. There are two major steps in these reactions (Fig. 10.2). First, electrons are removed from water and transferred to a pool of plastoquinone (PQ) molecules via a series of enzymes collectively termed photosystem II (PS II). Light absorbed by chlorophyll in a second complex of enzymes termed photosystem I (PS I) catalyzes the removal of electrons from the plastoquinone pool and the transfer of these electrons to NADP with the concomitant synthesis of ATP.

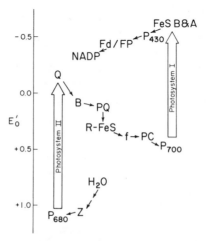

**Figure 10.2**  Series of components that catalyze noncyclic electron transport in the photosynthetic light reactions. Light energy absorbed by PS II pigments is transferred to the reaction center chlorophyll $P_{680}$; the PS II reaction results in reduction of a plastoquinone, Q, and oxidation of a primary electron donor, Z (a strong oxidant that removes electrons from water via a $Mn^{2+}$-containing enzyme). $Q^-$ is subsequently oxidized by a protein-bound plastoquinone, B. These reactions occur in a structural complex of polypeptides collectively termed photosystem II (see the inset diagram of Figs. 10.1$b$ and 10.4).

A pool of lipid-soluble bulk-phase plastoquinone (PQ) accepts pairs of electrons from $B^{2-}$, and protons from the stroma; the resultant hydroquinone ($PQH_2$) acts as a shuttle to interconnect the structurally separate PS II and PS I complexes (see inset of Fig. 10.1$b$). $PQH_2$ is reoxidized near the inner membrane surface (with proton release internally). The electron acceptor from $PQH_2$ is a Rieske iron-sulfur protein (R-FeS) (see ref. 36); electron transfer then occurs to cytochrome f, a copper-containing enzyme called plastocyanin (PC) and finally to $P_{700}$, the reaction center chlorophyll of PS I. Light absorbed by PS I results in donation of electrons from $P_{700}$ to two closely associated, membrane-bound iron-sulfur proteins designated as FeS B and A. These transfer electrons to soluble ferredoxin (Fd), which reduces NADP via a flavoprotein (FP) called Fd-NADP oxidoreductase.

At the heart of each photosystem is a chlorophyll molecule in a specialized environment (presumably created within a protein constituent of the chloroplast membranes). This "reaction center" chlorophyll molecule is uniquely located such that it is capable of losing an excited-state electron to an adjacent acceptor molecule. In addition, within this environment there is also an electron donor capable of transferring electrons back to the oxidized reaction center chlorophyll molecule. The essence of the photosynthetic process is that light energy, captured by an array of light-harvesting pigments, is transferred to the special reaction center chlorophyll, thus creating an excited-state electron. The loss of this "excited electron" from the chlorophyll to an acceptor molecule, and the re-reduction of the chlorophyll by its associated donor, give rise to the charge separation which is the fundamental step in all photosynthetic energy conversion.

## 3.1 The Photosystem II Reaction Center

The reaction center chlorophylls for the two photosystems are designated as $P_{680}$ or $P_{700}$ for PS II and PS I, respectively. The nomenclature indicates the absorption maximum for the specialized pigments serving in the reaction center complex. Both reaction centers contain chlorophyll $a$; the spectral differences are presumably determined by different polypeptides housing the reaction center complexes and thereby creating different microenvironments.

Within PS II, an enzyme that has $Mn^{2+}$ as a cofactor receives electrons from water. These electrons are then transferred to $P_{680}$ via Z, a theoretical unidentified electron carrier. It is now widely accepted that the primary electron acceptor for PS II is a specialized plastoquinone molecule called "Q." Because of its unique microenvironment, this quinone acts as an one-electron carrier. It can be reduced only as far as to the semiquinone anion state.[3]

Many early studies of photosynthesis inhibitors have provided evidence for a target site near the reducing side of PS II. As the conclusions of following sections focus on the second electron carrier as the specific site of herbicide action, we devote an extensive review to this membrane constituent.

### 3.1.1 The Reducing Side of Photosystem II

A current understanding of the second electron carrier on the reducing side of PS II has largely come about through biophysical analyses of electron transport. Spectral studies have clearly shown that a charge separation catalyzed by the PS II reaction center results in a one-electron reduction of Q. Analysis of how these electrons are transferred to PS I was initially investigated by using short flashes of light to induce single turnover events of the PS II reaction center.[4] By kinetic analysis, it was established that after one turnover of a dark-adapted PS II complex, the electron initially transferred to Q by $P_{680}$ was unable to pass through the electron transport chain to $P_{700}$. In contrast, after two very short flashes (indicating two turnovers of the PS II reaction center and two sequential reductions of Q), a pair of electrons was passed through the electron transport chain and were made available to $P_{700}$. The conclusion of these kinetic studies[4-6] was that a "gating" mechanism exists at the level of B, an electron carrier in the PS II complex. It has been concluded that this "gating" component must be physically isolated from similar membrane components or other carriers in the electron transport chain since it was stable in a singly reduced state (B$^-$) for relatively long periods of time.[4,5] Subsequent spectral analyses indicated that a plastoquinone molecule (the cofactor of B) undergoes conversion to a semiquinone anion after one flash and that this is stable for many seconds.[7,8] This redox behavior was considered unusual; semiquinones in the bulk lipid phase of chloroplast membranes, or in solutions, rapidly undergo disproportionation and protonation. The spectral studies,[7,8] therefore, lead to the conclusion that PS II contains two unique plastoquinone molecules. The first is the cofactor of the primary electron acceptor Q, as mentioned above. The second bound plastoquinone is the cofactor of the electron carrier, which we now identify as B. It seems highly probable that the properties influencing the quinone cofactor of B can only be ascribed to a unique environment created by the apoprotein of this electron transport component (for further discussion, see ref. 9).

Upon completion of two sequential transfers of electrons from Q to B, giving a fully reduced quinone (B$^{2-}$), the pair of electrons is rapidly donated to another bulk-phase plastoquinone molecule. For each electron transport chain, there are approximately three to six plastoquinones which act as mobile electron transport components.[10] The reduction of one of these molecules by a pair of electrons from B is accompanied by protonation to a fully reduced hydroquinone (PQH$_2$) utilizing protons removed from the external aqueous phase near the membrane.[11] The function of the reduced, bulk-phase PQ is to transfer electrons between the structural complexes of PS II and PS I (see diagrammatic indication in Fig. 10.1$b$). Urea and $s$-triazine-type herbicides are considered to act on the reducing side of PS II.

## 3.2  Photosystem I

The reduced, bulk-phase plastoquinone (PQH$_2$) transfers electrons to $P_{700}$ (the PS I reaction center chlorophyll) via a Rieske iron-sulfur center (R-FeS), cytochrome f, and the copper-containing enzyme, plastocyanin (PC), which ultimately serves as the electron donor to the special reaction center chlorophyll of PS I ($P_{700}$). Two nonheme iron-sulfur centers (FeS B and A) act as the primary electron acceptors for $P_{700}$. These transfer electrons to NADP via ferredoxin (Fd) and a flavoprotein (FP), functioning as Fd/NADP oxidoreductase. Bipyridilium-type herbicides are considered to operate on PS I.

## 3.3  Localization of Herbicide Receptor Sites on Thylakoids

A long-range goal of biochemical studies of herbicide effects on chloroplast membranes is the identification and isolation of the membrane constituent which is the actual herbicide receptor. This endeavor requires an understanding of the nature of the receptor and its interactions with other membrane components. One approach to this problem has entailed the use of enzymatic modification of chloroplast membrane polypeptides; the proteolytic digestion of thylakoids by trypsin was found to result in the selective alteration of membrane proteins.[12-14] The trypsin treatment resulted in the loss of action of many PS II inhibitors.[13,14]

The fact that trypsin diminishes herbicide activity in isolated chloroplasts has been exploited by several investigators in an attempt to focus on the identity of the inhibitor-binding component. An experimental approach used to localize selective effects of trypsin has been to analyze kinetically changes in the protein content and functional activities of isolated thylakoids. Using low trypsin concentrations and short incubation times, it has been observed that an alteration (or removal) of surface-exposed polypeptides reduces PS II-dependent electron transport activity over a time course when PS I partial reactions remain unaffected.[15-17] The binding of radioactive atrazine to the isolated, trypsinized thylakoids was lost, with a parallel loss of PS II-dependent dichlorophenolindophenol (DCPIP) photoreduction[18] (Fig. 10.3). It has been concluded. that a specific component of the PS II complex is exposed on the surface of the thylakoids and it is accessible to degradation by trypsin.[16,18] Modification of

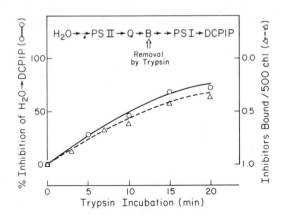

**Figure 10.3** Concurrent loss of PS II activity and triazine binding upon trypsinization. Trypsin incubation of isolated spinach chloroplasts causes a parallel reduction of PS II-mediated electron transport and loss of inhibitor binding sites due to selective digestion of the apoprotein of B. Stroma-free thylakoid membranes (100 $\mu$g Chl/ml) were incubated with 2 $\mu$g/ml trypsin for 0 to 20 min at room temperature in Na phosphate, pH 7.2 buffer containing 100 mM sorbitol, 5 mM MgCl$_2$. The action of the trypsin was stopped at various points by the addition of a 20-fold excess of trypsin inhibitor, centrifugation, and subsequent washing in trypsin-free medium. For details of trypsin incubation conditions, spectrophotometric determination of PS II-mediated reduction of DCPIP, and determinations of inhibitor binding, see refs. 15 and 30. (Data adapted from Steinback et al.[18])

this component results in a block of PS II-mediated electron flow into the PQ pool and a simultaneous loss of herbicide binding. These data are evidence for the involvement of a PS II protein in creating the PS II : inhibitor binding site. This protein may either be the apoprotein of B or another polypeptide of the PS II complex, which is very closely associated with B such that it sterically influences the function of the bound plastoquinone.

The proteins that comprise the reaction centers and electron carriers for both photosystems I and II are thought to be organized in separate functional complexes.[19] There are now a variety of techniques involving either charged or uncharged detergents to solubilize the lipid phase of chloroplast membranes. This results in the release of detergent-coated hydrophobic protein complexes which can be then characterized for photochemical activity and herbicide sensitivity. It is thus possible to obtain highly purified complexes which will carry out the reactions of either PS II or PS I. These preparations have been used to analyze polypeptide content of both photosystems.

A schematic representation of the structural organization of proteins in the PS II complex is presented in Fig. 10.4. This model emphasizes that at least six polypeptides are complexed into a functional PS II unit; these include pigment-binding proteins as well as electron carriers. The entire complex spans the lipid bilayer, which forms the fluid matrix of the membrane. The model indicates an exposed surface of the apoprotein of B and that a herbicide receptor region (binding site) exists within this protein. Implicit in the concept of a herbicide binding site at B is the idea that binding results in an inactivation of the function of this component. The nature of herbicide-induced inactivation can be characterized by a number of biochemical and biophysical techniques. These include the measurement of electron transport by biochemical assays and analysis of PS II photochemistry via chlorophyll fluorescence—these techniques are described below.

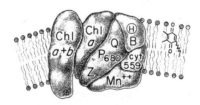

**Figure 10.4** Structural model for the organization of functional components in the photosystem II complex. The pigment protein and enzymes catalyzing PS II electron flow are thought to be localized in a structural complex that spans the lipid bilayer of the membrane. Surface-exposed segments of some polypeptides of the complex confer susceptibility to externally added proteolytic enzymes (see the text). A possible location for the PS II-herbicide binding site is indicated by "H." Background information relating to the development of this model was presented in ref. 1.

## 3.4  Herbicide Inhibition of PS II Electron Transport in Susceptible and Resistant Weed Biotypes

As described in the preceding section, a series of enzymes, as well as a lipid-soluble pool of plastoquinone, catalyze the transfer of electrons from water to the terminal electron acceptor NADP in the *in vivo* noncyclic electron transport pathway. By using various artificial electron donors and/or acceptors, it is possible to monitor the activity of PS II and PS I. These partial reactions can be used to localize the region of the electron transport chain where a specific inhibitor exerts its mode of action or to quantify the efficiency of the inhibitor.

Various assays are used to measure electron transport dependent upon PS II activity, and the concomitant release of oxygen due to oxidation of water, the so-called Hill reactions. The most widely used assays measure reduction of ferricyanide or the monitoring of reduction of a dye such as DCPIP. Either reaction is highly dependent upon turnover of the PS II reaction center. There is now abundant evidence that the primary mode of action of inhibitors such as diuron and atrazine is to block PS II.[20,21]

Photoreduction of DCPIP was used to test the activity of herbicides affecting PS II in triazine-resistant and triazine-susceptible chloroplasts (see Chapter 9). The goal in such assays was to quantify differences in resistance to various chemical families of inhibitors of PS II in the chloroplasts of newly discovered triazine-resistant weed biotypes. An example of the differences has already been shown in Fig. 9.4. In each case, the triazines showed $I_{50}$ concentrations over the range of $3 \times 10^{-7}$ to $4 \times 10^{-8}$ M in susceptible chloroplasts, with approximately 1000-fold higher concentrations needed to achieve similar inhibition in the chloroplasts from resistant biotypes.

To determine the degree of resistance that has occurred in the resistant chloroplasts, the $I_{50}$ values for a givern herbicide in the susceptible and resistant plants were compared. This degree of resistance (the R/S value) is defined as the $I_{50}$ (resistant) divided by the $I_{50}$ (susceptible) for the same compound. Examples of the R/S values for the *s*-triazines for chloroplasts isolated from a variety of different weed species are presented in Fig. 10.5*a*. Similar findings were obtained using other techniques (see Table 3.1). It should be noted that there are slight differences among the species, but the general trend of the resistance phenomenon is consistent.

The effects of an ethylthio-*s*-triazine (dipropetryn) and an asymmetrical triazinone (metribuzin) on PS II electron transport in the susceptible and resistant chloroplasts from *Amaranthus* sp. were also measured. The R/S values values for these herbicides were 270 and 290, respectively. When metribuzin was tested with chloroplasts of other weed species, there were marked differences in the R/S values. Examples of R/S determinations are: *Chenopodium album*, 33; *Brassica campestris*, 250; *Ambrosia artimesiifolia*, 21. These differences probably indicate that the "biochemical phenotype" of resistant weeds is not always identical (the $I_{50}$ concentration for all susceptible chloroplasts were nearly identical). This may indicate that different modifications have

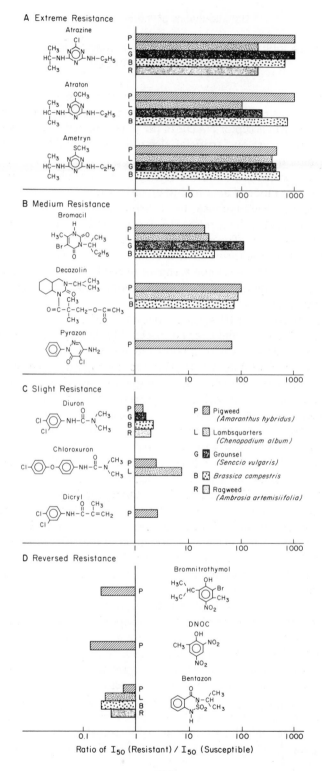

A Extreme Resistance

Atrazine

Atraton

Ametryn

B Medium Resistance

Bromacil

Decazolin

Pyrazon

C Slight Resistance

Diuron

Chloroxuron

Dicryl

P ▨ Pigweed
    (*Amaranthus hybridus*)

L ▨ Lambsquarters
    (*Chenopodium album*)

G ▨ Grounsel
    (*Senecio vulgaris*)

B ▨ *Brassica campestris*

R ▨ Ragweed
    (*Ambrosia artemisiifolia*)

D Reversed Resistance

Bromnitrothymol

DNOC

Bentazon

Ratio of I₅₀ (Resistant) / I₅₀ (Susceptible)

194

occurred in the binding site (see discussion below) in the different weeds. We emphasize, however, that the final result is always the same—a binding-site modification is responsible for herbicide resistance.

R/S values for *s*-triazines were all in the range 200 to 1000 (Fig. 10.5*a*). This indicates extreme herbicide resistance residing in the thylakoids. In a survey of other chemical families, it was found that bromacil (a uracil herbicide), decazolin (a quinazoline herbicide), and pyrazon or brompyrazon (pyridazinone herbicides) gave R/S values in ranges of 20 to slightly greater than 100 (Fig. 10.5*b*). These compounds fall in a group we designate as "medium resistance." In further tests, diuron and chloroxuron (urea herbicides) and dicryl (an amide herbicide) showed R/S values in the range 1.4 to 7.5 (Fig. 10.5*c*); we regard these as showing only "slight resistance." An unexpected trend appeared when various nitrophenol derivatives were characterized in the PS II assay system. For example, bromnitrothymol, DNOC, and the unrelated compound bentazon showed increased activity in the triazine-resistant chloroplasts; R/S values for the nitrophenol derivatives ranged from 0.1 to 0.6 (Fig. 10.5*d*). Similar results were observed using *Amaranthus* chloroplasts with dinoseb (R/S = 0.08), ioxynil (a nitrile herbicide; R/S = 0.6) and benzazin (a benzoxazinone herbicide; R/S = 0.8). We designate this pattern as examples of "reversed resistance." It has not yet been established whether trends of varying resistance to these individual herbicides or increased susceptibility to the nitrophenol-type herbicides will be detected when intact, resistant plants are compared to the susceptible controls. This is an interesting area of weed physiology where more research is needed.

## 4  CHLOROPHYLL FLUORESCENCE FOR DETERMINING SITES OF HERBICIDE ACTION

A goal of studies of herbicide effects on thylakoids is to understand the mode of action of the chemicals at the molecular level. For PS II-inhibiting herbicides, this is greatly facilitated by using chlorophyll fluorescence as a direct monitor of PS II primary photochemistry.

Absorption of a photon by a chlorophyll molecule results in the appearance of an "excited state" as the electron in the pigment molecule attains a higher energy level. This excited state is unstable; the electron is either lost from the molecule or, after very short times, the electron returns to its ground state. The most likely fate of the high-energy electron in PS II is transfer to the primary acceptor Q. This results in the storage of the absorbed light energy in a

---

**Figure 10.5** Summary of herbicide resistance patterns in various weed species as determined from PS II-mediated DCPIP photoreduction by isolated chloroplasts of susceptible and resistant biotypes. The R/S values expressed in the figure were calculated from inhibition curves such as shown in Fig. 9.4.

useful chemical form (a reduced enzyme, $Q^-$). If the reaction center chlorophyll of PS II is excited to the high-energy state under conditions in which Q is reduced (i.e., unable to accept another electron), the excited electron can return to its ground state by release of heat or by reemission of a photon of a lower energy level as fluorescence. Measurement of the reemitted light is a tool for direct analysis of the normal functioning of the PS II reaction centers. The chlorophyll fluorescence at room temperature arises almost entirely from PS II[22]; this fortuitous feature of the organization of the two photosystems makes chlorophyll fluorescence analysis a highly selective procedure for solely analyzing the PS II photochemical processes.

## 4.1 Measurement of Fluorescence Induction Transients

Blue light is used for chlorophyll excitation and a light detector senses the red fluorescence near 685 nm. Following onset of illumination of dark-adapted chloroplasts, chlorophyll fluorescence rises immediately to a level called $F_0$. This $F_0$ fluorescence is considered to arise from either inactive photochemical reaction centers which have a very high fluorescence yield or from the antenna pigment itself. Following attainment of $F_0$, there is a time-dependent increase in fluorescence to a maximal level termed $F_M$. The attainment of the $F_M$ level of fluorescence indicates that all PS II primary electron acceptors are fully re-

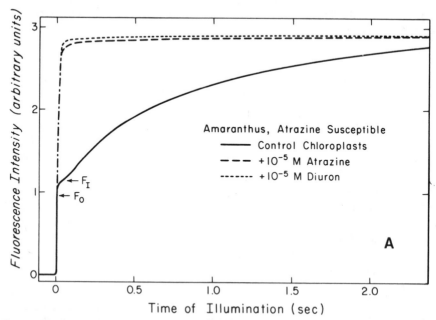

**Figure 10.6** Chlorophyll fluorescence transient changes observed upon illumination of dark-adapted chloroplasts isolated from susceptible (10.6a) and resistant (10.6b) *Amaranthus hybridus* biotypes. $F_O$ levels were determined within 1 msec after complete shutter opening, and the $F_I$ level after 50 msec of illumination. The higher $F_I$ level in uninhibited resistant chloroplasts as compared to the

duced. The time it takes for fluorescence to rise to a maximum level gives information about the electron transport steps occurring prior to complete accumulation of $Q^-$.[23]

The addition of herbicides affecting PS II to a suspension of isolated chloroplasts dramatically alters the characteristics of a chlorophyll fluorescence induction (the time-dependent change in fluorescence emission occuring after onset of illumination of dark-adapted chloroplasts). As is shown in Fig. 10.6a both atrazine and diuron dramatically stimulate the fast portion of the fluorescence induction rise. These experiments, conducted with susceptible chloroplasts, can be interpreted by assuming that the inhibitors block electron flow very near the reducing site of PS II (i.e., as the inhibitors block electron transfer from $Q^-$ to other carriers, a rapid accumulation of $Q^-$ occurs). By calculating the areas above the induction curves (with and without herbicides), one can determine the relative site of action of various inhibitors in the comparative set of assays. The usefulness of this approach for characterizing the similarity in herbicide site of action has been discussed in greater detail (see ref. 24).

When chloroplasts were isolated from resistant weed biotypes and the samples were subjected to chlorophyll fluorescence induction transient analysis, it was observed that diuron, but not atrazine, could alter the transient (see Fig. 10.6b). This is consistent with a lack of inhibition of photosynthesis by atrazine in the isolated resistant chloroplasts. We found that the initial, rapid-

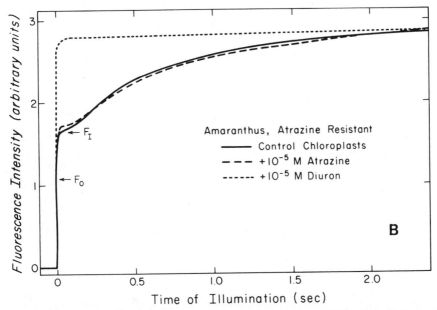

susceptible controls (Fig. 10.6b vs. 10.6a) indicates a reduced rate of $Q^-$ to B electron transport (discussed in ref. 25 and text). A stimulation of the fluorescence rise by herbicide addition is a direct indication of a block in electron transfer in the PS II complex. (Data adapted with permission from Arntzen et al.[25])

increase portion of the fluorescence induction curve ($F_i$ level) occurring in the absence of herbicides was larger in the herbicide-resistant plants (Fig. 10.6; see ref. 25 for further details). The interpretation of this observation is that the primary electron transfer steps on the reducing side of PS II are altered in the herbicide-resistant chloroplasts. This innate characteristic of the resistant thylakoids indicates a reduced rate of electron transfer from $Q^-$ to B, even in the absence of added inhibitors. This will be discussed more extensively below.

## 4.2  Measurement of Fluorescence Following Short Flashes

In Section 4.1, techniques were described that allow the analysis of changes in chlorophyll fluorescence during continuous illumination. These procedures utilize a measuring beam that is actinic and also induces changes in the oxidation–reduction state of the electron transport chain. As an alternative procedure, it is possible to use a very low-intensity measuring beam and still detect the level of chlorophyll fluorescence from the sample. In these experiments, the goal is to add a secondary perturbation of the system and monitor changes in fluorescence of a sample that has a largely oxidized electron transport chain. Experiments using short, intense flashes have been of great value in determining characteristics of the Q to B portion of the electron transport chain.[6,9,29,27] In addition, the data described below provide evidence pertaining to the molecular mechanism of action of inhibitors of PS II.

Figure 10.7 shows characteristic changes in fluorescence which occur in the presence and absence of inhibitors of photosynthesis following flash illumination. Isolated chloroplasts were illuminated with a single, intense flash that resulted in a single turnover of PS II reaction centers. By monitoring the fluorescence of the sample with a very weak measuring beam, changes in the redox state of Q could be monitored. If a single turnover event of PS II occurred in the absence of herbicide, there is no net change in the level of fluorescence (Fig. 10.7a). This indicates that the electron transferred to Q during the single intense flash was subsequently transferred to the rest of the electron transport chain in a time period faster than that resolved by our instrumentation. In contrast, if atrazine or diuron (Fig. 10.7b, c) were added to the sample prior to the single turnover flash, a very large increase in fluorescence occurred. This indicates that the site of action of these herbicides is immediately after Q; that is, a single turnover flash on the PS II reaction center can reduce Q completely, resulting in an increase in fluorescence. The electrons could not be removed from $Q^-$ by the remaining portion of the electron transport chain.

The effect of adding PS II-herbicides to dark-adapted or preilluminated chloroplasts was previously investigated extensively.[5,6] We have repeated similar experiments using *Amaranthus* chloroplasts to monitor the site of action of diuron and atrazine[7] (Fig. 10.8). With chloroplasts illuminated by a very weak measuring beam, the addition of diuron to a dark-adapted sample resulted in little or no change in fluorescence (i.e., Q remained oxidized). In contrast, if the chloroplasts had been subjected to one intense, 5-$\mu$ sec flash shortly

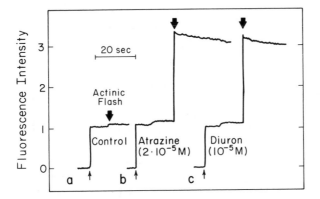

**Figure 10.7** Changes in chlorophyll fluorescence level in chloroplasts from susceptible *Amaranthus* sp. induced by a single saturating actinic flash (indicated by the downward arrow). Fluorescence was detected by a very low intensity measuring beam (turned on at small upward arrow) which, in itself, does not result in changes in the redox state of Q. The fact that one flash (a single PS II turnover) gave a maximal fluorescence increase when herbicides were present, but not in their absence, indicates that atrazine and diuron act to block electron flow immediately after Q. Herbicides were added to the sample while the weak measuring beam was turned on but before the actinic flash; time of addition is indicated by arrows.

before diuron addition, to cause a single turnover of the PS II reaction center, there was a marked increase in fluorescence intensity caused by the herbicide. We emphasize that this change occurs in the dark; it does not reflect photoreduction of electron transport components. The diuron-stimulated increase in fluorescence ($\Delta F$) showed binary oscillations with respect to the number of prior actinic flashes (Fig. 10.8*b, c*). Earlier studies[1,6] indicated that atrazine causes similar effects on fluorescence in susceptible chloroplasts. It is to be emphasized that diuron effects on the "dark" fluorescence showed the same flash dependence in both susceptible and resistant chloroplasts (Fig. 10.8). This indicates that the inhibitor has the same mechanism of action in both samples.

### 4.3 Fluorescence to Determine the Molecular Mode of Action of PS II Herbicides

The interpretation of "dark" effects of photosynthesis inhibitors (shown in Fig. 10.8) are schematically indicated in Fig. 10.9. In dark-adapted chloroplasts, Q and the majority of B are in the oxidized state. If diuron is added in the "dark" (the weak measuring beam is disregarded here as it has no actinic effect), the herbicide can have no effect on the redox state of Q. However, when the dark-adapted chloroplasts are exposed to one intense actinic flash, an electron is transferred from $P_{680}$ to Q and then to B. This electron transfer event,

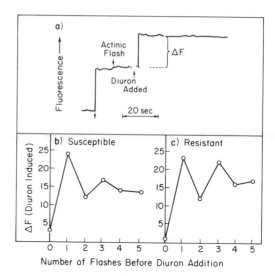

**Figure 10.8** (*a*) Stimulation of "dark" chlorophyll fluorescence from *Amaranthus* sp. chloroplasts by herbicide addition. The fluorescence was measured in a very weak measuring beam turned on at the upward arrow. After about 10 sec of measurement, a single intense actinic flash was used to cause a single electron transfer through PS II. Within 10 sec thereafter, diuron (to $10\,\mu$M final concentration) was added to the sample. An increase in fluorescence was detected, as indicated by $\Delta F$.

(*b,c*) Plot of the $\Delta F$ induced by diuron as described in (*a*) except that the sample was subjected to 0 to 5 intense actinic flashes prior to diuron addition. Chloroplasts were isolated from susceptible (*b*) and resistant (*c*) biotypes of *Amaranthus* sp. Similar data for both samples indicate an identical mode of action of diuron in both sources of chloroplasts. Similar data were observed with atrazine but only in the susceptible chloroplasts. Further details relating to the binary oscillations of $\Delta F$ are described in Fig. 10.9 and the text. (Reproduced with permission from Pfister and Arntzen.[1])

on the slow time scale of our recordings in Fig. 10.8, does not affect the fluorescence level, as $Q^-$ is quickly reoxidized and the detecting system "sees" only the oxidized Q. The stored charge on the relatively stable semiquinone $B^-$ does not affect the fluorescence yield. When diuron was added in the "dark" after an actinic flash, a fluorescence increase results from a back reaction in which the electron from $B^-$ is transferred to Q.[1,5,6] The resultant $Q^-$ formation causes a corresponding increase in fluorescence yield. Our interpretation of these data, in line with Velthuys' hypothesis,[6] is that reversed electron flow from $B^-$ to Q occurs because of inhibitor binding to B. In this interpretation, binding results in a decrease of the redox potential of the bound quinone of B with respect to that of Q (see Fig. 10.9). A decrease in the redox potential of B would, of course, result in the imposition of an energetic barrier in that portion of the electron transport chain.

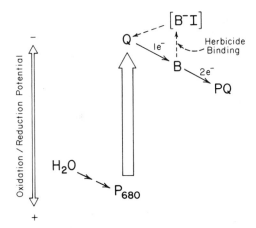

**Figure 10.9** Interpretation of the effects of herbicides on "dark" chlorophyll fluorescence as described in Fig. 10.8. The primary electron acceptor Q is believed to be a single electron carrier, whereas B accepts electrons singly but transfers these electrons to the next carrier only in pairs. In dark-adapted chloroplasts, both Q and B are oxidized. After one actinic flash, the electron donated to Q by $P_{680}$ will quickly be transferred to B, giving a stable semiquinone $B^-$. If two sequential flashes occur, two electrons arrive at B and a rapid transfer of the electron pair to PQ leaves both Q and B oxidized. Herbicide binding to PS II is suggested to alter the redox properties of the bound quinone forming the cofactor of B; the relative shift in redox coupling between Q and B results in a reversed electron flow from $B^-$ to Q, thus giving $Q^-$ in the "dark"–this increases the chlorophyll fluorescence of the sample, as shown by $\Delta F$ in Fig. 10.8. A prediction of this model is that an increase in $\Delta F$ will occur only when the second carrier is in the $B^-$ state; this is consistent with the binary oscillations in $\Delta F$ in Fig. 10.8b and c. The damping of these oscillations is due to incomplete excitation of all PS II centers during the flashes and/or inadequate times in these experiments to allow complete electron transfer between Q, B, and PQ, thus giving a mixed population of B, $B^-$. (Reproduced with permission from Pfister and Arntzen.[1])

## 4.4  Analysis of Electron Transfer from Q to B

Another important way of using chlorophyll fluorescence as a probe in herbicide research is to monitor the rate of electron transfer from $Q^-$ to B using very fast fluorimetry techniques.[9,26,27] The analytical procedures currently employed utilize a short (a few nanoseconds) laser pulse to cause a single turnover of PS II in isolated chloroplasts. At various times, beginning just a few microseconds after the laser pulse, a low-intensity measuring beam is used to detect chlorophyll fluorescence from the sample. The data obtained show a rapid decay in the yield of fluorescence at increasing times after the flash (Fig. 10.10). This decay monitors the kinetics of reoxidation of $Q^-$ by B.[26] The effects of atrazine on the fluorescence decay in susceptible chloroplasts of *Ama-*

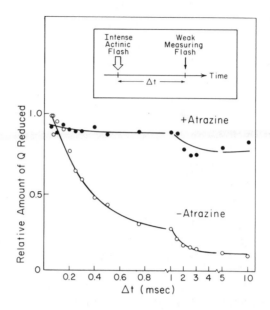

**Figure 10.10** Effect of atrazine on fluorescence decay following a single laser flash in chloroplasts of susceptible *Amaranthus* sp. The experimental protocol is defined in the inset; the chlorophyll fluorescence yield measured during the weak intensity flash is directly proportional to the amount of Q⁻ present. A time-dependent decrease in fluorescence therefore monitors the rate of electron transfer from Q⁻ to B. In the presence of atrazine, this process is blocked. (Data adapted from Bowes et al.[9])

*ranthus* is presented in Fig. 10.10; while the decay is half completed in approximately 0.4 msec in untreated samples, negligible decay in fluorescence occurred over a 10-msec duration in the atrazine-treated chloroplasts. These data are direct indications that herbicides such as atrazine do not affect the reduction of Q, but strongly inhibit Q⁻ reoxidation by the next component in the electron transport chain.

We have also used the flash-fluorimetry decay technique to monitor Q⁻ to B electron transfer in untreated samples of chloroplasts from resistant and susceptible *Amaranthus*. The data, shown in Fig. 10.11 indicate that the half-time for Q⁻ decay was approximately 0.4 to 0.5 msec in susceptible chloroplasts. In contrast, the fluorescence decay was more than 10-fold slower in chloroplasts isolated from resistant plants. The binary oscillations in the rate of Q⁻ decay were different in the resistant chloroplasts.[9] The data were interpreted to indicate that the apoprotein of B was altered in the PS II complex of the resistant plants, giving rise to a change in the microenvironment around the quinone cofactor of B. The less accessible quinone would, therefore, be thermodynamically more difficult to reduce.[9]

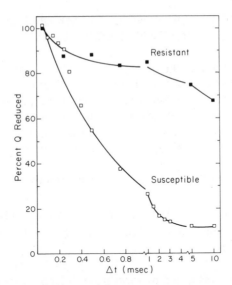

**Figure 10.11**  Analysis of electron transfer from $Q^-$ to B in chloroplasts of susceptible and resistant *Amaranthus* sp. Procedures were as defined in Fig 10.10. The data indicate a much slower electron transfer in the resistant chloroplasts. This experiment was conducted with no added inhibitors; the decrease in reaction rates is an innate property of the resistant membranes.

## 4.5  Summary of Studies Using Fluorescence as an Assay

The use of techniques to monitor chlorophyll fluorescence changes have been instrumental in understanding the exact site of action of some herbicides affecting photosynthesis, as well as providing a clue to their molecular mechanism of action. Fluorescence induction transients (Fig. 10.6) combined with single flash turnover experiments (such as shown in Fig. 10.7) clearly indicate that the site of electron transport inhibition caused by inhibitors such as diuron and atrazine is on the reducing side of PS II. The data demonstrate that Q can be reduced in diuron-inhibited chloroplasts, but cannot be reoxidized by B, the second component in the electron transport chain.

Two lines of evidence from the fluorescence experiments indicate that the target site for these herbicides affecting photosynthesis is at B. First, if B is in the semiquinone anion state ($B^-$) in the dark, addition of a herbicide such as diuron or atrazine can cause reversed electron flow to Q. This is interpreted as indicating that herbicide binding results in a change in the redox potential of B (see Fig. 10.9). In addition, with respect to the triazine-resistant weed biotypes, $Q^-$ to B electron transfer is slower in the resistant chloroplasts. This property is indicative of a modified microenvironment near B, causing it to be more difficult to reduce. This trait is characteristic of the triazine-resistant chloroplasts.

## 5  MEASUREMENT OF HERBICIDE BINDING IN SUSCEPTIBLE AND RESISTANT CHLOROPLASTS

A cursory examination of the data presented above may suggest an apparent contradiction. It has been argued from several lines of evidence that both diuron and atrazine act at the same site in the electron transport chain in susceptible chloroplasts. In spite of this, it has been shown that atrazine no longer functions in the resistant chloroplasts, whereas diuron is still effective. Three possible explanations for this apparent contradiction are:

1  Atrazine is selectively excluded from the membrane in the resistant chloroplasts.

2  Atrazine binds to chloroplast membranes of resistant plants but is inactive.

3  The specific atrazine binding domains are selectively lost in the herbicide-resistant chloroplasts.

The first possibility was tested by measuring the inhibitory activity of various concentrations of atrazine or terbutryn added to chloroplasts with short or long (10 min) incubation times prior to assay.[1] Terbutryn was used in these experiments because it is an extremely active triazine; this allowed direct determination of the $I_{50}$ concentration even in triazine-resistant chloroplasts. The results (see Table 10.1) show that no large change in sensitivity of either sample occurred after increasing incubation time. In a separate study, no change in

Table 10.1  $I_{50}$ Concentrations for the Inhibition of PS II in Chloroplasts Isolated from Triazine-Susceptible and Triazine-Resistant Biotypes of *Amaranthus hybridus*

| Sample | Preincubation Time (min) | $I_{50}$ Concentration ($\mu$M) | |
|---|---|---|---|
| | | Terbutryn | Atrazine |
| *Susceptible* | | | |
| Chloroplasts | 0 | 0.032 | 0.40 |
| Chloroplasts | 10 | 0.019 | 0.42 |
| PS II Particles | 0 | 0.025 | 0.35 |
| *Resistant* | | | |
| Chloroplasts | 0 | 5.6 | >100 (est. 400)[a] |
| Chloroplasts | 10 | 7.5 | >100 (est. 400)[a] |
| PS II Particles | 0 | 7.0 | >100 (est. 400)[a] |

*Note:* Preincubations with inhibitors were done at room temperature in the dark. Submembrane particles were prepared with digitonin as described in ref. 1. Activity was assayed in the diphenylcarbazide to DCPIP system.

[a]Because of the low solubility of atrazine, an $I_{50}$ could not be achieved.

atrazine activity was found even after 3-hr incubations of *Senecio vulgaris* chloroplasts with inhibitor.[28] As a further check on this point, detergent-derived PS II submembrane fragments were prepared.[7] These particles, in which penetration barriers should have been removed, possess the same triazine resistance as whole membranes (Table 10.1). Atrazine exclusion from PS II can therefore be ruled out as a mechanism of herbicide resistance (see also refs. 1, 28, and 29).

To test the alternative explanations, we have used direct measurements of binding of radioactively labeled herbicides to isolated chloroplast membranes.[30] The technique involved incubation of isolated, stroma-free thylakoids with radiolabeled herbicide. Following a time interval needed for equilibration, centrifugation techniques were used to separate the thylakoids from the suspension media. Distribution of bound and unbound herbicide was then determined. An example of our data using chloroplasts from triazine-susceptible and triazine-resistant *S. vulgaris* are shown in Figs. 10.12 and 10.13. The data are pre-

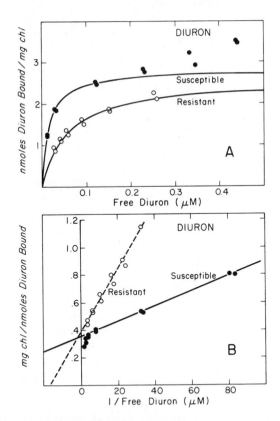

**Figure 10.12** (a) Binding curves of [14]C-diuron to isolated chloroplast membranes of susceptible and resistant *Amaranthus* sp.

(b) Data of Fig. 10.12a plotted in double reciprocal form. (Reproduced with permission from Pfister et al.[30])

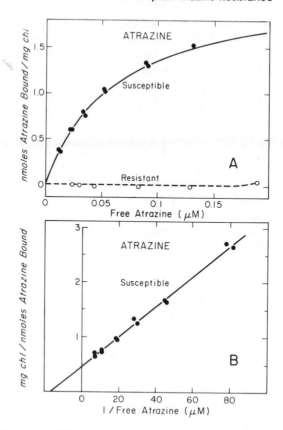

**Figure 10.13** Binding of $^{14}C$-atrazine to chloroplast membranes of *Amaranthus* sp.; data presented as in Fig. 10.12. (Reproduced with permission from Pfister et al.[30])

sented in two forms. When the amount of bound herbicide was plotted as a function of free herbicide remaining in solution, a hyperbolic binding curve was obtained which deviates from ideal form at high herbicide concentrations due to the onset of secondary, nonspecific binding. (For more details of non-specific binding, see refs. 30 and 31.) When the binding data are expressed in a double reciprocal form, the data allow a quantitative evaluation of the binding constant (abscissa intercept) as well as the number of chlorophylls per binding site (ordinate intercept). A summary of our data for both *Senecio* and *Amaranthus* chloroplasts is presented in Table 10.2. Clear evidence for a high-affinity binding site for both diuron and atrazine was observed for susceptible chloroplasts of both species. In contrast, no evidence for an atrazine binding site could be detected in the resistant chloroplasts. The values obtained for the amount of bound inhibitor on a chlorophyll basis agree well with the photosynthetic unit size measurements for similar sample preparations (unpublished data). We have previously reported Hill plots for diuron and atrazine inhibition of photosynthetic electron transport; these demonstrate that one inhibitor molecule binds per active site (i.e., per electron transport chain).[30]

Table 10.2  Calculated Binding Constants ($K$) and Number of Chlorophylls per One Bound Inhibitor (CHI/I) for Diuron and Atrazine in Chloroplasts of Susceptible and Resistant Biotypes

| | Susceptible | | Resistant | |
|---|---|---|---|---|
| | K | Chl/I | K | Chl/I |
| *Amaranthus* sp. | | | | |
| Diuron | $1.6 \times 10^{-8}$ | 410 | $3.3 \times 10^{-8}$ | 450 |
| Atrazine | $7.1 \times 10^{-8}$ | 440 | No binding detected | |
| *Senecio vulgaris* | | | | |
| Diuron | $1.4 \times 10^{-8}$ | 420 | $5 \times 10^{-8}$ | 500 |
| Atrazine | $4 \times 10^{-8}$ | 450 | No binding detected | |

*Note:* The values were determined by analysis of double reciprocal plots of data obtained from binding of $^{14}$C-diuron or $^{14}$C-atrazine to chloroplast membrane samples (see refs. 1 and 30 for experimental details).

In total, our studies of herbicide binding to isolated chloroplasts have provided definitive evidence for a nearly total loss of affinity of triazine herbicides for chloroplast membranes isolated from triazine-resistant weed biotypes. This lack of binding explains the inability of triazines to inhibit photosynthesis (and growth) in the resistant weed biotypes. We conclude that there has been a modification in a component of the chloroplast membranes that determines the biochemical specificity of the site, resulting in a highly selective alteration in the activity of herbicides directed against these photosynthetic membranes.

## 5.1  A Model of Modified Binding Sites to Explain Triazine Resistance

Based on the data presented in previous sections, we can reach the following conclusions:

1  There is one binding site for each herbicide group affecting photosynthesis in each electron transport chain (Figs. 10.12, 10.13; Table 10.2; refs. 1 and 30).
2  A single herbicide molecule binds at each binding site.[30]
3  Binding of a herbicide results in an alteration in the redox properties of the Q/B complex and a concomitant onset of inhibition of electron transport (Fig. 10.9; refs. 1, 5, and 6).
4  Trypsin can prevent herbicide binding to thylakoids; this results in a block in electron flow at the level of B (Fig. 10.3; refs. 16 and 18).

With respect to the triazine-resistant weed biotypes, we can add an additional list of thylakoid traits which have been characterized through studies of many different weed species over recent years.

1  Chloroplasts from triazine-resistant biotypes have an altered electron transport response to other herbicides, with a range of specificities being expressed for different chemical families (including a reversal of the resistance in some cases) (see Section 3.4).

2  Herbicide binding has been selectively modified in chloroplasts from all triazine-resistant weed species analyzed to date (see Section 3.4). The only exception to this is *Echinochloa* (see Table 3.1).

3  Chloroplasts from triazine-resistant weeds contain a PS II complex which is modified from those in susceptible chloroplast samples. This difference correlates with a reduction in the rate of electron transport from $Q^-$ to B even in the absence of all added inhibitors (see Section 4).

Based on these conclusions, we can state that the weed biotypes that evolved triazine resistance via a chloroplast membrane alteration must contain a thylakoid component that has a subtle alteration responsible for reduced triazine binding affinity. We believe that this herbicide binding component is a functional constituent of the PS II complex and therefore cannot be expected to be absent or even highly disorganized, as diuron and several other herbicides are still effective inhibitors, and the PS II complex, per se, is photochemically active. If such a conclusion is to be reached, it is necessary that we have a model to explain the mechanism for similar inhibitory action but variability among chemical classes of inhibitors in determining the actual binding process.

Herbicide research has extensively used structure/activity analyses of the various chemical families involved in PS II electron transport inhibition. It has been suggested that there is a common essential element which is shared by PS II inhibitors.[21,32] This consists of an electron deficient $sp^2$-carbon which is adjacent to a nitrogen with a lone electron pair. In addition to this essential element, it has been concluded that all inhibitors of PS II must contain various hydrophobic side chains which determine the efficiency of action of the various chemical classes.[32-35] These two chemical features, an essential element plus specificity-determining hydrophobic substituents, are recognized in a new model for inhibitor binding (Fig. 10.14).

The model in Fig. 10.14 defines the various domains that exist within and/or around the proposed binding sites for PS II inhibitors. We suggest that the binding sites must contain a domain that responds to the essential element of all PS II-inhibiting herbicides; this element may interact specifically with a portion of the herbicide binding site that is essential for electron transport. In addition, however, the hydrophobic side chains surrounding the essential element of the herbicide must occupy various other parts of the surrounding domain. It is known that the hydrophobic side chains for various PS II inhibitors display elaborate steric diversity and variable hydrophobicity properties. No common portion of any of these side chains is consistently required for binding and/or activity among different chemical families. We suggest that a variety of amino acids in a protein (or closely associated proteins) of PS II create domains around the site occupied by the essential element. These specify

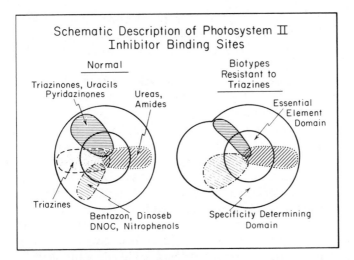

**Figure 10.14** Model of the binding sites for PS II inhibitors. The existence of two domains that determine binding properties is the major feature. The essential element domain is a region of the site shared by all herbicides affecting PS II; this may be near a portion of the quinone cofactor of B which controls redox properties of this electron carrier. The specificity-determining domain is a region of the binding site, with steric and biochemical features determined (probably by amino acid side chains), which determines the "fit" of various herbicide classes. The model suggests that any change in the binding site configuration can cause alterations in the specificity-determining domain, thus conferring altered binding properties to the various herbicides. (Reproduced with permission from Pfister and Arntzen.[1])

the selectivity of the hydrophobic regions of the inhibitor. Successful inhibition of electron transport requires interaction of herbicides with both binding site domains.

With the model of Fig. 10.14 in mind, an explanation for triazine resistance becomes apparent. By genetically modifying selective regions of the "specificity-determining domains" of the herbicide binding site, a chloroplast can selectively detect and respond to various classes of chemicals. In the model of Fig. 10.14 we have indicated that diuron and atrazine respond to markedly different portions of the specificity-determining domain. The justification for this is that R/S values for the triazines (see Section 3.4) were very large, whereas the R/S values for ureas were only slightly modified in herbicide-resistant plants. We interpret these data as indicating that some amino acids of the herbicide binding site have been altered such that atrazine occupation of the specificity-determining domain is severely limited, but these same alterations in amino acids have only marginal affects upon the diuron binding capacity. As R/S values of pyrazon, bromacil, and metribuzin were not as large as the triazines, but higher than the ureas, we interpret this as indicating that these chemical families occupy only part of the triazine specificity domain. Similarly, increasing activities of nitrophenols in the triazine-resistant plants indicate that alterations of the specificity-determining domain which cause limited triazine binding actually create a more favorable microenvironment for nitrophenols.

## 5.2  Implications of the Herbicide Binding Model

Analysis of triazine-resistant chloroplasts has lead us to conclude that some biological diversity exists within the PS II complex and that this confers differential herbicide sensitivity. The resistance that has developed alters the response to all *s*-triazines. We emphasize, however, that data obtained with isolated chloroplasts indicate that resistance should also be evident when certain triazinones, pyridazinones, and uracils are applied. Conversely, bentazon and some nitrophenol herbicides have enhanced inhibition in the resistant chloroplasts. These compounds may be more efficient in the field; the information derived in laboratory studies should be of value in design of herbicide testing programs in areas where triazine resistance is already a problem.

A second implication of the herbicide binding model of Fig. 10.14 relates to the design of chemicals that might act as protectants against PS II inhibitors. A useful protectant would be a compound that could be applied to a desirable crop plant and would then protect that plant from subsequent damage by a PS II-inhibiting herbicide. Ideally, such a compound should occupy the inhibitor binding site within the PS II complex, but should not result in interruption of electron flow. If seems highly probable that the "essential element" is the component of the herbicide molecule which actually influences electron transport and is thereby in control of the mode of action of the herbicides. The extensive diversity in other aspects of the molecular structure of the inhibitors affecting PS II is a reflection of the differences in the microenvironment of the specificity-determining domain. We suggest that a search for potential protectants against PS II-inhibiting herbicides should focus on those compounds that can occupy the specificity-determining domain but which do not contain the essential element; these compounds could, in theory, compete with the binding of active herbicides but would not in and of themselves have herbicidal activity. To our knowledge, this concept has not been tested. Because very simple *in vitro* screening systems can be devised with isolated chloroplasts, it is possible that such a program can be developed to test for potential protectants against PS II-inhibiting herbicides.

## EPILOGUE

Several papers that appeared after this review was completed focus on the identity of the herbicide-receptor protein(s) of the photosystem II complex. These have utilized three major approaches: proteolytic modification of the binding proteins, isolation of photosystem II particles, and the use of photoaffinity herbicides.

Trypsin-induced removal of a 34–32 kilodalton polypeptide, coincident with the loss of high-affinity binding of triazine and urea herbicides but not phenol-type herbicides, was observed in isolated chloroplast thylakoids.[18] In isolated photosystem II particles, trypsin treatment also eliminated urea and triazine activity with concomitant alteration of a 32 kilodalton protein constituent of the particles.[38,39] Extraction of photosystem II particles with urea/cholate caused the selective extraction of a 32 kilodalton polypeptide of the complex. This resulted in loss of triazine and urea herbicide inhibition with little effect on the inhibitory activity of a nitrophenol herbicide.[39]

The use of photoaffinity-labeled herbicide derivatives was introduced with 2-azido-4-ethylamino-6-isopropylamino-s-triazine (azido-atrazine). This compound replaces atrazine at the triazine binding site with equal inhibitory activity. UV-irradiation of the azido-[14]C-atrazine, when bound to thylakoid membranes, resulted in the production of a highly reactive nitrene which became covalently attached to the binding protein.[40,41] Subsequent analysis of thylakoid polypeptides by SDS polyacrylamide gel electrophoresis revealed covalent association of the radiolabeled herbicide with polypeptides of 34–32 kilodaltons only in triazine-susceptible thylakoids.[41] The 34–32 kilodalton polypeptide was also tagged by the azido-atrazine label in isolated photosystem II particles.[39]

Azido-i-dinoseb (4-nitro-2-azido-6-[2',3'-[3]H] isobutylphenol) has also recently been synthesized. It binds covalently to several thylakoid polypeptides; labeling of a polypeptide in the 40 kilodalton size class was partially blocked by the presence of an unlabeled inhibitor which would compete for the same binding site.[42] This protein was not equivalent to the polypeptide tagged by azido-atrazine (K. Steinback and A. Trebst, unpublished observations). It can be concluded from these data and the analysis of photosystem II particles[39] that at least two polypeptides create an herbicide-binding domain in the photosystem II complex.

The 34–32 kilodalton polypeptides, which determine the triazine-receptor site and which are tagged by azido-atrazine, have been shown to be identical with a chloroplast-encoded polypeptide synthesized by isolated chloroplasts[43,44] This is consistent with the observation of maternal inheritance of triazine-receptor protein properties in reciprocal crosses of susceptible and resistant weed biotypes.[45,46] It has been suggested that this protein is a rapidly turned over component of the photosynthetic membranes.[47]

## REFERENCES

1   K. Pfister and C. J. Arntzen, The mode of action of photosystem II—specific inhibitors in herbicide-resistant weed biotypes, Z. Naturforsch., **34c**, 996 (1979).

2   K. Wright and J. R. Corbett, Biochemistry of herbicides affecting photosynthesis, Z. Naturforsch., **34c**, 966 (1979).

3   J. Amesz and L. N. M. Duysens, Primary and associated reactions of system 2, *In Primary Processes in Photosynthesis,* J. Barber (Ed.), Elsevier, New York, 1977, pp. 149-185.

4   B. Bouges-Bocquet, Electron transfer between the two photosystems in spinach chloroplasts, *Biochim. Biophys. Acta,* **314,** 250 (1973).

5   B. R. Velthuys and J. Amesz, Charge accumulation at the reducing side of system 2 of photosynthesis, *Biochim. Biophys. Acta,* **333,** 85 (1974).

6   B. R. Velthuys, Charge accumulation and recombination in system 2 of photosynthesis, Thesis, University of Leiden, The Netherlands, 1976.

7   M. P. J. Pulles, H. J. van GorKom, and J. Gerben-Willemsen, Absorbance changes due to the charge-accumulating species in system 2 of photosynthesis, *Biochim. Biophys. Acta,* **499,** 536 (1976).

8   P. Mathis and J. Haveman, Analysis of absorption changes in the ultraviolet related to charge-accumulating electron carriers in photosystem II of chloroplasts, *Biochim. Biophys. Acta,* **461,** 167 (1977).

9   J. Bowes, A. R. Crofts, and C. J. Arntzen, Redox reactions on the reducing side of photosystem II in chloroplasts with altered herbicide binding properties, *Arch. Biochem. Biophys.,* **200,** 303 (1980).

10   J. Amesz, Plastoquinone, *In Encyclopedia of Plant Physiology,* Vol. 5; *Photosynthesis,* A. Trebst and M. Avron (Eds.), Springer, New York, 1977, pp. 238–246.

11.   H. T. Witt, Energy conversion in the functional membrane of photosynthesis, Analysis by light-pulse and electric pulse methods, *Biochim. Biophys. Acta,* **505,** 355 (1979).

12   G. Regitz and I. Ohad, Changes in the protein organization in developing thylakoids of *Chlamydomonas reinhardii* y-1 as shown by sensitivity to trypsin, *Biochim. Biophys. Acta,* **505,** 355 (1974).

13   G. Renger, Studies on the structural and functional organization of system II of photosynthesis, The use of trypsin as a structurally selective inhibitor at the outer surface of the thylakoid membranes, *Biochim. Biophys. Acta,* **440,** 287 (1976).

14   G. Renger, K. Erixon, G. Doring, and C. Wolff, Studies on the nature of the inhibitory effect of trypsin on the photosynthetic electron transport of system II in spinach chloroplasts, *Biochim. Biophys. Acta,* **440,** 278 (1976).

15   K. E. Steinback, J. J. Burke, and C. J. Arntzen, Evidence for the role of surface-exposed segments of the light-harvesting complex in cation-mediated control of chloroplast structure and function, *Arch. Biochem. Biophys.,* **195,** 546 (1979).

16   A. Trebst, Inhibition of photosynthetic electron transport by phenol and diphenylether herbicides in control and trypsin-treated chloroplasts, *Z. Naturforsch.,* **34c,** 986 (1979).

17   P. Boger and K.-J. Kunert, Differential effects of herbicides upon trypsin-treated chloroplasts, *Z. Naturforsch.,* **34c,** 1015 (1979).

18   K. E. Steinback, K. Pfister, and C. J. Arntzen, Trypsin-mediated removal of herbicide binding sites within the photosystem II complex, *Z. Naturforsch.,* **36c,** 98 (1981).

19   C. J. Arntzen, Dynamic structural features of chloroplast lamellae, *In Current Topics in Bioenergetics,* D. R. Sanadi and L. P. Vernon (Eds.), Academic Press, New York, 1978, pp. 112-160.

20   S. Izawa, Inhibitors of electron transport, *In Encyclopedia of Plant Physiol.,* Vol. 5, *Photosynthesis,* A. Trebst and M. Avron (Eds.), Springer, New York, 1977, pp. 266-282.

21   D. E. Moreland, Mechanisms of action of herbicides, *Ann. Rev. Plant Physiol.,* **31,** 597 (1980).

22   Govindjee and R. Govindjee, Introduction to photosynthesis, *In Bioenergetics of Photosynthesis,* Govindjee (Ed.), Academic Press, New York, 1975, pp. 1-50.

23   L. N. M. Duysens and H. E. Sweers, Mechanisms of two photochemical reactions in algae as studied by means of fluorescence, *In Studies in Microalgae and Photosynthetic Bacteria,* J. Ashida (Ed.), University of Tokyo Press, Tokyo, 1963, pp. 353-372.

24  P. E. Brewer, C. J. Arntzen, and F. W. Slife, Effects of atrazine, cyanazine and procyazine on the photochemical reactions of isolated chloroplast, *Weed Sci.,* **27,** 300 (1979).

25  C. J. Arntzen, C. L. Ditto, and P. E. Brewer, Chloroplast membrane alteration in triazine-resistant *Amaranthus retroflexus* biotypes, *Proc. Natl. Acad. Sci. U.S.A.,* **76,** 278 (1979).

26  A. Joliot, Fluorescence rise from $36\mu$ s on following a flash at low temperature ($+2°$ − $60°$), *In Proc. 3rd Int. Congr. Photosynth. Res., Rehovot,* M. Avron (Ed.), Elsevier, Amsterdam, 1974, pp. 315-322.

27  J. Bowes and A. R. Crofts, Binary oscillations on the acceptor side of photosystem II, *Biochim. Biophys. Acta,* **590,** 373 (1980).

28  S. R. Radosevich, K. E. Steinback, and C. J. Arntzen, Effect of photosystem II inhibitors on thylakoid membranes of two common groundsel (*Senecio vulgaris*) biotypes, *Weed Sci.,* **27,** 216 (1979).

29  V. Sousa-Machado, C. J. Arntzen, J. D. Bandeen, and G. R. Stephenson, Comparative triazine effects upon system II photochemistry in chloroplasts of two common lambsquarters (*Chenopodium album*) biotypes, *Weed Sci.,* **26,** 318 (1978).

30  K. Pfister, S. R. Radosevich, and C. J. Arntzen, Modification of herbicide binding to photosystem II in two biotypes of *Senecio vulgaris,* L., *Plant Physiol.,* **64,** 995 (1979).

31  W. Tischer and H. Strotmann, Relationship between inhibitor binding by chloroplasts and inhibition of photosynthetic electron transport, *Biochim. Biophys. Acta,* **460,** 113 (1977).

32  A. Trebst and W. Draber, Structure activity correlations of recent herbicides in photosynthetic reactions, *In Advances in Pesticide Science,* Vol. 2, H. Geissbuhler (Ed.), Pergamon Press, New York, 1979, pp. 223-234.

33  N. C. Hansch, Theoretical considerations of the structure-activity relationship in photosynthesis inhibitors, *In Prog. in Photosynth. Res.,* Vol. 3, H. Metzner (Ed.), Tubingen, 1969, pp. 1685-1692.

34  D. E. Moreland, Inhibitors of chloroplast electron transport: Structure activity relations, *In Prog. in Photosynth. Res.,* Vol. 3, H. Metzner (Ed.), Tubingen, 1969, pp. 1693-1711.

35  A. Trebst, and E. Harth, Herbicidal N-alkylated-ureas and ring closed N-acylamides as inhibitors of photosystem II, *Z. Naturforsch.,* **29c,** 232 (1974).

36  J. Whitmarsh and W. A. Cramer, Photooxidation of the high-potential iron-sulfur center in chloroplasts, *Proc. Natl. Acad. Sci. U.S.A.,* **76,** 4417 (1979).

37  K. Sauer, Photosynthesis—The light reactions, *Ann. Rev. Phys. Chem.,* **30,** 155 (1979).

38  E. Croze, M. Kelly, and P. Horton, Loss of sensitivity to diuron after trypsin digestion of chloroplast photosystem II particles, *FEBS Lett.,* **103,** 22 (1979).

39  J. E. Mullet and C. J. Arntzen, Identification of a 32–34 kilodalton polypeptide as a herbicide receptor protein in photosystem II, *Biochim. Biophys. Acta,* **635,** 236 (1981).

40  G. Gardner, Azidoatrazine: photoaffinity label for the site of triazine herbicide action in chloroplasts, *Science,* **211,** 937 (1981).

41  K. Pfister, K. E. Steinback, G. Gardner, and C. J. Arntzen, Photoaffinity labeling of an herbicide receptor protein in chloroplast membranes, *Proc. Natl. Acad. Sci. U.S.A.,* **78,** 981 (1981).

42  W. Oettmeier, K. Masson, and U. Johanningmeier, Photoaffinity labeling of the photosystem II herbicide binding protein, *FEBS Lett.,* **118,** 267 (1980).

43  K. E. Steinback, L. McIntosh, L. Bogorad, and C. J. Arntzen, Identification of the 32 kdalton triazine binding polypeptides of thylakoid membranes as a chloroplast gene product, *Plant Physiol. Suppl.,* **67,** 64 (1981).

44  K. E. Steinback, K. Pfister, and C. J. Arntzen, Identification of the receptor site for triazine herbicides in chloroplast thylakoid membranes, *In Biochemical Responses Induced by Herbicides,* ACS Symposium Series 181, D. E. Moreland, J. P. St. John and F. D. Hess (Ed.), 1982, pp. 37-55.

**45** V. Souza-Machado, J. D. Bandeen, G. R. Stephenson, and P. Lavigne, Uniparental inheritance of chloroplast atrazine tolerance in *Brassica campestris, Can. J. Plant Sci.,* **58**, 977 (1978).

**46** S. Darr, V. Souza-Machado, and C. J. Arntzen, Uniparental inheritance of a chloroplast photosystem II polypeptide controlling herbicide binding, *Biochim. Biophys. Acta,* **634**, 219 (1981).

**47** A. K. Matoo, U. Pick, H. Hoffman-Falk, and M. Edelman, The rapidly metabolized 32,000-dalton polypeptide of the chloroplast is the "proteinaceous shield" regulating photosystem II electron transport and mediating diuron herbicide sensitivity, *Proc. Natl. Acad. Sci. U.S.A.,* **78**, 1572 (1981).

# CHAPTER ELEVEN

# Tolerance to Bipyridylium Herbicides

**B. M. R. HARVEY and D. B. HARPER**

Faculty of Agriculture and Food Science
The Queen's University of Belfast
Belfast, Northern Ireland

## 1 INTRODUCTION

Paraquat, normally employed as its dichloride salt, and diquat, normally employed as its dibromide salt, are the bipyridylium herbicides finding most widespread use in agriculture. They are nonselective contact herbicides with a rapid desiccant action. Although transported to a limited extent in the xylem,[1-3] neither of the compounds is metabolically degraded by plants.[4-6] Significant photochemical degradation can occur on the surface of the plant in daylight.[5,7] Irreversible adsorption of the herbicides on soil colloids and clay minerals ensures that residues in the soil are not normally available for root uptake. However, some uptake can occur in peat and other soils where the herbicides are initially weakly adsorbed to organic components and strong adsorption on inorganic soil components occurs more slowly.[8] Although diquat shows slightly greater toxicity to dicotyledonous species as compared with monocotyledonous species, this selectivity is not sufficient to be of any practical application. A third member of this class of herbicide, morfamquat, displays differential toxicity of an order sufficient to allow its application in postemergence weed control in cereals.[9] However, this compound is no longer commercially available.

It was noted early in the study of these compounds that herbicidal activity was confined to quaternary salts of 2,2'-and 4,4'-bipyridyl with the two pyridine rings in the same plane and possessing redox potentials [$E_o'$] between $-300$ and $-500$ mV and which can be reduced to water-soluble free radicals by the addition of one electron.[10] Paraquat ($E_o' = -446$ mV), diquat ($E_o' = -349$ mV), and morfamquat ($E_o' = -305$ mV) all share these properties, which enable them to compete for electron flow from the primary electron acceptor of photosystem I in photosynthetic electron transport,[11,12] as shown in Fig. 11.1.

Paraquat

Diquat

morfamquat

The bipyridylium free radical formed on reduction is instantly reoxidized by molecular oxygen, with the production of the superoxide radical anion, $O_2^{\cdot-}$,[13-15] which can undergo dismutation to oxygen and hydrogen peroxide.[16] Oxidants of this nature and free radicals derived from their interaction with each other and other cell components[17] can rapidly initiate a chain reaction resulting in peroxidation of unsaturated fatty acids, which are essential constituents of cell membranes.[18,19]

Thus the first observable effect of the bipyridylium herbicides on the ultrastructure of plant tissues is membrane damage involving either the plasmalemma,[20] the tonoplast,[21] the chloroplast envelope,[22] or the thylakoid membranes.[23] At the molecular level there is a concomitant increase in malondialdehyde,[24] a characteristic breakdown product of the unsaturated fatty acid hydroperoxides which are formed in the chloroplast membranes.[25] This sequence of events explains the finding that the rapid bleaching and desiccation, which are the most typical of the phytotoxic effects of this class of herbicide, require not only the presence of chlorophyll but also light and oxygen.[26,27] The importance of the superoxide ion in the initiation of the phytotoxic effects of bipyridylium herbicides is demonstrated by the recent observation that a copper chelate of D-penicillamine protects photosynthetic tissues against paraquat damage.[28] This compound, which has high activity in the dismutation of superoxide ion,[29] was found to reduce the loss of both chlorophyll and carotenoid pigments and to prevent the oxidation of unsaturated fatty acids in paraquat-treated cotyledon leaves of flax.

Paraquat damage occurs more slowly in nonphotosynthetic tissues. The mechanism of toxicity in such tissues has not been fully elucidated, but injury is widely attributed to the reoxidation of free radicals arising from reduction of the bipyridylium compound by the respiratory electron transport chain.[21,26,30] This may be analogous to the action of paraquat in animal tissues. Bipyridylium herbicides are reduced by microsomal preparations from rat liver[31] and

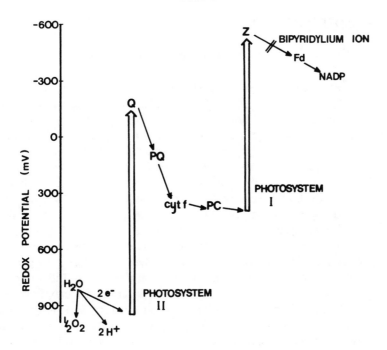

**Figure 11.1** Reduction of bipyridylium cation in photosynthetic electron transport. Q, primary electron acceptor of photosystem II; PQ, plastoquinone; cyt, cytochrome; PC, plastocyanin; Z, primary electron acceptor of photosystem I; Fd, ferredoxin.

lung.[32] The lungs are apparently the principal site of paraquat damage in animals.[33] *In vitro* reduction by the mitochondrial electron transport chain has also been demonstrated.[31,34] The relevance of the latter observation *in vivo* is questionable, as bipyridylium herbicides do not readily penetrate the inner membrane of intact mitochondria.[31,34]

The alleviation of the toxic effects of paraquat in animals by treatment with free radical scavengers[35] or superoxide dismutase[36,37] suggests that, as in green plants, damage is due to the formation of superoxide ion on reoxidation of the bipyridylium free radical. However, some evidence that paraquat toxicity is not due to formation of superoxide ion has been obtained from experiments with *Escherichia coli* by Simonds et al.[38] and with rat liver microsomes by Talcott et al.[39] The latter workers have also demonstrated that paraquat free radical can be reoxidized by ferric pyrophosphate *in vitro* and have proposed that *in vivo,* the resultant ferrous ion may cleave lipid hydroperoxides to generate alkoxy radicals, which then could initiate a chain reaction of lipid peroxidation. Such a sequence of events could account for the observation by Merkle et al.[27] that even in a nitrogen atmosphere, paraquat can damage leaf membranes without loss of chlorophyll.

## 2 DIFFERENCES IN TOLERANCE TO BIPYRIDYLIUM HERBICIDES

### 2.1 Interspecific Differences

Perennial plants with underground storage organs, although susceptible to top-kill by the bipyridylium herbicides, generally regrow vigorously from the base.[40] This can be ascribed not only to the inherently lower toxicity of these compounds in nonphotosynthetic tissues but also to their very limited movement out of the foliage.[1-3] Translocation would appear the more important factor, as conditions which increase translocation apparently improve control by paraquat in *Cyperus* spp.[41] and *Agropyron repens*.[42] Although annual plants are usually highly susceptible to bipyridylium herbicides, differences between annual species have been recorded. In field trials Stubbs[43] found that most broad-leaved annuals were killed by diquat applied at 1.1kg/ha, but a few weed species (e.g., *Papaver rhoeas* and *Chrysanthemum segetum*) exhibited tolerance and flax recovered from diquat applied at 2.2 kg/ha. It is not clear whether these species differ biochemically in susceptibility to bipyridylium herbicides. As no wetting agent was used, the effect may have merely involved differences in spray retention, as determined by leaf morphology or leaf surface characters, such as epicuticular wax. In the absence of wetting agent, Thrower et al.[2] found abrasion of leaf surface wax to be necessary to obtain toxic effects in soybean using a diquat treatment which killed unabraded leaves of broad and French beans.

The extent of adsorption of bipyridylium herbicides on the leaf surface may also be important in determining the amount of herbicide available for uptake into the cytoplasm. Brian[44] has defined three phases of uptake of bipyridylium herbicides. An initial rapid adsorption at or near the leaf surface lasting about 30 sec is followed by adsorption into the less accessible Donnan free space, which continues up to 2 hr. Finally, there is a slow accumulation, presumed to be within the cell membrane. Adsorbed bipyridylium herbicides are not readily desorbed[44] and hence differences between species in surface adsorption capacity can influence the amount of herbicide available for uptake into the cytoplasm. Lignified or tannin-containing tissue can strongly adsorb bipyridylium ions,[45] thus greater lignification may reduce the susceptibility of a species to these herbicides. Indeed, the use of paraquat for postemergence weed control in corn is possible principally because the lignified outer leaf sheaths strongly adsorb the herbicide.[46] Similarly, the total inactivation of paraquat on application to mature bark allows the herbicide to be used for weed control in tree plantations and orchards.[47]

Leaves of some perennial species have been shown to differ markedly in susceptibility to damage by paraquat.[48] Thus *Ilex vomitoria* and *Quercus virginiana* exhibited much less necrosis than *Propis juliflora* when treated with a comparable amount of the herbicide. These species also appeared to differ in the rate of paraquat uptake; washing leaves 1 hr after treatment halved necrosis in *Q. virginiana* and *I. vomitoria,* but washing only 20 min after treatment did not reduce necrosis in *P. juliflora*. However, Davis et al.[49] observed that uptake of 2,4,5-T and picloram were much lower in leathery-leaved species such as *Q. virginiana* and *I. vomitoria* than in *P. juliflora*. Therefore, their tolerance to paraquat must be considered in this context.

## 2.2 Intraspecific Differences

### 2.2.1 Genetic Factors

Varieties of a given species may differ in susceptibility to bipyridylium herbicides, and in some species tolerance has been developed by selection. In the absence of any direct selection pressure for development of tolerance to bipyridylium herbicides, varieties of both Italian and perennial ryegrass[50] exhibit differences in susceptibility to paraquat. Fourfold differences were also found when 280 varieties of wheat were screened.[50a] However, as wheat is very susceptible to paraquat, differences of this magnitude were not considered useful in development of tolerant varieties which would permit the application of selective methods of weed control.

Selection for resistance to bipyridylium herbicides can be used to identify photosynthesis mutants, and known photosystem II mutants of corn have been demonstrated to be resistant to diquat.[51] Similarly, albino mutants[52] and etiolated seedlings[26] exhibit tolerance to bipyridylium herbicides. Although the resistance of photosynthesis mutants to bipyridylium herbicides has considerable significance in the elucidation of the electron transport processes of photosynthesis, it is of little practical value since such plants are obviously not competitive under natural conditions. A tolerant line of *Poa annua* has arisen under unusually strong selective pressure as a result of continued use of paraquat as the sole method of controlling annual weeds in a market garden (Chapter 3, Section 6.1). Normal *P. annua* is killed by 0.1 to 0.2 kg/ha paraquat, but more than 0.8 kg/ha is required to kill the tolerant strain.[50a] No information is yet available concerning the mode of tolerance in this strain. Paraquat tolerance has also arisen under strong selection pressure in perennial ryegrass[53] and has been exploited in the development for agricultural purposes of paraquat-tolerant lines of perennial ryegrass (see Chapter 12). These lines exhibit tolerance at all stages of the life cycle but the degree of tolerance is dependent on growth stage, growth conditions, and method of herbicide treatment. The results of experiments which compare three closely related tolerant lines and a number of normal susceptible varieties subjected to a range of treatments with paraquat are summarized in Table 11.1.

The physiological basis of variation within species in susceptibility to bipyridylium herbicides has been investigated in detail for perennial ryegrass,[23,57,58] where certain bred lines are tolerant to paraquat and diquat but susceptible to a wide range of other herbicides.[56] Tolerance to a herbicide could arise as a result of reduced uptake by the plant. However, experiments using [14]C-methyl labeled paraquat applied to the leaf surface or supplied to the cut ends of excised leaves have demonstrated that uptake is similar in tolerant lines and normal susceptible varieties.[57] Washing the treated leaves with [12]C-paraquat displaced similar amounts of [14]C-paraquat from both tolerant and susceptible varieties; hence tolerance would not appear to be due to greater adsorption of the herbicide in the free space.

Herbicide tolerance is frequently due to metabolic detoxification in the plant, but neither the evolution of [14]C-volatiles nor the presence of [14]C-labeled breakdown products of paraquat was detected in plants that had been treated with the labeled herbicide and maintained under conditions not conducive to paraquat photooxidation.[57]

Differences in the pattern or extent of translocation could also be responsible for plants showing different degrees of susceptibility to a herbicide. Although qualitative and quantitative assessments of [14]C-paraquat movement showed very great variability within each variety, neither the total herbicide translocated out of the treated leaf nor the pattern of transport to other tillers and the root system could be related to the degree of herbicide tolerance.[57]

Uptake, translocation, and metabolic stability of paraquat appear similar in tolerant and normal varieties. Thus tolerance probably arises from fundamental differences related to the mode of action of the herbicide. Furthermore, as chloroplasts isolated from paraquat-tolerant and normal varieties of perennial ryegrass display equal sensitivity to the herbicide, tolerant and normal genotypes cannot differ in the interaction of paraquat with photosystem I. These observations contrast with results of some studies on the basis of atrazine resistance, which revealed resistance resulting from alterations in chloroplast thylakoid membranes (Chapter 9 and 10). Although investigations appear to preclude such a possibility as a basis for paraquat tolerance, it is not inconceivable that paraquat-tolerant and normal varieties differ in permeability of the chloroplast envelope. This hypothesis has not been tested, as it has not yet proved practicable to prepare perennial ryegrass chloroplasts with undamaged envelopes. Nevertheless, a mode of tolerance directly related to the photosynthetic process appears unlikely, as tolerance is apparent as early as the germination stage (Table 11.1) and also in root tissue (Table 11.2).

As paraquat-induced damage is caused by superoxide ion and hydrogen peroxide, an increased capacity to detoxify these oxidants could reduce membrane damage resulting from their presence and thus provide a basis for tolerance. However, because the herbicidal action of paraquat is extremely rapid, it is unlikely that induction of detoxifying enzymes by the herbicide would provide a basis for tolerance. Instead, a biochemical mechanism of this type would require the necessary enzymes to be present in the plant prior to herbicide treatment. Activities of superoxide dismutase (E.C.1.15.1.1), catalase (E.C.1.11.1.6), and peroxidase (E.C.1.11.1.7) were therefore compared in untreated leaves of 11 normal varieties and four tolerant lines of perennial ryegrass (Table 11.3). Mean activities of superoxide dismutase (SOD), catalase, and peroxidase were respectively 56%, 32%, and 35% higher in herbage of paraquat-tolerant lines than in herbage of normal susceptible varieties and

Table 11.1  Paraquat Tolerance in Perennial Ryegrass

| Stage of Growth | Environment | Tolerant Line | Normal Variety | Herbicide Treatment | Degree of Tolerance[a] | Reference |
|---|---|---|---|---|---|---|
| Seed germination | Darkness in incubator | PRP VII | Barlenna | Supplied in hydroponics | > 10* | 54 |
| Germination/ emergence | Greenhouse | PRP VII | Aberystwyth S101 | Sprayed on soil (10.5% organic matter) prior to sowing | Approx. 3† | 54 |
| Germination/ emergence | Greenhouse | PRP VII | Aberystwyth S101 | Sprayed on microswards prior to reseeding | Approx. 9† | 55 |
| Two-leaf seedings | Greenhouse | PRP II | Barlenna | Foliar spray | 5.6* | 56 |
| Established sward | Field | PRP VII | Talbot | Foliar spray | Approx. 4† | 56a |
| Excised leaves | Growth cabinet | PRP IX | Barlenna, Kent Indigenous | Through cut end of leaf | Aprox. 3† | 57 |

[a]Comparison of paraquat toxicity to tolerant and normal genotypes calculated as either *a ratio of $ED_{50}$ values estimated by probit analysis, or †a ratio of concentrations at which equivalent toxicity is observed in tolerant and normal varieties.

221

Table 11.2  Paraquat Toxicity to Roots of Normal and Tolerant
Perennial Ryegrasses

| Paraquat Treatment (mM) | Root Tip Browning (%) | | Solute Leakage (mhos x $10^{-5}$) | |
|---|---|---|---|---|
| | Normal[a] | Tolerant[b] | Normal[a] | Tolerant[b] |
| 0 | 1.5 | 0 | 7.0 | 6.3 |
| 0.15 | 56.9 | 1.7 | 10.5 | 6.7 |
| 0.30 | 73.1 | 1.6 | 16.1 | 7.1 |

*Source:* Data from Faulkner, Lambe and Harvey.[57a]

*Note:* Root systems were excised from 8-week-old seedlings. Each treatment comprised triplicate samples of four roots incubated in 20 ml of aqueous paraquat solution. After 8 hr, conductivity of the herbicide solution with and without roots was measured to assess solute leakage from the roots and the percentage of nodal root tips exhibiting browning was scored.

[a]Normal (susceptible) cultivar Kent Indigenous.
[b]Tolerant line PRP VII, supplied by J. S. Faulkner.

these differences were statistically significant ($p < 0.001$, $p < 0.001$, and $p < 0.05$, respectively). Some susceptible varieties had elevated levels of one of these enzymes alone, but only the tolerant lines showed consistently high activities of both SOD and catalase, and in most instances, high activity of peroxidase as well. As destruction of superoxide ion by SOD generates hydrogen peroxide, an increase in SOD activity would only confer tolerance to paraquat if catalase and peroxidase levels were adequate to detoxify the hydrogen peroxide formed (see Fig. 11.2). A prerequisite of increased activities of several different enzymes for tolerance is consistent with the polygenic nature of inheritance of paraquat tolerance in these lines of perennial ryegrass (see Chapter 12).

Elevated activities of SOD, catalase, and peroxidase should provide a basis for paraquat tolerance not only in the green parts of plants but also in non-photosynthetic tissues, although studies have been confined to the former. As the paraquat-tolerant lines of perennial ryegrass exhibit this character during early germination (radicle-protrusion), enzymes for detoxification of superoxide ion and hydrogen peroxide must be present in the ungerminated seed or must rapidly become active following imbibition. Giannopolitis and Ries[59] demonstrated that ungerminated corn, oats, and pea seeds possessed high levels of SOD, although the activity of the enzyme did not increase during an imbibition period of 25 hr. Catalase and peroxidase activities have been demonstrated in wheat grains,[60] and Do Quy Hai et al.[61] have reported SOD and peroxidase activity in imbibed seeds of a wide range of species. Thus enzymes for detoxification of both the superoxide ion and peroxides appear to be present in ungerminated and imbibing seeds, and it is conceivable that their activities are higher in the seeds as well as in herbage of paraquat-tolerant perennial ryegrass.

Table 11.3  Superoxide Dismutase, Catalase, and Peroxidase Activities in Normal and Paraquat Tolerant Perennial Ryegrasses

| | Soluble Protein | Superoxide Dismutase | Catalase | Peroxidase |
|---|---|---|---|---|
| *Normal varieties* | | | | |
| Barlenna | 100 | 100 | 100 | 100 |
| Cropper | 129 | 75 | 114 | 55 |
| Petra | 132 | 65 | 77 | 61 |
| Aberystwyth S23 | 121 | 118 | 95 | 94 |
| Aberystwyth S321 | 77 | 81 | 76 | 56 |
| Gremie | 108 | 104 | 98 | 117 |
| Melle | 90 | 103 | 90 | 108 |
| Kent Indigenous | 87 | 98 | 99 | 104 |
| Talbot | 116 | 72 | 101 | 103 |
| Taptoe | 111 | 74 | 107 | 58 |
| Hora | 103 | 97 | 99 | 107 |
| Mean | 107 | 90 | 96 | 88 |
| *Tolerant lines* | | | | |
| PRP II | 114 | 127 | 122 | 105 |
| PRP VI | 107 | 146 | 143 | 132 |
| PRP VII | 122 | 146 | 117 | 121 |
| PRP IX | 117 | 140 | 124 | 118 |
| Mean | 115 | 140 | 127 | 119 |

*Source:* Data from Harper and Harvey.[58]

*Note:* Batches of six varieties, always including Barlenna and at least one tolerant line, were grown by hydroponics. Foliage was clipped when 8 weeks old and regrowth harvested 3 to 4 weeks later. Soluble proteins and enzyme activities were assayed and activities expressed relative to Barlenna. Tolerant lines were supplied by J. S. Faulkner.

Whether or not the higher enzyme activities alone can account for the degree of paraquat tolerance observed is difficult to ascertain, particularly in view of the lack of information on the proportion of the total photosynthetic electron flow which must be diverted into the production of superoxide ion before obvious toxic effects are observed. Asada et al.[62] estimate that the steady-state concentration of superoxide ion is approximately $6 \times 10^{-9}$ M in illuminated chloroplasts under normal conditions. This calculation assumes that between 5 and 10% of the total electron flow is diverted into the formation of the ion. Presupposing that all the reductant generated in the chloroplasts transfers its electrons to the herbicide, the formation of superoxide ion should increase 10- to 20-fold, giving a final concentration of 6 to $12 \times 10^{-8}$ M. Clearly, to main-

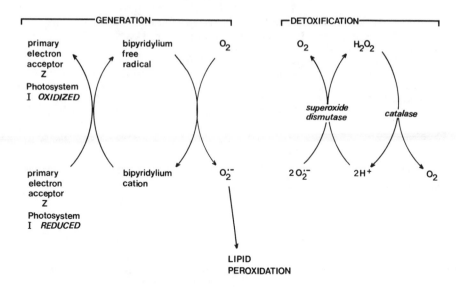

**Figure 11.2** Generation and detoxification of superoxide ion in paraquat-tolerant perennial ryegrass.

tain a steady-state concentration of superoxide ion similar to that in untreated chloroplasts, an increase in SOD concentration of a similar magnitude would be required unless a change in the kinetics of the enzyme system is proposed. However, it is unlikely that the total photosynthetic electron flow is diverted into formation of superoxide ion at concentrations of paraquat that are at the toxicity threshold (i.e., the concentrations used to assess the degree of tolerance of ryegrass varieties). Possibly, a comparatively small increase in the proportion of electron flow diverted into additional superoxide ion formation could produce toxic effects. If such is the case, the 50% increase in SOD activity in tolerant ryegrass lines may be sufficient to provide tolerance.

The first detectable effect of paraquat on perennial ryegrass is inhibition of net carbon dioxide uptake.[23] Damage to thylakoid membranes occurs within 1 hr of paraquat treatment at high light intensity. Presumably, the damage is due to rapid formation of superoxide ion and hydrogen peroxide within the organelle. The effect of the herbicide on carbon dioxide uptake has been attributed to diversion of photosynthetic electron flow from NADP to reduction of paraquat.[63] However, low concentrations of hydrogen peroxide can cause a reversible inhibition of carbon dioxide fixation by isolated chloroplasts.[64,65] As there is little or no catalase or peroxidase activity present in chloroplasts,[66-68] the rapid inhibition of carbon dioxide uptake by paraquat may be due to accelerated formation of hydrogen peroxide. When tolerant ryegrass plants are treated with the minimum dose of paraquat that is toxic to normal plants, there is little inhibition of net carbon dioxide uptake and no ultrastructural damage to thylakoid membranes.[23] Although SOD activity in chloroplasts of tolerant plants may be adequate to convert all additional superoxide ion to hydrogen peroxide, the elevated activities of catalase and peroxidase are not located within these organelles.[58] Therefore, either hydrogen peroxide diffuses very rapidly out of the chloroplast to be detoxified by the enhanced extra-chloroplastic activities of catalase and peroxidase or, more probably, the chloroplasts have effective endogenous systems for detoxification of hydrogen peroxide, such as those involving ascorbate and glutathione.[69,70]

Tolerance to paraquat has arisen in populations of an annual weed, *Conyza linifolia*, growing in citrus and vine plantations where paraquat has been applied up to eight times annually for 9 years (Chapter 3). Under greenhouse conditions, these populations show tolerance to paraquat at application rates up to the equivalent of 10 kg/ha, thus displaying an approximately fivefold greater tolerance to the herbicide than do normal plants of the species. Youngman and Dodge have recently reported that only a slight inhibition of net $CO_2$ fixation occurs when tolerant plants are treated with a concentration of paraquat which totally inhibits net $CO_2$ fixation in normal biotypes.[71] Furthermore, the paraquat-tolerant biotype possesses two electrophoretically distinct cyanide-insensitive superoxide dismutase enzymes in addition to the two cyanide-sensitive enzymes detectable in both the normal and tolerant biotypes.[71] Total SOD activity is almost threefold greater in the tolerant than in the normal biotype, and this suggests a mode of tolerance analogous to that which has arisen in ryegrass. However, other studies have revealed that paraquat adsorption on leaf tissue is somewhat greater in tolerant than in normal plants.[71a] Thus it is possible that several different factors are responsible for tolerance to the herbicide. Further investigations are necessary to determine the relative contributions of the various factors involved in paraquat tolerance in *C. linifolia*.

Paraquat-resistant cell lines of a number of species have been obtained by *in vitro* selection procedures (see Chapter 14). Hughes[72] isolated a soybean cell line whose growth was not inhibited by $10^{-3}$ M paraquat, a concentration that totally inhibited growth of normal lines. Resistance in this line appears related to herbicide uptake, which is less than half that of normal lines.[72a] Miller and Hughes[73] have also isolated paraquat-resistant lines from callus and cell suspension cultures of tobacco. When plantlets were successfully regenerated from these resistant lines and retested for paraquat resistance by floating leaf disks on herbicide solution, only 10% appeared to have retained the character. Nevertheless, when callus cultures were reestablished from the regenerated plants, all but one exhibited partial or complete paraquat resistance. They suggest that callus cells possess a sufficiently increased capacity to detoxify both superoxide ion and hydrogen peroxide to provide adequate protection against paraquat damage, under *in vitro* conditions, given that these cells are pale green or yellow and are grown in low light intensity. The enhanced detoxifying capacity may be insufficient to prevent damage resulting from the far greater rates of production of superoxide ion and hydrogen peroxide that occur when the regenerated leaf disks, which are fully photosynthetic, are exposed to the herbicide under higher light intensity. One of the resistant lines has recently been reported to have higher activities of catalase and peroxidase than control lines, but activities of superoxide dismutase were similar in normal and resistant cell lines.[74]

## 2.2.2 Nutritional Factors

The nutritional status of plants of the same genotype can cause variations in susceptibility to bipyridylium herbicides. Increasing the supply of nitrogen increased the susceptibility of *Agropyron repens* to paraquat, probably because at higher nitrogen levels, fewer rhizome buds remained dormant prior to spraying and the treated plants therefore had a lower capacity for regrowth.[75] Lutman et al.[76] did not find an effect of nitrogen supply on susceptibility of *A. repens* to paraquat but demonstrated that seedlings of cereals and several varieties of ryegrass were more susceptible to paraquat when grown under a high nitrogen regimen. They attributed this to greater spray interception by the expanded area of foliage and increased spray retention due to altered leaf surface characters.[77] However, in established swards of perennial ryegrass, susceptibility to paraquat appeared to be somewhat lower under conditions of high nitrogen availability,[76] even though well fertilized plants might be expected to have an increased photosynthetic rate and consequently an accelerated rate of reduction of paraquat. Furthermore, when microdroplets of paraquat were applied to excised leaf segments, plants grown in soils of high nitrogen status were considerably more tolerant to paraquat than those grown in soils of comparatively low nitrogen status.[78] This method of paraquat application eliminated possible effects arising from differences in spray interception and retention. By vacuum infiltration of paraquat into excised leaf segments, changes in susceptibility due to effects of nutritional status on cuticle thickness and permeability were also minimized. Again, plants grown under a high nitrogen regimen manifested significantly less paraquat damage than plants grown under conditions of low nitrogen availability.[78] Thus differences in nutrition may possibly influence susceptibility to bipyridylium herbicides at the enzymatic as well as at the morphological level.

Nutrient availability can also affect paraquat toxicity to *E. coli*. Cultures growing in a nutrient-rich medium had a greater tolerance to paraquat than those grown in minimal medium.[79] Introduction of paraquat into cultures grown in nutrient-rich medium resulted in a very large increase in the activity of SOD,[80] an enzyme that is normally induced by exposure to increased concentrations of oxygen[81,82] in a number of prokaryotic and eukaryotic organisms. When puromycin was added to cultures to inhibit protein synthesis, the lethal effects of paraquat were greatly enhanced. This suggested that when synthesis of SOD was prevented, cells became more vulnerable to the superoxide-mediated toxic effects of paraquat.[80] Conversely, pretreatments which elevated intracellular SOD levels protected against the toxic effects of subsequent exposure to paraquat in the presence of puromycin.[79]

Whether these findings are relevant to the reduced paraquat susceptibility of some higher plants grown under conditions of high fertility is debatable, as the effects of nutrition on SOD activity are not well documented. However, effects of leaf aging on SOD levels and on susceptibility to paraquat have been investigated separately. SOD activity apparently declined with age in leaves of pepper and poplar,[83,84] while tolerance to paraquat increased during leaf aging in oats.[78]

Despite the observations that paraquat induces synthesis of SOD in *E. coli*, little information is currently available on whether sublethal doses of paraquat can induce synthesis of additional SOD in higher plants. Low concentrations of paraquat were reported to increase SOD activity and growth of callus cultures of two *Pinus* sp., whereas in a third species the same concentration of paraquat inhibited growth and had no effect on SOD activity.[84] An investigation of the effects of sublethal doses of bipyridylium herbicides on the activities of SOD and other oxidative enzymes in plants differing in susceptibility to these herbicides, whether due to their genotype or nutritional status, would be of considerable interest.

## 3 TOLERANCE OF MONOCOTYLEDONOUS SPECIES TO MORFAMQUAT

Monocotyledonous plants are more tolerant to diquat and much more tolerant to morfamquat than are dicotyledonous species. Rates of morfamquat application required to control most dicotyledonous weeds (1.1 to 1.6 kg/ha) only caused a transient scorch of cereal crops, and grain yield was not depressed.[9] Brian[85] investigated the physiological basis of selectivity and using simple washing procedures demonstrated that uptake of morfamquat into the "cytoplasm + vacuole fraction" is much lower than that of paraquat. However, both dicotyledonous and monocotyledonous species exhibit lower permeability to morfamquat and hence it appears unlikely that this can account for the tolerance of monocotyledons. Brian[85] suggested, but did not test the following possibilities: (*a*) morfamquat is activated in dicotyledon or detoxified in monocotyledons, (*b*) morfamquat is acccumulated in the nonphotosynthetic cells of leaves of monocotyledons, and (*c*) morfamquat fails to penetrate the chloroplast envelope in monocotyledons. A further possibility is that bipyridylium compounds of low redox potential compete with natural electron acceptors of the photosynthetic electron transport chain less successfully in monocotyledons than in dicotyledons. However, any mode of tolerance based simply on prevention of morfamquat interaction with the photosynthetic electron transport chain in monocotyledons appears improbable, since these exhibit tolerance to morfamquat even at the early stages of germination in darkness (Fig. 11.3). Metabolic transformation either to a more active form

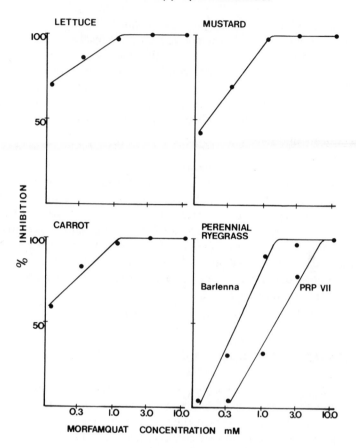

**Figure 11.3** Inhibition by morfamquat of etiolated shoot grown in germinating seeds of dicotyledonous and monocotyledonous species. Duplicate samples of 100 seeds of each species were germinated for 5 days in darkness with 5 $cm^3$ of aqueous herbicide solution. Hypocotyl length was measured for lettuce, mustard, and carrot and coleoptile length for perennial ryegrass. PRP VII was supplied by J. S. Faulkner. (Previously unpublished data of B. M. P. Harvey and D. B. Harper.)

in dicotyledons or alternatively to a nontoxic compound in monocotyledons would seem to be the most probable mode of selectivity.

It is perhaps significant that photosynthetic and nongreen tissues of paraquat tolerant ryegrass have a considerably greater tolerance to morfamquat than normal varieties (Fig. 11.3). This tolerance to a bipyridylium herbicide with a structure quite distinct from diquat or paraquat is consistent with tolerance in perennial ryegrass arising, as previously suggested, from fundamental differences relating to the mode of action of the bipyridylium herbicides, rather than merely from differences in the uptake, translocation, or metabolism of these compounds.

## 4 CONCLUDING REMARKS

We have briefly reviewed the extensive literature on the mode of action of bipyridylium herbicides, listed some instances of inter- and intraspecific differences in susceptibility to these herbicides, and discussed the genetic and environmental factors that may give rise to these differences. However, several areas of uncertainty remain. Firstly, the mode of action of bipyridylium herbicides in nonphotosynthetic plant tissues has not yet been established unequivocally, although it is generally assumed to be analogous to the mode of action of these compounds in animal tissues. Secondly, the basis of selectivity of morfamquat and, to a lesser extent, diquat between monocotyledons and dicotyledons has not been elucidated. Thirdly, it is not yet certain whether a superior ability to detoxify superoxide radical anion and hydrogen peroxide can alone account for the tolerance of tissues of lines of certain species to bipyridylium herbicides, or whether additional mechanisms are also involved.

Tolerance or resistance to bipyridylium herbicides is now known to have arisen within a number of species. Investigation of the physiological basis of this tolerance may provide us with further information not only on the mode of action of these herbicides but also, perhaps more importantly, on the fundamental process of photosynthesis.

## REFERENCES

1   B. C. Baldwin, Translocation of diquat in plants, *Nature (Lond.)*, **198**, 872 (1963).

2   S. L. Thrower, N. D. Hallam, and L. B. Thrower, Movement of diquat (1,1'-ethylene-2,2'-dipyridylium) dibromide in leguminous plants, *Ann. Appl. Biol.*, **55**, 253 (1965).

3   P. Slade and E. G. Bell, The movement of paraquat in plants, *Weed Res.*, **6**, 267 (1966).

4   H. H. Funderburk and J. M. Lawrence, Mode of action and metabolism of diquat and paraquat, *Weeds*, **12**, 259 (1964).

5   P. Slade, The fate of paraquat applied to plants, *Weed Res.*, **6**, 158 (1966).

6   A. A. Akhavein and D. L. Linscott, The dipyridylium herbicides, paraquat and diquat, *Residue Rev.*, **23**, 97 (1968).

7   H. H. Funderburk, N. S. Negi, and J. M. Lawrence, Photo-chemical decomposition of diquat and paraquat, *Weeds*, **14**, 240 (1966).

8   R. G. Burns and L. J. Audus, Distribution and breakdown of paraquat in soil, *Weed Res.*, **10**, 49 (1970).

9   H. M. Fox and C. R. Beech, Bipyridylium herbicides. Field trials with PP407 and PP745 as selective herbicides for weed control in cereals, *Proc. 7th Br. Weed Control Conf.*, 108 (1964).

10  R. F. Homer, G. C. Mees, and T. E. Tomlinson, Mode of action of dipyridyl quaternary salts as herbicides, *J. Sci. Food Agric.*, **11**, 309 (1960).

11  G. Zweig and M. Avron, On the oxidation–reduction potential of the photoproduced reductant of isolated chloroplasts, *Biochem. Biophys. Res. Commun.*, **19**, 397 (1965).

12  C. C. Black, Chloroplast reactions with dipyridyl salts, *Biochim. Biophys. Acta*, **120**, 332 (1966).

13  T. C. Stancliffe and A. Pirie, The production of superoxide radicals in reactions of the herbicide diquat, *FEBS Lett.*, **17**, 297 (1971).

14  J. A. Farrington, M. Ebert, E. J. Land, and K. Fletcher, Bipyridylium quaternary salts and related compounds. V. Pulse radiolysis studies of the reaction of paraquat radical with oxygen. Implications for the mode of action of bipyridyl herbicides, *Biochim. Biophys. Acta*, **314**, 372 (1973).

15  C. N. Giannopolitis and S. K. Ries, *In vitro* production of superoxide radical from paraquat and its interactions with monuron and diuron, *Weed Sci.*, **25**, 298 (1977).

16  B. Halliwell, Superoxide dismutase, catalase and glutathione peroxidase: solutions to the problems of living with oxygen, *New Phytol.*, **73**, 1075 (1974).

17  B. Halliwell, Biochemical mechanisms accounting for the toxic action of oxygen on living organisms: the key role of superoxide dismutase, *Cell Biol. Int. Rep.*, **2**, 113 (1978).

18  E. W. Kellogg and I. Fridovich, Superoxide, hydrogen peroxide, and singlet oxygen in lipid peroxidation by a xanthine oxidase system, *J. Biol. Chem.*, **250**, 8812 (1975).

19  B. A. Svingen, F. O. O'Neal, and S. D. Aust, The role of superoxide and singlet oxygen in lipid peroxidation, *Photochem. Photobiol.*, **28**, 803 (1978).

20  J. R. Baur, R. W. Bovey, P. S. Baur, and Z. El-Seify, Effects of paraquat on the ultrastructure of mesquite mesophyll cells, *Weed Res.*, **9**, 81 (1969).

21  N. Harris and A. D. Dodge, The effect of paraquat on flax cotyledon leaves: changes in fine structure, *Planta*, **104**, 201 (1972).

22  J. D. Dodge and G. B. Lawes, Some effects of the herbicides diquat and morfamquat on the fine structure of leaf cells, *Weed Res.*, **14**, 45 (1974).

23  B. M. R. Harvey and T. W. Fraser, Paraquat tolerant and susceptible perennial ryegrasses: effects of paraquat treatment on carbon dioxide uptake and ultrastructure of photosynthetic cells, *Plant Cell Environ.*, **3**, 107 (1980).

24  N. Harris and A. D. Dodge, The effect of paraquat on flax cotyledon leaves: physiological and biochemical changes, *Planta*, **104**, 210 (1972).

25  R. L. Heath and L. Packer, Photoperoxidation in isolated chloroplasts: 1. Kinetics and stoichiometry of fatty acid peroxidation, *Arch. Biochem. Biophys.*, **125**, 189 (1968).

26  G. C. Mees, Experiments on the herbicidal action of 1,1'-ethylene-2,2'-dipyridylium dibromide, *Ann. Appl. Biol.*, **48**, 601 (1960).

27  M. G. Merkle, C. L. Leinweber, and R. W. Bovey, The influence of light, oxygen and temperature on the herbicidal properties of paraquat, *Plant Physiol.*, **40**, 832 (1965).

28  R. J. Youngman and A. D. Dodge, Mechanism of paraquat action: inhibition of the herbicidal effect by a copper chelate with superoxide dismutating activity, *Z. Naturforsch.*, **34C**, 1032 (1979).

29  E. Lengfelder and E. F. Elstner, Determination of the superoxide dismutating activity of D-penicillamine copper, *Hoppe-Seyler's Z. Physiol. Chem.*, **359**, 751 (1978).

30  D. M. Stokes and J. S. Turner, The effects of the dipyridyl diquat on the metabolism of *Chlorella vulgaris*: III. Dark metabolism: effects on respiration rate and the path of carbon, *Aust. J. Biol. Sci.*, **24**, 433 (1971).

31  J. C. Gage, The action of paraquat and diquat on the respiration of liver cell fractions, *Biochem. J.*, **109**, 757 (1968).

32  J. S. Bus, S. D. Aust, and J. E. Gibson, Superoxide and singlet oxygen-catalyzed lipid peroxidation as a possible mechanism for paraquat (methyl viologen) toxicity, *Biochem. Biophys. Res. Commun.*, **58**, 749 (1974).

33  A. P. Autor (Ed.), *Biochemical Mechanisms of Paraquat Toxicity,* Academic Press, New York, 1977, p. ix.

34  K. Kopaczyk-Locke, *In vitro* and *in vivo* effects of paraquat on rat liver mitochondria, in *Biochemical Mechanisms of Paraquat Toxicity,* A. P. Autor (Ed.), Academic Press, New York, 1977, p. 93.

35  J. S. Bus, S. D. Aust, and J. E. Gibson, Lipid peroxidation: a possible mechanism for paraquat toxicity, *Res. Commun. Chem. Pathol. Pharmacol., 11,* 31 (1975).

36  A. P. Autor, Reduction of paraquat toxicity by superoxide dismutase, *Life Sci., 14,* 1309 (1974).

37  B. Wasserman and E. R. Block, Prevention of acute paraquat toxicity in rats by superoxide dismutase, *Aviat. Space Environ. Med., 49,* 805 (1978).

38  R. S. Simons, P. S. Jackett, M. E. W. Carroll, and D. B. Lowrie, Superoxide independence of paraquat toxicity in *Escherichia coli, Toxicol. Appl. Pharmacol., 37,* 271 (1976).

39  R. E. Talcott, H. Shu, and E. T. Wei, Dissociation of microsomal oxygen reduction and lipid peroxidation with the electron acceptors, paraquat and menadione, *Biochem. Pharmacol., 28,* 665 (1979).

40  R. C. Brian, R. F. Homer, J. Stubbs, and R. L. Jones, A new herbicide 1,1'-ethylene-2,2'-dipyridylium dibromide, *Nature (Lond.), 181,* 446 (1958).

41  G. H. Wood and J. M. Gosnell, Some factors affecting the translocation of radioactive paraquat in *Cyperus* species, *Proc. S. Afr. Sugar Technol. Assoc., 40,* 286 (1966).

42  A. A. Akhavein and D. L. Linscott, Effects of paraquat and light regime on quackgrass growth, *Weed Sci., 18,* 378 (1970).

43  J. Stubbs, The herbicidal properties of 1,1'-ethylene-2,2'-dipyridylium dibromide, *Proc. 4th Br. Weed Control Conf.,* 251 (1958).

44  R. C. Brian, The uptake and adsorption of diquat and paraquat by tomato, sugar beet and cocksfoot, *Ann. Appl. Biol., 59,* 91 (1967).

45  M. Damanakis, D. S. H. Drennan, J. D. Fryer, and K. Holly, The adsorption and mobility of paraquat on different soils and soil constituents, *Weed Res.* 10, 264 (1970).

46  J. Hill, L. S. Lloyd, and A. F. J. Wheeler, Post-emergence weed control in maize with paraquat, *Outlook Agric., 7,* 227 (1973).

47  W. R. Boon, Diquat and paraquat—new agricultural tools, *Chem. Ind., 19,* 782 (1965).

48  R. W. Bovey and F. S. Davis, Factors affecting the phytotoxicity of paraquat, *Weed Res., 7,* 281 (1967).

49  F. S. Davis, R. W. Bovey, and M. G. Merkle, The role of light, concentration, and species in foliar uptake of herbicides in woody plants, *For. Sci., 14,* 164 (1968).

50  C. E. Wright, A preliminary examination of the differential reaction of perennial and Italian ryegrass cultivars to grass-killing herbicides, *Proc. 9th Br. Weed Control Conf.,* 477 (1968).

50a A. F. Hawkins, ICI, Jealott's Hill, England, personal communication (1979).

51  C. D. Miles, Selection of diquat resistance photosynthesis mutants from maize, *Plant Physiol., 57,* 284 (1976).

52  R. W. Bovey and F. R. Miller, Phytotoxicity of paraquat on white and green hibiscus, sorghum and alpinia leaves, *Weed Res., 8,* 128 (1968).

53  B. Lovelidge, "Rogue" grass defies paraquat, *Arable Farming,* 1(9), 9 (1974).

54  J. S. Faulkner and B. M. R. Harvey, Paraquat tolerant *Lolium perenne:* effects of paraquat on germinating seedlings, *Weed Res.* 21, 29 (1981).

55  J. S. Faulkner, The effects of paraquat and glyphosate residues in sprayed herbage on the development of seedlings of a normal and a paraquat tolerant variety of *Lolium perenne, Grass Forage Sci.* 35, 311 (1980).

56  J. S. Faulkner, A paraquat tolerant line in *Lolium perenne, Proc. 1st Eur. Weed Res. Soc. Symp.,* Paris, 349 (1975).

**56a** J. S. Faulkner, Northern Ireland Plant Breeding Station, Loughgall, Armagh, U.K. personal communication (1979).

**57** B. M. R. Harvey, J. Muldoon, and D. B. Harper, Mechanism of paraquat tolerance in perennial ryegrass: 1. Uptake, metabolism and translocation of paraquat, *Plant Cell Environ.,* **1,** 203 (1978).

**57a** J. S. Faulkner, C. B. Lambe, and B. M. R. Harvey, Towards an understanding of paraquat tolerance in *Lolium perenne, Proc. 1980 Crop Prot. Conf.—Weeds,* 445.

**58** D. B. Harper and B. M. R. Harvey, Mechanism of paraquat tolerance in perennial ryegrass: II. Role of superoxide dismutase, catalase and peroxidase, *Plant Cell Environ.,* **1,** 211 (1978).

**59** C. N. Giannopolitis and S. K. Ries, Superoxide dismutases 1. Occurrence in higher plants, *Plant Physiol.,* **59,** 309 (1977).

**60** I. Hagima, V. Alexandrescu, and Z. Cseresnyes, Peroxidases and catalases of dormant, able to germinate and germinated wheat seeds, *Rev. Roum. Biochim.,* **15,** 273 (1978).

**61** Do Quy Hai, K. Kovacs, I. Matkovics, and B. Matkovics, Properties of enzymes: X. Peroxidase and superoxide dismutase contents of plant seeds, *Biochem. Physiol. Pflanz.,* **167,** 357 (1975).

**62** K. Asada, M. Takahashi, K. Tanaka, and Y. Nakano, Formation of active oxygen and its fate in chloroplasts, in *Biochemical and Medical Aspects of Active Oxygen,* O. Hayaishi and K. Asada (Eds.), University Park Press, Baltimore, Md., 1977, p. 45.

**63** A. D. Dodge, The mode of action of the bipyridylium herbicides, paraquat and diquat, *Endeavour,* **30,** 130 (1971).

**64** W. Kaiser, The effect of hydrogen peroxide on $CO_2$ fixation of isolated intact chloroplasts, *Biochim. Biophys. Acta,* **440,** 476 (1976).

**65** J. F. Allen and F. R. Whatley, Effects of inhibitors of catalase on photosynthesis and on catalase activity in unwashed preparations of intact chloroplasts, *Plant Physiol.,* **61,** 957 (1978).

**66** N. E. Tolbert, Microbodies—peroxisomes and glyoxysomes, *Ann. Rev. Plant Physiol.,* **22,** 45 (1971).

**67** R. W. Parish, The intracellular location of phenol oxidases, peroxidase and phosphatases in the leaves of spinach beet (*Beta vulgaris* L. subspecies *vulgaris*), *Eur. J. Biochem.,* **31,** 446 (1972).

**68** J. F. Allen, Effects of washing and osmotic shock on catalase activity of intact chloroplast preparations, *FEBS Lett.,* **84,** 221 (1977).

**69** C. H. Foyer and B. Halliwell, The presence of glutathione and glutathione reductase in chloroplasts: a proposed role in ascorbic acid metabolism, *Planta,* **133,** 21 (1976).

**70** D. Groden and E. Beck, $H_2O_2$ destruction by ascorbate-dependent systems from chloroplasts, *Biochim. Biophys. Acta,* **546,** 426 (1979).

**71** R. J. Youngman and A. D. Dodge, On the mechanism of paraquat resistance in *Conyza* sp., in *Proc. 5th Int. Congr. Photosynth.* G. Akoyunoglou (Ed.), Balaban International Science Services, 2242 Mt. Carmel Ave., Glenside, Pa. 19038 (1981).

**71a** M. Parham, ICI, Jealott's Hill, England, personal communication (1980).

**72** K. W. Hughes, Isolation of a herbicide-resistant line of soybean cells, in *Plant Cell and Tissue Culture: Principles and Applications,* W. R. Sharp, P. O. Larson, E. F. Paddock, and V. Raghavan (Eds.), Ohio State University Press, Columbus, Ohio, 1979, p. 874.

**72a** K. W. Hughes, University of Tennessee, Knoxville, personal communication (1980).

**73** O. K. Miller and K. W. Hughes, Selection of paraquat-resistant variants of tobacco from cell cultures, *In Vitro,* **16,** 1085 (1980).

**74** K. W. Hughes and R. W. Holton, Levels of superoxide dismutase, peroxidase and catalase in a tobacco cell line selected for herbicide resistance, *In Vitro,* **17,** Abstr. 44 (1981).

**75** A. R. Putnam and S. K. Ries, Factors influencing the phototoxicity and movement of paraquat in quackgrass, *Weed Sci.,* **16,** 80 (1968).

76  P. J. W. Lutman, G. R. Sagar, C. Marshall, and D. W. R. Headford, The influence of nutrient
    status on paraquat activity, *Weed Res.,* **14,** 355 (1974).

77  P. J. W. Lutman and G. R. Sagar, The influence of the nitrogen status of oat plants (*Avena
    sativa* L.) on the interception and retention of foliar sprays, *Weed Res.,* **15,** 217 (1975).

78  P. J. W. Lutman, G. R. Sagar, C. Marshall, and D. W. R. Headford, The influence of nitro-
    gen status on the susceptibility of segments of cereal leaves to paraquat, *Weed Res.,* **15,** 89
    (1975).

79  I. Fridovich and H. M. Hassan, Paraquat and the exacerbation of oxygen toxicity, *Trends
    Biochem. Sci.,* **4,** 113 (1979).

80  H. M. Hassan and I. Fridovich, Regulation of the synthesis of superoxide dismutase in
    *Escherichia coli.* Induction by methyl viologen, *J. Biol. Chem.,* **252,** 7667 (1977).

81  E. M. Gregory and I. Fridovich, Induction of superoxide dismutase by molecular oxygen, *J.
    Bacteriol.,* **114,** 543 (1973).

82  E. M. Gregory, S. A. Goscin, and I. Fridovich, Superoxide dismutase and oxygen toxicity in
    a eukaryote, *J. Bacteriol.,* **117,** 456 (1974).

83  B. Matkovics, Effects of plant and animal tissue lesions on superoxide dismutase activities,
    in *Superoxide and Superoxide Dismutases,* A. M. Michelson, J. M. McCord and I. Fridovich
    (Eds.), Academic Press, New York, 1977, p. 501.

84  K. Tanaka and K. Sugahara, Role of superoxide dismutase in defense against $SO_2$ toxicity and
    an increase in superoxide dismutase activity with $SO_2$ fumigation, *Plant Cell Physiol.,* **21,**
    601 (1980).

85  R. M. Barni, Banding patterns and induction of superoxide dismutase in three species of
    southern pines, *In Vitro,* **16,** Abstr. 112 (1980).

86  R. C. Brian, Observations on the physiological basis for the selectivity of morfamquat to
    graminaceous plants, *Pestic. Sci.,* **3,** 409 (1972).

# CHAPTER TWELVE

# Breeding Herbicide-Tolerant Crop Cultivars by Conventional Methods

### J. S. FAULKNER

Northern Ireland Plant Breeding Station
Loughgall, Armagh, United Kingdom

## 1  INTRODUCTION

### 1.1  Why Breed Tolerant Cultivars?

The main aim of plant breeders is to produce new cultivars that are superior in yield or quality. In principle, selection of plants for yield and quality is likely to be more successful if the plants are grown in an environment similar to that of commercial practice. The principle is implicitly recognized by conducting selection trials in the geographical areas where the new cultivars will be grown. The principle applies not only to environmental factors such as climate that are beyond the control of the farmer or grower, but also to environmental factors that are determined by management practices. Plant breeders do not invariably copy the currently prevailing management practices, however, because new cultivars may themselves be a catalyst of change. Well-known examples include the development of short-strawed cereals adapted to high levels of nitrogenous fertilizer and of cultivars of many crops adapted to mechanical harvesting. Similarly, the breeding of cultivars tolerant to specific herbicides may lead to changes in herbicide usage.

Every field crop has associated weeds. In the more advanced agricultural systems of the world, weeds are now mainly controlled by herbicides. The range of herbicides available is extensive; for example, 83 herbicides were approved for use by farmers and growers in the United Kingdom in 1979.[1]

Nevertheless, there remain many weeds that are technically difficult or expensive to control in particular crops. In addition, new problems are continually arising as a result of changes in weed populations caused by the selective pressure of the herbicides themselves. These problems arise both from the spread of inherently tolerant weed species, and from the evolution of tolerance within originally susceptible species. The former has been the more marked change hitherto (see Chapter 4), but the latter is likely to become more serious as particular herbicides are used repeatedly for longer periods[2] (see Chapter 17).

The standard response to these problems is to search for new herbicides or new formulations of old herbicides, but it is a search that is subject to progressively diminishing returns. Even if a suitable herbicide can be discovered, the cost of developing it is very high because of the need for field trials, toxicological tests, manufacturing plant, and so on. As a result of increases in costs, in the stringency of safety requirements, and in the number of compounds that have to be screened to discover new herbicides, there is a trend for fewer new herbicides to be introduced each year, and this trend is expected to continue.[3]

The costs of developing a new cultivar are, in contrast, relatively small, perhaps in the range 1 to 5% of the costs of a new herbicide. Thus it would be more economical to breed a cultivar tolerant to a herbicide known to kill the major weeds of that crop than to develop a new selective herbicide. The breeding of herbicide-tolerant cultivars could extend the range of herbicides available for use on the crop, thus permitting the control of a wider range of weed species and helping to delay the evolution of tolerance to individual herbicides.

There are various methods of breeding herbicide-tolerant cultivars. In this chapter, conventional plant breeding methods are considered, and in Chapters 13 and 14 specialized methods are described.

## 1.2  The Genetics of Tolerance

Since the concept of selective herbicides was introduced, it has been accepted that plant species differ in their responses to herbicides. Species belonging to the same genus or family are more likely to respond in the same way than are species belonging to different families; thus members of the Gramineae are mostly resistant to 2,4-D, and annual *Veronica* spp. are tolerant to paraquat. At first it was satisfactory to regard all members of one species as a homogenous group for practical weed control purposes. Later, just as mutants with specific resistance to antibiotics and insecticides had arisen, reports began to appear of plant populations evolving tolerance to herbicides.[4,5] Intraspecific variation in herbicide sensitivity has now been reported for many herbicides and in weed and crop species from diverse families (see Appendix Table A1).

In some cases, herbicide tolerance is inherited simply due to genes at a single locus. These major genes for herbicide tolerance may be either dominant[6,7] or recessive.[8] In other cases, genetic analyses have shown a clear-cut segregation of tolerant and susceptible genotypes under the control of more than one pair of genes.[9,10] Examples of maternal inheritance of tolerance are described in Chapter 13. The majority of reports of differential herbicidal sensitivity involve differences between clones, ecotypes, or cultivars which follow a more or less continuous pattern of variation (see Appendix Table A1). Although few of these cases have been genetically investigated, it is probable that their responses to herbicide are most often controlled polygenically.

In polygenically controlled characters, the concept of heritability—a measure of the amount of genetic as opposed to environmental control of variation—is of major importance as a determinant of the response of a population to natural or artificial selection. In the progeny of a cross between a selected atrazine-tolerant line and a susceptible cultivar of flax, Comstock and Andersen[11] found broad-sense heritabilities of 0.29 and 0.34. Similar rather low levels were found for MCPA tolerance in flax by Stafford et al.[12]

Higher levels of heritability were found for tolerance of chloramben methyl ester in cucumber by Miller et al,[13] studying the progeny of crosses between two tolerant and two susceptible selections. Using three criteria of tolerance, they obtained figures from 0.49 to 0.93 for broad-sense heritability and 0.36 to 0.87 for narrow-sense heritability. In a variable outbreeding population of *Lolium perenne* L., Faulkner[14] studied the narrow-sense heritability of paraquat tolerance by analyzing half-sib families and by comparing progeny with parents; on most criteria, the narrow-sense heritability estimates lay within the range of 0.51 to 0.72.

Heritability estimates usually have a high standard error and strictly are applicable only to the population on which they were carried out. Levels of heritability of the order found in flax would be interpreted by breeders as indicating that they should concentrate their efforts on finding more tolerant parents (see Section 2.1). A satisfactory response to recurrent selection might be expected with the higher levels of heritability found in cucumber and perennial ryegrass.

Although the mechanisms of herbicide tolerance are diverse and of great interest, from the narrow viewpoint of the breeder trying to select tolerant plants, the physiological mechanisms of tolerance are normally irrelevant. Provided that the mechanism functions in field conditions and does not adversely affect the yield or quality of the crop, it matters little whether tolerance depends on reduced uptake, impaired translocation, lack of biochemical sensitivity, or enhanced rate of degradation of the herbicide. Exceptions to this generalization may arise if selection in cell cultures is contemplated, since certain types of tolerance are only expressed in intact plants (Chapter 14).

## 1.3  Choice of Herbicide

The breeding of herbicide-tolerant cultivars is a refinement in weed control technology. The breeder works on a long time scale within a framework provided by the herbicide manufacturers and with constraints imposed by environmental and patent regulations. The breeder's role is to adjust the genotype of the crop so that the selectivity of the herbicide between crop and weeds is maximized. It is improbable that breeding will have much influence, at least in the short term, on which herbicides are manufactured, and what their costs are to the farmer, so it is rational to give careful thought to the choice of herbicide before embarking on a selection program.

The foremost requirement of the herbicide is that it should control the most important weeds of the crop at an economic dosage. Tolerance to a broad-spectrum herbicide would normally be more valuable than to a narrow-spectrum herbicide. The appropriate dosage for the projected tolerant cultivar may well be different from the recommended dosage of the herbicide used for weed control in other situations. For example, 1.1 to 1.7 kg/ha of paraquat is recommended for killing grass swards,[15] but 0.2 kg/ha is sufficient for controlling weed seedlings in young stands of paraquat tolerant perennial ryegrass.[16] A higher-than-standard dosage might be appropriate for other crop-weed problems, but the breeder should beware of creating a means of weed control that is too expensive to use. In practice, it will only be possible to determine the precise optimum dosage of the herbicide empirically after the herbicide-tolerant crop has been bred.

A second factor to be borne in mind is the inherent tolerance of the crop species to the herbicide before selection for increased tolerance. If the crop is more tolerant than the majority of its associated weed flora, the amount of improvement in tolerance required to permit selective control of the weeds would be smaller than if it were less tolerant. Thus it was the knowledge that *Lolium perenne* in pastures was more tolerant than *Agrostis* spp. and *Poa trivialis* L. to dalapon[17] that led to the choice of dalapon as a herbicide to which the tolerance of the *Lolium* should be increased.[18] When little is known about the relative tolerance of the crop to herbicides, a preliminary investigation is required. Fisher and Faulkner[19] investigated the responses of 12 grass species to 12 herbicides, in order to identify promising combinations of herbicide and agricultural or amenity grass species for selective breeding. Several combinations were suggested and some of these, such as *Festuca arundinacea* and simazine,[20] are now being followed up.

Although the existence of heritable genetic variation in response to a herbicide within the crop species is a prerequisite of a successful selection program, there will usually be little information on this subject to influence the initial choice of herbicide. Preliminary studies, for example on variation among cultivars, may be valuable as a means of revealing the extent of variation, and may incidentally help to identify promising parent plants (see Section 2.1).

Plant breeding is slow. There is a long interval between the start of a selection program and the eventual widespread use of any new cultivars that may be developed. Breeders therefore aim to produce cultivars suitable for use one to two decades ahead of their selection work. Forecasting the future is especially important for breeders of cultivars with tolerance to specific herbicides, as herbicides themselves can become outdated, either because they are superseded by more effective ones, or because of changes in safety or environmental regulations. It would therefore be prudent of the breeder to eschew herbicides that seem to be of limited agronomic value or are notably toxic to human beings or wildlife. Thus Lupton and Oliver[10] concluded that selection for metoxuron tolerance in winter wheat, although possible, was unjustifiable because of the introduction of isoproturon, a herbicide that controls the same weeds as metoxuron without severely damaging any cultivars of winter wheat. Breeding turf grasses with increased resistance to endothall, despite its effectiveness against the weed grass *Poa annua* L.,[21] would be inadvisable because of the high mammalian toxicity of this herbicide.

## 1.4  Genetic Strategy

Three alternative courses of action can be envisaged for creating a herbicide-tolerant cultivar that is as good as the best other cultivars in yield and quality.

1  Find alleles for herbicide tolerance and combine them with alleles for general agronomic traits.
2  Choose a superior but susceptible cultivar and improve its tolerance by intravarietal selection.
3  Use mutagenesis to increase tolerance in an existing cultivar.

The first course is applicable to any species, regardless of the breeding system, but is likely to take a long time unless, by good fortune, the tolerant genotypes happen to be agronomically superior as well. The need to select for agronomic characters, rather than herbicide tolerance, could impose the main restrictions to progress.

This course is being followed in the breeding of paraquat tolerant cultivars of *Lolium perenne* at my station.[22] As paraquat tolerance can be assessed rapidly on a large number of plants, highly tolerant segregate lines were readily bred from crosses between tolerant parents and agronomically superior parents. This was done by applying two annual cycles of selection to the $F_2$ and $F_3$ generations.[22a] Selection for yield and persistence in the pasture, which is still in progress, requires a generation interval of 3 or more years and a far greater expenditure of resources for every plant screened. A possible pitfall in this breeding strategy is that the herbicide tolerance may be due to pleiotropic genes that have a deleterious effect on agronomic performance. These effects would undermine the value of the breeding program, but would not be discovered until the later stages.

The second course, selection within cultivars, is applicable to cross-fertilized species. If the tolerant alleles promoted by this means are indissolubly associated with inferior agronomic performance through pleiotropy, the association will at least be detectable relatively early in the selection program. However, there are two other risks with intravarietal selection: if the amount of heritable variation present in the population is too low, the level of resistance achieved by recurrent selection within a cultivar will reach a plateau below the level required; and inbreeding may depress the agronomic value of the cultivar. Both of these eventualities could be countered only by crossing with unrelated populations, but they could be anticipated by basing the selection program on more than one cultivar.

Some success with intravarietal selection was reported by Warwick,[23] who developed a line of seed rape with increased tolerance to simazine by three cycles of recurrent selection in the cultivar Rigo. The tolerant line had larger seeds and greater seedling vigor than Rigo, but its agronomic performance was not tested and the degree of tolerance achieved was not sufficient to be commercially valuable. Recurrent selection is likely to produce slower results in seed rape, which is an allotetraploid, than it would in comparable diploid species. Intravarietal selection was also practised by Fisher and Wright[24] and Lee and Wright[24a] using cultivars of three species of lawn grass; increased tolerance to amitrole was obtained, but the lawn qualities of the selected tolerant lines were not investigated.

The third course, mutagenesis, is especially attractive for self-fertilized crops because a single tolerant mutant could be the foundation of a new cultivar. The vast majority of induced mutations are between neutral and deleterious. Depending on the mutagenic dose, many treated plants may carry more than one mutation. If desirable mutations are to be found, it is necessary to screen a large number of treated plants, a practice that would be easier for herbicide tolerance than for many other characters.

An attempt has been made to produce herbicide-tolerant plants of two self-fertilized crops by exposing seeds to the chemical mutagen ethyl methane sulfonate (EMS).[25] In a sample of 50,000 third-generation descendants of EMS-treated seed of a wheat cultivar, 588 seedlings apparently tolerant to terbutryn were found. Similarly, in 20,000 third-generation descendants of treated seed of a tomato cultivar, there were 120 individuals apparently tolerant to diphenamid. Lines selected from these apparently tolerant plants were shown to be significantly less affected by the respective herbicides than the original cultivars, but apparently not by a very wide margin. Genetic analysis of these lines and examination for genetic defects were not attempted. The proportion of apparently tolerant individuals found in these experiments (1.2% and 0.6%) was much higher than would be expected for truly tolerant mutants unfettered with other mutations. It is likely that many of them had escaped rather than tolerated the herbicide treatment. Interesting though this line of research may seem, it has not been followed up. Experimentation with a more rigorous screening procedure, larger samples, and varying mutagenic doses would be required to explore the potentialities of mutagenesis more fully.

Irrespective of the course employed, the breeder's dream will be to find a major gene for herbicide tolerance. If the differences between susceptible and tolerant plants are large and clear cut, the selection process consists essentially of retaining plants that survive treatment with the herbicide and rejecting those that are killed. Any hybridization program required to transfer major genes into agronomically superior genotypes would be straightforward, especially if the tolerance genes are dominant. Once homozygosity for the major gene has been established, there will be no problems of variability in herbicide response within the resulting cultivars. Unlike major gene resistance to disease, which is often specific to races of the pathogen,[26] major gene tolerance to a herbicide will not break down.

The problem with aiming to select for major gene tolerance to a particular herbicide is that the existence of this form of tolerance in the required crop will usually be a matter of conjecture. Although a few instances of major gene tolerance are known, it should not be assumed that it is present in most species to most herbicides. Even if major gene tolerance does exist, individuals phenotypically tolerant to a particular herbicide may be represented at such a minute frequency that there is little chance of finding them. In contrast, at least a small degree of polygenic variation in herbicide tolerance is probably ubiquitous. In practice, therefore, it may often be better for the breeder to adopt a selection strategy which is designed to accumulate polygenic tolerance through transgressive segregation (i.e. the occurrence in a segregating generation of genotypes that are more tolerant than any of the parent plants) but which would allow major gene tolerance to be detected if it is present. A strategy for polygenic tolerance requires evaluation of degrees of tolerance rather than simple discrimination between tolerance and susceptibility. It would therefore have more stringent technical requirements, thereby restricting the number of plants that can be screened with fixed resources. Given that the order of magnitude of the frequency of major genes for tolerance is unknown, it is unlikely that this restriction would make the difference between finding and not finding a major gene for tolerance. For example, if the number of plants that could be screened for polygenic tolerance is $10^6$ and for major gene tolerance is $10^7$, only if major gene-tolerant phenotypes occurred with a frequency in the order of $10^{-6}$ to $10^{-7}$ would there be a high probability of discovering them among the larger sample but not in the smaller one. If major gene-tolerant phenotypes occurred at a frequency greater than $10^{-6}$, both samples would be likely to contain them, and if they were much rarer than $10^{-7}$ (or nonexistent), neither sample would be likely to contain them.

Various considerations affect the relative merits of selecting for polygenic and major gene tolerance. In species with a long generation interval, the need for several generations of recurrent selection may make the accumulation of polygenic tolerance impractical. If there are positive indications that major gene tolerance might be present—for instance, the presence of major gene tolerance in a related species—selection for polygenic tolerance would be a relatively unattractive proposition. In cell cultures also, because of the enormous numbers of individuals that can be screened, it may be more realistic to aim to find major genes rather than polygenes for tolerance (see Chapter 14).

## 2  BREEDING FOR RESISTANCE

### 2.1  Source of Breeding Material

Herbicide-tolerant mutants of weeds are normally detected on field sites which have been treated with a herbicide, sometimes repeatedly over a period of years (Chapters 2, 3, and 17). Because herbicides are not commonly applied to susceptible crop species, there is less opportunity for tolerant mutants of crops to appear in this way. Nevertheless, some cultivated species regularly occur as rogues in other crops and are then exposed to herbicides to which they are susceptible. After 9 successive years of paraquat application for direct drilling arable crops, Lovelidge[27] reported that a problem was encountered with controlling rouge plants of *Lolium perenne*. Alleles for paraquat tolerance obtained from this site contributed to the paraquat tolerant lines of *Lolium* bred at my station.[28]

In practice, the main source of genetic material for selection for herbicide tolerance is existing cultivars and breeders' stocks. Screening cultivars to detect different responses to herbicides has been carried out in many species, often with the primary objective of detecting any cultivars to which the herbicide could be safely applied, rather than exploring the cultivars as a source of material for breeding.

More than 30 years ago, it was recognized that there was variation in tolerance of 2,4-D among cultivated stocks of the turfgrass *Agrostis stolonifera* L.[29] Most of the differences between cultivars in their responses to herbicides are small, but exploitable degrees of tolerance have been found in some species. For example, the tolerance of 12 winter wheat cultivars to chlortoluron and isoproturon has been examined.[30] With chlortoluron a clear distinction was found between six tolerant and six susceptible cultivars, but with isoproturon there was a continuous range of tolerance levels. For weed control purposes, it was concluded that isoproturon was safer on some cultivars, chlortoluron on others. Although there appears to have been no intention to follow up this work by breeding for increased herbicide tolerance, potentially valuable information for this objective was obtained.

The effect of dalapon on 35 cultivars of *Lolium perenne* was studied by Faulkner.[31] Significant differences between cultivars were found, and it was concluded that dalapon would have a better selective action on grass swards that had been sown with one of the more tolerant cultivars of ryegrass than with one of the more susceptible ones. Tolerance levels varied in a continuous manner and were not related to other characteristics of the cultivars, except that most tetraploid cultivars were relatively tolerant. Plants selected out of the more tolerant diploid cultivars were used in the development of Rathlin, a cultivar bred for high dalapon tolerance.[18]

Preparatory to selection for increased tolerance, Fisher[32] compared the response of 12 cultivars of *Festuca rubra* L. to paraquat and glyphosate. She found marked differences between cultivars which corresponded largely to their subspecific and cytogenetic groupings and a significant correlation between responses to the two herbicides. Cultivars of *F. rubra* ssp. *commutata* (hexaploid) were relatively susceptible to both herbicides, octaploid cultivars of *F. rubra* ssp. *rubra* were intermediate, and hexaploid cultivars of *F. rubra* ssp. *rubra* were relatively tolerant. The most tolerant cultivar to both herbicides was Dawson, which was therefore chosen as the subject for a selection experiment.[24]

Even the most extensive screenings of crop species do not necessarily reveal many genotypes with a substantial measure of tolerance. A search for atrazine tolerance in the flax collection of the U.S. Department of Agriculture, amounting to 1541 samples, revealed only one line that was about twice as tolerant as the standard cultivar.[33]

Herbicides are a very recent selective factor in the time scale of crop evolution. Thus there is no reason to suppose that the alleles for herbicide tolerance are much more common among modern cultivars than among primitive or wild races, or related wild species. Primitive or wild relatives, which are often more genetically varied than cultivated forms, are therefore a potential source of herbicide tolerance.

Triazine resistance under cytoplasmic control has been discovered in a wild genotype of *Brassica campestris* L.[34] and the transfer of this tolerance into cultivated varieties of *B. campestris* (Polish rape) and *B. napus* (seed rape) by backcrossing is described in Chapter 13. De Gournay et al.[35] considered the possibility of investigating primitive wheats and related species for herbicide tolerance but did not undertake this project. The problem would be that primitive or wild varieties are not broadly adapted to current agricultural or horticultural practice. After finding genes for herbicide tolerance, the breeder would be faced with the difficult task of incorporating them into a cultivar that fulfills modern needs. The screening of wild or primitive relatives for herbicide tolerance would therefore be a last resort in crops such as cereals, which have been evolving in cultivation for thousands of years, but a more practical proposition for recently domesticated species, such as the agricultural and amenity grasses.

## 2.2 Techniques for Selection Trials

The concept of selecting for herbicide resistance is very simple. An obvious technique is to sow a large area of field with the crop, spray it with a herbicide to kill all susceptible plants, and select the survivors. This technique may work, but there is a host of possible reasons why the results may be less than perfect.

1  Unevenness of spray application—because of drift, imperfect nozzle spacing or height, asymmetrical nozzles, and so on.
2  Unevenness in soil conditions—leading to greater vigor and thus more survival in one area than in another.
3  Inadequate or total mortality—because the calculations of the rate at which to apply the herbicide were based on inadequate premises, or because of unpredictable climatic events (e.g., rain) after spraying.
4  Avoidance of the herbicide (e.g., by delayed germination).
5  Protection of plants by neighbors or weeds from foliar application, or by stones or deep sowing from preemergence applications.
6  Differential interception of spray because of varying orientations of leaves at spraying.

The crop, the herbicide, the growth stage at application, and the facilities available must be taken into account in reaching a decision on the most appropriate technique of selecting for herbicide resistance. Selecting in the greenhouse at the seedling stage has been found to be the most satisfactory basic technique in breeding grasses at my station. Seedlings are preferred to mature plants because it is possible to screen a very much larger number of individuals in a shorter time. The greenhouse provides independence from the seasons, allowing selection to be carried out at any time of year, and independence from the weather, allowing precise timing of herbicide application and of the selection of survivors. Temperature and lighting in the greenhouse are controlled. Growth chambers are ideal for small-scale work,[37] but are not usually practical for large-scale screening.

For tolerance to foliar herbicides in grasses, seeds are sown in a fine soil-based compost, at regular spacings of 15 mm for the large-seeded species. In a good seed sample sown in optimal conditions, the majority of seeds germinate within a few days in one even flush, and the remainder germinate during the next 1 to 2 weeks. A laboratory sprayer especially designed to spray at repeatable and uniform rates is used to apply the herbicide when the main flush of seedlings has reached the two-leaf stage. Because the ability of seedlings to withstand these herbicides gradually increases with age, most of the younger seedlings are killed by dosages which are selective at the two-leaf stage. There is a small proportion of survivors that either had not emerged or had only a short vertical leaf at the time of spraying. These "false tolerants" are detected as one-leaf seedlings with no lesions by inspection at intervals after spraying. True tolerants have two or more leaves, and often have small herbicide induced lesions.

The greatest disadvantage of this method is the laboriousness of the manual sowing procedure. Regular spacing of seeds is, however, a very important feature of the method because it ensures that there is no variation in sheltering and competition effects (except at the borders, which can be discarded) and helps an observer distinguish between tillers of neighboring seedlings. The method permits discrimination not only between seedlings that do or do not survive the herbicide treatment but also between different degrees of tolerance. This is a very valuable feature. Suppose that it is intended to select the most tolerant 1% of seedlings: to try to choose a herbicide dosage that would kill exactly 99% of a population of seedlings would be unrealistic because the effects of a given dosage cannot be predicted so accurately. If the target adopted is 95% mortality, there is a high probability of a result within the range 90 to 99%, and it will be possible to select the required fraction as the least damaged seedlings among the surviving 1 to 10%. Close control of the proportion of seedlings selected is particularly important in a recurrent selection program in a cross-fertilized species, because a balance must be achieved between maximizing the genetic advance and avoiding inbreeding.

Soil-active herbicides are normally mixed into the compost before sowing rather than sprayed onto the soil surface. To ensure thorough mixing, the herbicide can be dissolved or suspended in water and sprayed into the compost in a revolving cement mixer. It is particularly important to ensure as even a supply of water to the seedlings as possible: clusters of survivors can frequently be observed, for example in sites that have been leached by drips of condensation or in positions directly above holes in the base of seed trays set on capillary beds. Incorporating the herbicide into the seedbed has been used in evaluating the relative tolerances of grass species and cultivars to soil-active herbicides.[19] A similar technique has been used in selecting for simazine tolerance in *Sinapis alba* L.[36] and seed rape.[23,37]

A theoretical objection to screening seedlings is that they may not respond to a herbicide in the same way as mature plants. This objection applies only to those herbicides that are envisaged as potentially useful on mature rather than seedling stands. Two parallel comparisions were made of the efficiency of selecting for paraquat or dalapon tolerance in *Lolium perenne* by spraying at successive stages of growth. With dalapon, which is active both as a foliar and soil herbicide, an additional technique involving the incorporation of the dalapon into the compost was also tried. The areas of greenhouse used for each method were equal, although fewer plants were sprayed at the older stages. After spraying, 25 "tolerant" plants were selected from each population and planted out in the field. Their tolerance was reassessed after a field spray (Table 12.1). Taking into account the length of the growing period before selection, the results indicated that screening at 2 weeks from sowing would give the maximum response for equivalent deployment of resources. A similar conclusion was reached by Fisher[32] working with *Agrostis tenuis* and amitrole. The alternative technique of incorporating dalapon into the seedbed compost did not give a good response in *Lolium* (Table 12.1). However, when *A. tenuis*

Table 12.1 Efficiency of Selection for Herbicide Tolerance at Different Growth Stages: Field Reassessment of Tolerance (0-40 Visual Scale) of *Lolium perenne* Plants Selected by Treatment in the Greenhouse at Successive Stages

| Stage of Treatment for Selection | Number of Plants Screened | Mean Field Tolerance | |
|---|---|---|---|
| | | Dalapon | Paraquat |
| Control (unselected) | — | 22.4 | 15.9 |
| Seedbed | 1200 | 23.6 (NS) | — |
| 2 weeks | 1200 | 26.6*** | 19.8*** |
| 4 weeks | 600 | 24.6 (NS) | 19.3*** |
| 6 weeks | 400 | — | 18.2* |
| 9 weeks | 400 | 23.8 (NS) | 17.7(NS) |
| 12 weeks | 400 | 24.4* | 19.2* |

*Source:* J. S. Faulkner, unpublished data.

*Note:* Significances of differences from the control were calculated by *t* tests: NS, not significant; *, $P < 0.05$; ***, $P < 0.001$.

was selected for tolerance to amitrole (which is also both foliar and soil active) by this means, a superior response was obtained compared with spraying at the two-leaf stage.[32] In these experiments with both *L. perenne* and *A. tenuis* it was evident that although selection in seedlings raised the overall level of tolerance, some of the plants selected were susceptible on retesting. A second screening at a later growth stage is therefore recommended for maximizing the response to selection.

When comparing the herbicide tolerance of populations, such as cultivars or breeders' lines, rather than selecting tolerant individuals within a segregating population, each population can be represented by many plants. The effects of unevenness in the environment can therefore be reduced by replication. Nevertheless, de Gournay et al.[35] encountered problems in screening winter wheat cultivars for atrazine tolerance by preemergence spraying in the field. Among the difficulties encountered were variation according to the source of the seed, variation from year to year in the performance of cultivars, lack of a definitive criterion for selection, clustered distribution of survivors, and edge effects. The dominant problem, however, seemed to be the low level of genetic variation in atrazine tolerance, not so much within each of the cultivars (which is expected in a self-fertilized crop), but in winter wheat as a whole.

Incorporation of the herbicide into a field soil was practiced by Andersen and Behrens[33] in screening flax lines for atrazine tolerance. Single rows of each line were sown into the treated soil and individual tolerant plants were selected at maturity. Pure-line progeny of most of the selections were found to be as susceptible as a control cultivar on retesting in the greenhouse, although one partially tolerant line was obtained.

Soil-less media and water culture represent the opposite extreme from field selection, but these techniques do not accurately simulate field conditions, especially in respect of herbicide uptake through the root system. Individual seedlings of *Festuca arundinacea* Schreb. growing between two vertical layers of filter paper in acrylic plastic tanks containing a suspension of simazine were selected by Johnston.[20] Progeny of the selected individuals, produced by poly-crossing, were shown to be more tolerant than the base population in the tank conditions, but they were not tested in soil. A vermiculite growing medium was used for progeny testing sugarcane for ametryn resistance,[38] and a petri dish technique has been suggested for selecting 2,4-D-resistant genotypes of bird's foot trefoil.[39] For genetic studies of tolerance to chloramben methyl ester, single cucumber plants were grown in flasks of nutrient plus herbicide solution,[13] but this method would hardly be practical for large-scale screening.

The use of cell culture techniques for selecting herbicide tolerant mutants is reviewed in Chapter 14.

## 2.3 Selection Criteria

After exposing seedlings to a herbicide, the simplest of all criteria—survival—can normally be used in selecting for tolerance. As described in the preceding section, it is sometimes advantageous to select within the survivors on grounds of vigor to increase the selection intensity.

In screening older plants, detection of degrees of tolerance is more important. Three quantitative measures of tolerance to chloramben methyl ester in cucumber were compared by Miller et al:[13] plant height, dry weight, and visual rating. All three characters were highly heritable, but it was concluded that visual rating was the best measure because it integrated various facets of tolerance, and because some of the variation in the other two characters was unrelated to tolerance.

Dry weight of herbage and visual rating were compared as criteria of paraquat tolerance in *Lolium perenne* by Faulkner.[14] Weight of sprayed as compared with unsprayed clonal replicates was regarded as the ideal criterion, but visual scores were preferred in practice because they were much quicker to record, did not require unsprayed replicates, and were almost as effective as a selection criterion.

Correlations between anatomical or morphological characters and herbicide resistance have been demonstrated in some weeds, but the effects of herbicides on plants are usually so obvious to the eye and so readily measured that the use of correlated characters as criteria for selection in crops is unnecessary. Characters such as cuticular wax[40] or lemma color and pubescence[41] are more difficult to measure or only partially associated with herbicide response. Indirect selection may, however, be appropriate for certain clonally propagated crops in which the provision of growing plants for special herbicide trials would be troublesome. For example, Gawronski et al.[42] devised a laboratory sinking leaf-disk test for metribuzin tolerance in potatoes which requires only a few leaves (see also Chapter 7).

Characters correlated with herbicide tolerance are, perhaps, more likely to be a snag than an asset. Pesticide-resistant microorganisms, insecticide-resistant insects, and herbicide-resistant weeds tend to have a lower evolutionary fitness than their normal counterparts in the absence of chemical treatment (see Chapters 9, 15, and 17). Although fitness is not identical with agronomic value, there is a strong relationship between them. Thus it is possible that many herbicide tolerance genes in crop species have pleiotropic deleterious effects on yield or quality. There are indications to this effect in the poor agronomic qualities of the winter wheats found to have tolerance to atrazine[35] and the indifferent performance of the earliest paraquat-tolerant cultivars of perennial ryegrass.[43] However, it should not be expected that the earliest herbicide-tolerant lines of crop species, produced after a few years of selection, would necessarily be as good agronomically as the best modern cultivars, which are the outcome of decades or centuries of crop evolution.

If differences in fitness under conditions of repeated herbicide application could be used to favor tolerance genes, it might be possible to dispense with selection criteria altogether. Grignac[44] made 24 applications of metoxuron over 7 years to a natural population of the weed grass *Poa annua*, and found that the ability of the population to survive metroxuron treatment was increased markedly (e.g., from 2.5% to 68.5% survival at 3.0 kg/ha of metoxuron). However, considering that 7 years represents up to 28 generations in *P. annua,* the rate of genetic advance achieved would be too slow for crop-breeding purposes. Applications of 2,4-D to three successive generations of three cultivars of flax were not found to have altered the degree of tolerance,[45] but since flax is predominantly self-fertilized, a genetic response within cultivars would have been unlikely. The generally rather slow rate of appearance of herbicide-tolerant weeds in agriculture[46] also indicates that methods based on natural selection would probably be too slow for deliberate production of herbicide-tolerant genotypes. Indeed, were natural selection generally able to deliver tolerant populations within a few generations, the effectiveness of the herbicide against weeds would very soon be eroded.

## 2.4 The Scope for Herbicide Tolerance Breeding

Breeding herbicide-tolerant crops is an alternative to developing new herbicides. Since the potential return to the manufacturer of a new herbicide is related to the area on which the herbicide may be applied, much of the research and development effort on herbicides is directed toward the weed problems of large-area crops. In developing a herbicide-tolerant cultivar, the commercial motivations are more complicated. The breeders may, for example, be in business to make money directly out of new cultivars through royalties or by selling seed. They may, on the other hand, be employed on behalf of the growers and consumers to improve the efficiency of food production. They may conceivably be employed by a herbicide manufacturer with the intention of developing a wider market for the herbicide. Thus there can be inducements to breed a crop with tolerance to a specific herbicide, even though the potential usage of herbicides on the crop would not be of sufficient magnitude to justify developing a new herbicide. Hence herbicide tolerance could be valuable in any crop, regardless of the area of cultivation, if there are weed problems. In large-area crops, the prizes are high but the competition from new herbicides is stiffer; in small-area crops, the prizes are lower but there is less competition from new herbicides.

The breeder of herbicide-tolerant cultivars is not only in competition with the developers of new herbicides but also with breeders of conventional cultivars. Unless the weed problems of the crop are exceptionally severe, it is pointless to breed tolerant cultivars unless they are as good as existing cultivars in other characters. In the short term, at least, it is this restriction that is likely to determine the crop species in which herbicide tolerance breeding will be successful. In most of the major crops of temperate areas, modern cultivars are highly specialized genotypes resulting from a long history of selective breeding and from the use of increasingly advanced breeding methods. These cultivars often have a narrow genetic base within their respective species, and are agronomically greatly superior to their primitive or wild relatives. In these crops, the breeding of new cultivars that are as good as the existing ones is inevitably more difficult than in crops that have been bred less intensively.

Another factor that influences the potential value of herbicide-tolerant cultivars is the nature of the weed flora associated with the crop. It is relatively easy to find herbicides that will discriminate between a crop and weeds that are evolutionarily unrelated. The more recent the evolutionary divergence of crop and weed, the less likely it is that genetic drift would have brought about a differential reaction to a herbicide. Thus there are a large number of phenoxy acid-type herbicides that act selectively against dicotyledonous weeds in cereals. The more intractable weed problems in cereals are now represented by other species of Gramineae such as *Agropyron repens* (L.) Beauv., *Alopecurus myosuroides* Huds., and *Avena* spp.

The number of crops in which conventional breeding for specific herbicide tolerance has been attempted is small (Table 12.2) and it is too soon to predict on empirical grounds the most likely fields of commercial success. However, it is interesting that the longest catalog of problems emanated from the work on atrazine tolerance in winter wheat.[35] More promising results have been obtained in some forage plants, *Brassica* crops, and amenity grasses, which have a shorter history of intensive selective breeding than does wheat.

In bird's-foot trefoil, a forage legume, a line has been selected that recovers from dosages of 2,4-D normally used for weed control,[39] although the agronomic value of the line with increased resistance has not been demonstrated, and no commercial variety has been released.

In *Lolium perenne,* it has been shown that both dicotyledonous weeds and other grasses can be controlled by paraquat in seedling swards of a cultivar selected for resistance to this herbicide.[16,47] The cultivar on which these tests were done is not suitable for forage use, but was entered on the U.K. National List in 1980 for amenity use. Other paraquat-tolerant lines have since been produced and their forage traits are being investigated.[22]

Also in *L. perenne,* a cultivar Rathlin has been bred with increased tolerance of dalapon, a herbicide that was previously known to be slightly effective in favoring ryegrass in competition with other grasses.[18] The greater selectivity of dalapon between Rathlin and the "weed grasses" renders the use of dalapon for grassland improvement more attractive. Rathlin has good agronomic characteristics and was admitted to the U.K. National List in 1980.

Broad-leaved weeds, including wild members of the *Brassiceae* tribe, are major weeds in crops of the *Brassiceae*. Interest in herbicide tolerance in the *Brassiceae* has been focused on the triazine group of herbicides. Little progress has been achieved by recurrent selection,[37] perhaps because the scale of the selection work has been too small. Triazine-resistant cultivars of *Brassica campestris* and *B. napus* are being bred by transferring cytoplasmic resistance genes from a wild genotype of *B. campestris,* although it remains to be seen whether there will be any agronomic penalties (Chapter 13).

Table 12.2    Attempts to Select for Increased Herbicide Tolerance in Crop Species by Conventional Methods

| Herbicide | Species | Method | References |
|-----------|---------|--------|------------|
| Amitrole | *Agrostis tenuis* | Recurrent Selection | 24, 24a |
| Amitrole | *Agrostis stolonifera* | Recurrent Selection | 24a |
| Amitrole | *Festuca rubra* | Recurrent Selection | 24a |
| Asulam | Seed rape | — | J. E. Flack and P. D Putwain, personal communication |
| | Barley | — | J. E. Flack and P. D. Putwain, personal communication |
| Dalapon | Perennial ryegrass | Recurrent selection | 18 |
| Diphenamid | Tomato | Chemical mutagenesis | 25 |
| Glyphosate | Perennial ryegrass | Recurrent selection | 22a |
| Paraquat | *Festuca rubra* | One cycle of selection | 24 |
| | Perennial ryegrass | Recurrent selection and selection in intervarietal hybrids | 22, 28, 22a |
| | *Trifolium repens* | One cycle of selection | 48 |
| | Wheat | Search for tolerant parents | A. F. Hawkins, personal communication |
| Terbutryn | Wheat | Chemical mutagenesis | 25 |
| Triazine | Polish rape | Transfer from wild genotype | Chap. 13 |
| | Seed rape | Recurrent selection | 23, 37 |
| | Seed rape | Transfer from *B. campestris* | Chap. 13 |
| | Cabbage | One cycle of selection | 37 |
| | *Festuca arundinacea* | One cycle of selection | 20 |
| | White mustard | Recurrent selection | 36, 37 |
| | Wheat | Search for tolerant parents | 35 |
| 2,4-D | Bird's-foot trefoil | Recurrent Selection | 39 |

*Note:* Several of the selection programs are still in progress.

In the amenity grass *Agrostis tenuis,* selection for tolerance to the herbicide amitrole resulted in a line that was estimated to have sufficient resistance to allow the use of this herbicide for selectively removing the unsightly weed grass *P. annua* from closely mown turf.[24] The value of this particular line has not yet been tested in practice, but it represents a particularly interesting application of herbicide tolerance breeding. In crop plants there is commonly a minor delay in growth after the application of a selective herbicide, which has to be offset against the benefits of weed control. In amenity grasses and other ornamental plants, a loss of growth is of little or no consequence because yield is not an objective, although discoloration is undesirable. Thus the breeder of herbicide-tolerant ornamental species has somewhat different and perhaps less stringent requirements. In some perennial crops also, such as soft fruits, growth delays may be unimportant at certain times of year, and could even be beneficial in spring—to delay growth until the danger of frost has past.

## 3  CONCLUDING REMARKS

### 3.1  Secondary Benefits and Problems of Herbicide-Tolerant Cultivars

Although the normal reason for developing herbicide-tolerant cultivars would be to permit a herbicide to be used for weed control in a hitherto susceptible crop, or to improve the selectivity of a herbicide already in use, there are various other possible benefits to be derived.

In some crop rotations, a residual herbicide applied for weed control in one crop may cause damage to a succeeding crop. This problem could be overcome by sowing a more tolerant cultivar of the second crop. For example, in France stands of winter wheat are sometimes damaged by residues of atrazine that has been applied to the preceding corn crop. The main purpose of the search for atrazine tolerance in winter wheats[35] was to develop cultivars that would be consistently tolerant of these residues.

Some crops are commonly grown with a companion species, or are undersown with seed of a succeeding crop. This practice places restrictions on the use of herbicides, even nonresidual ones. White clover, for instance, is commonly included in grass seed mixtures for forage, because of its high nutritive value and nitrogen-fixing root nodules. The herbicide ethofumesate is very effective at controlling broad-leaved weeds and undesirable grasses in young grass leys, but also eliminates any white clover that is present. A tolerant cultivar of white clover would solve this problem. However, from preliminary investigations at my station, it seems that the amount of variation in white clover is insufficient for tolerant lines to be bred, at least using conventional selection methods.

Weeds can be serious in ordinary crops, but they are doubly so in seed crops. Seed certification regulations require a high standard of purity in general and of freedom from off-types and closely related species in particular. A seed crop that fails to meet the requirements may be valueless. The manual removal of rogues is very expensive or virtually impossible, but the grower of normal cultivars has no chance of roguing his crop by blanket spraying with herbicides. In cultivars with tolerance to specific herbicides, however, the removal of contaminant rogues would be possible,[49] at least until such time as the herbicide tolerance genes had been transferred to the potential rogue genotypes. Paraquat has proved valuable in the production of clean seed of paraquat-tolerant *Lolium perenne*.[47]

In conjunction with varietal registration and breeders' rights schemes for plant cultivars, there has developed a requirement that new plant cultivars should be distinct, uniform, and stable. To fulfil the distinctness requirement, a new plant cultivar must be consistently different from existing cultivars in at least one morphological or physiological character. Distinctness in yield, the most important of breeders' selection criteria, is not admissible because of the inherent instability of this character. In outbreeding crops, where the maintenance of genetic heterogeneity within the cultivar is a necessary insurance against inbreeding depression, the requirement for distinctness can be a problem. Breeders may resort to selection for trivial characters, such as ligule pubescence in grasses, in order to establish the distinctness of their cultivars. It has been suggested that herbicide tolerance could be used as a criterion of varietal distinctness.[50] In general, the differences between cultivars are probably too small for herbicide tolerance to be useful as a routine criterion in distinguishing cultivars. However, cultivars selected specifically for herbicide tolerance should be readily and consistently distinguishable.

Although herbicide-tolerant cultivars should fulfil the distinctness requirement without difficulty, the uniformity and stability requirement may present a quandary. Clonally propagated or inbred cultivars should be genetically uniform and stable between cycles of multiplication, and the use of the herbicide to maintain varietal purity should not cause any genetic changes. In outbred cultivars with polygenic herbicide tolerance, some variation in degree of tolerance within the cultivar will be unavoidable. Application of the herbicide to maintain the purity of the cultivar during multiplication may cause it to become even more tolerant. On the other hand, failure to apply the herbicide may lead to a gradual loss of tolerance if there is a negative correlation between herbicide tolerance and fitness (Chapters 9, 15, and 17). This quandary is best met with a compromise: the herbicide should be applied at a rate high enough to kill fully susceptible rogue plants, but not so high as to exert a strong selective pressure within the cultivar itself.

Herbicides are sometimes deliberately used to kill plants of crop species, such as volunteers in subsequent crops and grasses at the end of the ley phase of a crop rotation. Although tolerance to a herbicide may be valuable in the life of the crop, it could become a drawback thereafter. The case of paraquat-tolerant *Lolium perenne* illustrates this point well. *L. perenne* is perhaps the most important sown pasture grass in the cool wet temperate areas of the world, and it also occurs naturally on roadside verges, waste ground, and so on. The use of a paraquat-tolerant cultivar in grassland precludes the eventual use of paraquat for sward destruction, except at about five times the normal dosage. Moreover, the escape of paraquat-tolerant genotypes may render the herbicide ineffective for controlling *L. perenne* as a weed. Fortunately, the species is of minor importance in this guise: its seeds have little capacity for remaining dormant in the soil, and it is easily controlled by other herbicides. There is one recorded case of a natural population of *L. perenne* with a high level of tolerance to paraquat giving rise to a weed problem in a field subjected to repeated applications of paraquat for direct drilling.[27] The problem was resolved by using another grass-killing herbicide, glyphosate.

The risk of herbicide-tolerant cultivars becoming new weeds is not serious enough to outweigh the advantages of breeding them, but this problem should not be dismissed from consideration. Any projected herbicide-tolerant cultivar should be known to be readily controllable by an alternative herbicide, and breeding for tolerance to two or more unrelated herbicides should be eschewed.

## 3.2  Summary: The Pros and Cons of Selection

1  Breeding for tolerance to herbicides is a cheaper alternative than developing new herbicides as a means of combating weed problems.

2  Ideally, tolerance should be sought to a herbicide that is safe, cheap, and toxic to a wide range of weeds.

3  Major gene tolerance to herbicides is known, and would be ideal from the breeder's viewpoint, but polygenic tolerance is more likely to be found by conventional systematic screening.

4  Existing cultivars and breeders' collections are the most likely sources of herbicide tolerance, but rogue plants or spent crops that have been exposed to the herbicide should be screened when available.

5  Herbicide-tolerant cultivars may be bred by combining genes for herbicide tolerance with genes for agronomic characters from different sources, by selection for herbicide tolerance within outbreeding cultivars, or by mutagenic treatment of existing cultivars.

6  Screening for herbicide tolerance at the seedling stage permits the use of large populations and high selection intensities.

7  Screening can take place in the field, in soil or soil-less media in the greenhouse or growth chamber, or in cell cultures. The relative advantages of each method depend on the crop, the herbicide, and the facilities available to the breeder.

8  In practice, the primary criterion of selection in seedlings is survival and in older plants is a visual rating of tolerance.

9  In field populations, selection by repeated herbicide application is unlikely to give a rapid increase in tolerance unless augmented by breeders' selection of tolerant plants.

10  It is possible that selection for herbicide tolerance will sometimes, perhaps usually, lead to reduced agronomic performance.

11  Herbicide-tolerant cultivars have been developed in some forage and amenity grasses. It is suggested that, in the short term at least, the more successful herbicide-tolerant cultivars will be of crops without a long history of intensive breeding.

12  Herbicide tolerance may confer benefits in flexibility of crop mixtures or rotations, roguing of seed crops, and testing varietal distinctness, as well as in conventional weed control.

13  Herbicide-tolerant crops could become difficult weeds. To reduce this risk, it is recommended that tolerance should be sought only to single herbicides and in species for which there is an effective alternative means of control.

## REFERENCES

1 Agricultural Chemicals Approval Scheme, *Approved products for farmers and growers,* Ministry of Agriculture, Fisheries and Food, U.K., 1979.

2 J. Gressel, Genetic herbicide resistance; projections on appearance in weeds and breeding for it in crops, in *Plant Regulation and World Agriculture,* T. K. Scott (Ed.), Plenum Press, New York, 1979, p. 89–109.

3 Royal Commission on Environmental Pollution, *Seventh Report: Agriculture and Pollution,* Her Majesty's Stationery Office, London, 1979, Chap. 3.

4 H. Linser, Unkrautbekämpfung auf hormonaler Basis, *Bodenkultur,* 5, 191 (1951).

5 N. S. Hanson, Dalapon for control of grasses on Hawaiian sugarcane lands, *Down to Earth,* 12(2), 2 (1956).

6 C. O. Grogan, E. F. Eastin, and R. D. Palmer, Inheritance of susceptibility of a line of maize to simazine and atrazine, *Crop Sci.,* 3, 451 (1963).

7 R. S. Chaleff and M. F. Parsons, Direct selection *in vitro* for herbicide resistant mutants of *Nicotiana tabacum, Proc. Natl. Acad. Sci. USA,* 75, 5104 (1978).

8 J. D. Hayes, R. K. Pfeiffer, and M. S. Rana, The genetic response of barley to DDT and barban, and its significance in crop protection, *Weed Res.,* 5, 191 (1965).

9 A. B. Schooler, A. R. Bell and J. D. Nalewaja, Inheritance of siduron tolerance in foxtail barley, *Weed Sci.,* 20, 167 (1972).

10 F. G. H. Lupton and R. H. Oliver, The inheritance of metoxuron susceptibility in winter wheat, *Proc. Br. Crop Prot. Conf. Weeds,* 473 (1976).

11 V. E. Comstock and R. N. Andersen, An inheritance study of tolerance to atrazine in a cross of flax (*Linum usitatissimum* L.), *Crop Sci.,* 8, 508 (1968).

12 R. G. Stafford, V. E. Comstock, and J. H. Ford, Inheritance of tolerance in flax (*Linum usitatissimum* L.) treated with MCPA, *Crop Sci.,* 8, 423 (1968).

13 J. C. Miller, L. R. Baker, and D. Penner, Inheritance of tolerance to chloramben methyl ester in cucumber, *J. Am. Soc. Hort. Sci.,* 98, 386 (1973).

14 J. S. Faulkner, Heritability of paraquat tolerance in *Lolium perenne* L., *Euphytica,* 23, 281 (1974).

15 J. D. Fryer, and R. J. Makepeace, (Eds.), *Weed Control Handbook,* Vol. 2: *Recommendations,* 8th ed., Blackwell, Oxford, 1978, Chap. 5.

16 J. S. Faulkner, The use of a paraquat for controlling weeds in seedling swards of paraquat resistant *Lolium perenne* L., *Proc. Br. Crop Prot. Conf. Weeds,* 349 (1978).

17 G. P. Allen, The effect of July applications of dalapon on the growth and botanical composition of an *Agrostis/Lolium* pasture, *Weed Res.,* 8, 309 (1968).

18 J. S. Faulkner, Dalapon tolerant varieties—a possible basis for pure swards of *Lolium perenne* L., *Proc. Br. Crop Prot. Conf. Weeds,* 341 (1978).

19 R. Fisher and J. S. Faulkner, The tolerance of twelve grass species to a range of foliar-absorbed and root-absorbed grass-killing herbicides, *Proc. Eur. Weed Res. Soc. Symp. Status, Biol. Control Grassweeds Eur., Paris,* 204 (1975).

20 F. P. Johnston, Genetic and environmental variation associated with seedling establishment of Tall Fescue (*Festuca arundinacea* Schreb), Ph.D thesis, The Queen's University of Belfast, U.K., 1980.

21 P. McMaugh, A dessicant approach to *Poa annua* control, *J. Sports Turf Res. Inst. 1970,* 46, 63 (1971).

22 Anonymous, Breeding for tolerance to paraquat, *Annual Report on Research and Technical Work of the Department of Agriculture for Northern Ireland,* 1978, 104 (1979).

**22a** C. E. Wright and J. S. Faulkner, Effective selection for tolerance to grass-killing herbicides in perennial ryegrass (*Lolium perenne* L.), *Proc. XIV International Grassland Congress,* J. A. Smith and V. W. Hays, Ed., Westview Press, Boulder, Col., 1982, in press.

**23** D. D. Warwick, Factors contributing to the improved simazine resistance observed in oilseed rape variety Rigo after three cycles of selection, *Proc. Br. Crop Prot. Conf. Weeds,* 479 (1976).

**24** R. Fisher and C. E. Wright, The breeding of lines of *Agrostis tenuis* Sibth. and *Festuca rubra* L. tolerant of grass-killing herbicides, *Proc. 3rd Int. Turfgrass Res. Conf., Munich* 1977, **11**, (1980).

**24a** H. Lee and C. E. Wright, Effective selection for aminotriazole tolerance in *Festuca* and *Agrostis* turf grasses, *Proc. 4th Int. Turfgrass Res. Conf., Guelph* 1981, **41**, (1981).

**25** M. J. Pinthus, Y. Eshel, and Y. Shchori, Field and vegetable crop mutants with increased resistance to herbicides, *Science,* **177**, 715 (1972).

**26** J. E. Van der Planck, *Disease Resistance in Plants,* Academic Press, New York, 1968.

**27** B. Lovelidge, 'Rogue' grass defies paraquat, *Arable Farming,* 9 (Sept. 1974).

**28** J. S. Faulkner, A paraquat tolerant line in *Lolium perenne, Proc. Eur. Weed Res. Soc. Symp. Status, Biol. Control Grassweeds Eur., Paris,* 349 (1975).

**29** H. R. Albrecht, Strain differences in tolerance to 2,4-D in creeping bent grasses, *J. Am. Soc. Agron.,* **39**, 163 (1947).

**30** D. R. Tottman, J. Holroyd, F. G. H. Lupton, R. H. Oliver, R. T. Barnes, and R. H. Tysoe, The tolerance of chlortoluron and isoproturon by varieties of winter wheat, *Proc. Eur. Weed Res. Soc. Symp. Status, Biol. Control Grassweeds Eur., Paris,* 360 (1975).

**31** J. S. Faulkner, The effect of dalapon on 35 cultivars of *Lolium perenne, Weed Res.,* **14**, 405 (1974).

**32** R. Fisher, Herbicide tolerance in amenity grasses, Ph.D. thesis, The Queen's University of Belfast, U.K., (1975).

**33** R. N. Andersen and R. Behrens, A search for atrazine resistance in flax (*Linum usitatissimum* L.), *Weeds,* **15**, 85 (1967).

**34** V. Souza Machado, J. D. Bandeen, G. R. Stephenson, and P. Lavigne, Uniparental inheritance of chloroplast atrazine tolerance in *Brassica campestris, Can. J. Plant Sci.,* **58**, 977 (1978).

**35** X. de Gournay, J. Koller, and J. L. Dufour, Difficultes rencontrees dans la prospection de geniteurs pour la resistance a un herbicide; cas de l'atrazine et du ble d'hiver, *Proc. Eur. Weed Res. Soc. Symp. Status, Biol. Control Grassweeds Eur., Paris,* 388 (1975).

**36** A. Karim, and A. D. Bradshaw, Genetic variation in simazine resistance in wheat, rape and mustard, *Weed Res.,* **8**, 283 (1968).

**37** S. R. Sykes, Selection for herbicide resistance in cruciferous crop plants, Ph.D. thesis, University of Liverpool, U.K., 1980.

**38** R. V. Osgood and D. J. Heinz, Selecting sugarcane seedlings for ametryn tolerance, *Hawaiian Sugar Plant. Assoc. Exp. Stn., 1976 Annu. Rep.,* 44 (1977).

**39** T. E. Devine, R. R. Seaney, D. L. Linscott, R. D. Hagin, and N. Brace, Results of breeding for tolerance to 2,4-D in bird's-foot trefoil, *Crop Sci.,* **15**, 721 (1975).

**40** J. M. Hodgson, Lipid deposition on leaves of Canada thistle ecotypes, *Weed Sci.,* **21**, 169 (1973).

**41** D. J. Rydrych and C. I. Seely, Effect of IPC on selections of wild oats, *Weeds,* **12**, 265 (1964).

**42** S. W. Gawronski, R. H. Callihan, and J. J. Pavek, Sinking leaf-disk test for potato variety herbicide tolerance, *Weed Sci.,* **25**, 122 (1977).

**43** Anonymous, Breeding for tolerance to paraquat, *Annual Report on Research and Technical Work of the Department of Agriculture for Northern Ireland 1977,* 100 (1978).

44  P. Grignac, Deviations genetiques de biotypes de plantes adventices sous l'action repetee de traitements herbicides, *Proc. Eur. Weed Res. Soc. Symp. Status, Biol. Control Grassweeds Eur., Paris,* 340 (1975).

45  R. S. Dunham, Effects of herbicides on flax, *Proc. 9th N. Central Weed Control Conf.,* 4 (1952).

46  J. Gressel and L. A. Segel, The paucity of plants evolving genetic resistance to herbicides: possible reasons and implications, *J. Theor. Biol.,* **75**, 349 (1978).

47  J. S. Faulkner, A paraquat resistant variety of *Lolium perenne* under field conditions, *Proc. Br. Crop Prot. Conf. Weeds,* 485 (1976).

48  J. S. Faulkner, Variation in paraquat tolerance in seedlings of *Trifolium repens, Rec. Agric. Res. (Department of Agriculture for Northern Ireland)*, **28**, 27 (1980).

49  C. E. Wright, Differential reaction of *Lolium perenne* genotypes to certain herbicides, *Nature (Lond.)*, **210**, 327 (1966).

50  C. E. Wright, A preliminary examination of the differential reaction of perennial and Italian ryegrass cultivars to grass-killing herbicides, *Proc. 9th Br. Weed Control Conf.,* 477 (1968).

# CHAPTER THIRTEEN

# Inheritance and Breeding Potential of Triazine Tolerance and Resistance in Plants

V. SOUZA MACHADO

Department of Horticultural Science
University of Guelph
Guelph, Ontario, Canada

## 1 INTRODUCTION

Studies on the inheritance of herbicide tolerance in plants could provide weed scientists with basic knowledge needed to prevent gene exchange between plant populations in an attempt to control the spread of a weed. The introduction of herbicide-tolerant and herbicide-resistant genes into germ plasm could be utilized to promote economic crop husbandry practices as well.

Genetic variation in the herbicide tolerance of weeds can be used to modify the genotype of crop plants to accommodate herbicides or raise their level of genetic tolerance to herbicides, and thereby improve the selectivity of registered herbicides. Variation in weed populations is of potential use to plant breeders, not only to promote more effective weed control but also for the selection of other agronomic traits. In Chapter 12 the transfer of quantitative tolerance was discussed at length. In this chapter, attention is focused on the mode of inheritance of herbicide tolerance and resistance and the potential for breeding cytoplasmically inherited triazine resistance into economic crops.

Unfortunately, prospects of registering new herbicides are not encouraging, as described in Chapter 12. More stringent registration regulations pertaining to toxicological data and increasing costs of research and development have been cited as the main reasons. Herbicide manufacturers are placing their main emphasis on the weed control of major crops, because of their larger market potential. There arises a need not only to devise alternative methods to improve the use of proven registered herbicides, but also to widen their selectivity in crop and weed communities, particularly with minor crops.

There are several reports on the varietal differences in susceptibility or tolerance to individual herbicides (see Appendix Tables A1 and A2). Little use has been made of the inheritance of exploitable tolerances or attempts to increase varietal tolerance by plant breeding.[1] The failure of any of the studies to develop into active breeding programs may be a reflection of several factors. One of them is the high priority of plant breeders to increase disease resistance in crop cultivars, because of the rapid breakdown of disease resistance to a single race, when bred into them. Other factors involve the relatively few cases of simple Mendelian type of inheritance of tolerance to a particular herbicide, and unsatisfactory agronomic characteristics of some of the herbicide-tolerant cultivars. Also, it is no longer valid to assume that herbicides will have a short market life because of the rapid appearance of new chemicals. On the contrary, the need to improve the selectivity of our limited number of registered herbicides, is today a research field that merits further investigation.

## 2  GENETIC ASPECTS

### 2.1  Inheritance of Herbicide Tolerance

#### 2.1.1  Crop species

Hayes[2] was the first to report the inheritance of tolerance to a pesticide in a field crop. He reported that barley varietal differences in susceptibility to DDT were controlled by a single major gene with tolerance being recessive. Later, differences in the inheritance of different reactions in barley to barban were reported.[3] The differential chlorosis reaction of barley cultivars to barban was controlled by a single recessive gene independently inherited; resistance to apical inhibition was inherited quantitatively.

Wiese and Quinby[4] compared the inheritance of propazine tolerance in grain sorghum to the tolerance to 2,4-D. Of 60 cultivars tested with propazine, 'Martin' and 'Red Kafir' emerged as tolerant, and 'Caprock' and 'Pink Kafir' as most susceptible. Evaluation of the $F_3$ generation of plants from crosses among these sorghum cultivars indicated that the inheritance of tolerance to propazine injury was dominant, although more than one gene seemed to be involved. In contrast, tolerance to 2,4-D was dominant and controlled by a single gene.

The inheritance of atrazine susceptibility in corn was investigated by Grogan et al.[5] They used an unusually susceptible inbred line and found that its susceptibility was conditioned by a single recessive allele. Subsequently, they attempted to locate this gene in corn using translocation stocks, and concluded that the gene controlling the response to atrazine was located on the long arm of chromosome 8.[6]

Geadelmann and Andersen[7] investigated the inheritance of tolerance to diclofop-methyl in corn so as to estimate the likelihood of effective control of volunteer corn in soybean fields. They produced single cross hybrids of tolerant and susceptible inbreds. They concluded that several loci controlled tolerance to diclofop-methyl with good prospects to transfer tolerance to their progeny, namely broad-sense heritability in the 95% range. They cautioned that when dealing with segregating $F_2$ volunteer corn populations, control would be unsatisfactory if tolerant X tolerant or tolerant X susceptible single crosses were grown the previous year.

Comstock and Andersen[8] found that tolerance to atrazine in flax was a quantitative character. The heritability of the tolerance to atrazine was estimated to be about 30%. This relatively low heritability implied that selection for tolerance to atrazine in flax would be difficult with the procedures and criteria they used. Flax was also found to have a low heritability for tolerance to MCPA.[9]

Edwards et al.[10] found that a single recessive gene controlled susceptibility to metribuzin in a soybean cultivar. The inheritance of tolerance of tomatoes to metribuzin was investigated by Souza Machado.[11] 'Fireball' and 'Vision' were used as tolerant cultivars and 'Heinz 1706' as a susceptible cultivar. The parental, $F_1$, $F_2$, and backcross generations were studied. The parents and progeny were checked by a bioassay technique using a nutrient-herbicide solution culture in controlled environments. The bioassay was useful for classifying the segregating generations using visual phytotoxic scores, seedling height, and seedling dry weight. Good evidence from $F_2$ and backcross segregation ratios as well as progeny variance analyses indicated that the inheritance of tolerance to metribuzin in tomato was controlled by one major nuclear gene with modifiers. Broad-sense heritability values of about 65% indicated a potential for increasing tolerance in susceptible cultivars through breeding.

In conclusion, the inheritance of triazine tolerance in crops such as soybean,[10] corn[5] and tomato[11] is controlled by a single nuclear gene. Physiological studies in these crops have attributed triazine tolerance to differential metabolism[12-15] (see Chapter 8). Plant breeding programs to increase triazine tolerance in soybean and tomato, so as to control broadleaf weeds, might prove feasible. The development of 'Tracy M' soybean with good tolerance to metribuzin by Hartwig[16] is an example of this approach.

## 2.1.2  Weed species

There is evidence that weeds possess a wide diversity of modes of inheritance of herbicide tolerance. Schooler et al.[17] concluded that three dominant factors controlled the inheritance of siduron tolerance in *Hordeum jubatum*. Tolerance of *Avena fatua* to diallate was quantitatively inherited.[18] This indicates that in situations where nuclear inherited genes for herbicide tolerance existed in weed populations as dominant alleles, control would be very difficult, as cross pollination with weed populations from unsprayed areas would constantly enrich the tolerance of weeds in sprayed areas at rates discussed in Chapter 17. With self-pollinated weed species and where tolerance is inherited as a recessive character, control would be possible for longer periods.

## 2.2  Inheritance of Herbicide Resistance

Recently, the resistance of biotypes of annual weeds to triazines has been reported in *Senecio vulgaris, Amaranthus* spp., *Chenopodium* spp., *Ambrosia artemisiifolia, Brassica campestris, Solanum nigrum,* and several other species (see Chapters 2 and 3).

Unlike the crop tolerances, the mechanism of resistance in some of these weed species was not based on differential uptake, translocation, or metabolism of triazines between resistant or susceptible biotypes (see Chapter 8). The resistance was due to some modified function of the Hill reaction in the chloroplasts. This was considered to be a newly discovered mechanism of resistance to triazines centered in the chloroplasts. This mechanism afforded a much higher degree of resistance to atrazine than was described in Section 2.1. Differences at the level of the photosystem II complex in the thylakoid membranes were correlated to atrazine resistance (see Chapters 9 and 10).

Genetic investigations into the inheritance of the thylakoid-modified atrazine-resistant and unmodified susceptible biotypes of *Chenopodium album* were done.[19] *C. album* is hexaploid, mainly self-pollinated with some cross pollination, and has minuscule flowers that are laborious to hand-emasculate (Fig. 13.1). Simulated wind-pollinated progeny and parental populations were treated with atrazine, but the results were difficult to interpret. All the progeny responded similarly to their female parent, which might mean that either no cross pollination, some cross pollination with cross incompatibility, or cross pollination and uniparental inheritance were involved. Subsequent work by Warwick and Black[19a] confirmed uniparental inheritance through the female parent.

Wild *Brassica campestris,* a diploid species ($n = 10$), self-incompatible, cross pollinated with large flowers suited to hand emasculation in the bud stage, was chosen as a better species for determining the inheritance of triazine resistance (Fig. 13.2, Tables 13.1 and 13.2). Reciprocal crosses were made between susceptible and resistant biotypes and the $F_1$ progeny were treated with 3 kg/ha atrazine postemergence. All the $F_1$ seedlings with the atrazine-susceptible biotype as female parent had phytotoxic symptoms within a week and later died. None of the $F_1$ seedlings with the atrazine-resistant biotype as female parent had any symptoms of phytotoxicity (Fig. 13.3). Therefore, uniparental female inheritance (Fig. 13.4) for atrazine resistance was indicated.[20] A chlorotic cotyledon mutant was used to confirm that the $F_1$ seedlings were true hybrids derived from reduced gametes of both parents and not matromorphic plants derived from unreduced gametes. Chlorosis was controlled by a single partially dominant gene ($Cc$) in the $F_1$ progenies.[21] The atrazine-resistant biotype was crossed reciprocally with this atrazine-susceptible mutant. All $F_1$ progeny seedlings had chlorotic cotyledons, confirming that true hybrids were involved. These two progenies were next treated with 3 kg/ha atrazine. Resistance remained female uniparental controlled, as before.

An analysis of the segregation of triazine resistance in $F_2$ and backcross progeny was required to determine whether this uniparental effect was con-

**Figure 13.1** Influorescence of *Chenopodium album*. The close-up illustrates minuscule flower buds that are laborious to hand emasculate. Bud in the center has opened, showing anthers and pistil. (V. Souza Machado, previously unpublished.)

**Figure 13.2** Influorescence of wild *Brassica campestris*. The large flowers are suited to hand emasculation. (V. Souza Machado, previously unpublished.)

**Table 13.1 Nomenclature and Genome Relationships of Some *Brassica* Species and Related Genera**

| Species Name | Common Synomym | Chromosome Number (*n*) and Genome | Common Name |
|---|---|---|---|
| *Brassica campestris* L. | | | |
| ssp. *oleifera* (Metzg.) Sinsk. f. *annua* | *B. rapa* ssp. *oleifera* Metzg. | 10 AA | Summer turnip rape or Polish rape |
| ssp. *oleifera* (Metzg.) Sinsk. f. *biennis* | | | Winter turnip rape |
| ssp. *eu-campestris* (L.) Olsson | *B. campestris* L. *B. rapa* L. | | Bird rape; wild turnip rape |
| ssp. *sarson* Prain | | | Yellow and brown sarson |
| ssp. *dichotoma* (Roxb.) Olsson | *B. napus* ssp. *dichotoma* Prain | | Toria |
| ssp. *chinensis* (L.) Makino | *B. chinensis* L. | | Chinese mustard; Pe-tsai |
| ssp. *pekinensis* (Lour.) Olsson | *B. pekinensis* (Lour.) Rupr. | | Celery cabbage |
| ssp. *nipposinica* (Bailey) Olsson | *B. nipposinica* Bailey | | Curled mustard |
| ssp. *rapifera* (Metzg.) Sinsk. | *B. rapa* L. | | Turnip |
| *B. tournefortii* Gouan | | | Wild turnip |
| *B. nigra* (L.) Koch | | 8 BB | Black mustard |
| *B. oleracea* L. | | 9 CC | |
| ssp. *acephala* DC. | | | Kale |
| ssp. *botrytis* L. | | | Cauliflower |
| ssp. *capitata* L. | | | Cabbage |
| ssp. *gongylodes* L. | | | Kohlrabi |
| ssp. *italica* Plenck | | | Broccoli |
| ssp. *sylvestris* L. | | | Wild cabbage |
| ssp. *gemmifera* | | | Brussels sprouts |

| | | | |
|---|---|---|---|
| B. juncea (L.) Coss | B. cernua Forbes & Hemsl. | 18 AABB | Brown, Oriental, leaf, or Indian mustard |
| B. napus L. | | | |
| ssp. oleifera (Metzg.) Sinsk. f. annua | B. napus L. annua Koch | 19 AACC | Summer or Argentine rape |
| ssp. oleifera (Metzg.) Sinsk. f. biennis | B. napus L. biennis (Schuebl. Mart.) Reichenb. | | Winter rape |
| | B. napella Chaix | | |
| ssp. pabularis (DC.) Reichenb. | | | Rape-kale |
| ssp. rapifera (Metzg.) Sinsk. | B. napobrassica (L.) Mill. | | Rutabaga, swede |
| B. carinata Braun | | 17 BBCC | Abyssinian mustard |
| B. hirta Moench | Sinapis alba L. | 12 DD | White or yellow mustard |
| B. kaber (DC.) L.C. Wheeler | Sinapis arvensis L. | 9 SS | Charlock, wild mustard |
| Eruca sativa Mill. | | 11 EE | Rocket salad |
| Raphanus sativus L. | | 9 RR | Radish |
| Camelina sativa Crantz. | | 20 | Camelina, false flax |
| Crambe abyssinica | | 45 | Crambe, Abyssinian kale |

Source: Modified from Downey et al.[28]

Note: Genome designations according to Fig. 13.6.

Table 13.2  Ease of Crossing among Self-compatible (SC) and Self-incompatible (SI) Oilseed and Condiment Species and *Brassica* Related Species and Genera

| | | *B. campestris* | *B. napus* | *B. juncea* | *B. hirta* | *B. oleracea* | *B. carinata* | *B. nigra* | *B. kaber* | *B. tournefortii* | *B. fruticulosa* | *Raphanus sativus* | *Eruca sativa* | *Rapistrum rugosum* | *Diplotaxis tenuifolia* |
|---|---|---|---|---|---|---|---|---|---|---|---|---|---|---|---|
| *B. campestris* | ♀ | SI | 3[a] | 1 | 0 | 2 | 2 | 1 | 1 | 2 | 2 | 1 | 2 | — | — |
| | ♂ | SI | 5 | 4 | 0 | 1 | 2 | 0 | 1 | 1 | — | 1 | — | — | — |
| *B. napus* | ♀ | | SC | 2 | 1 | 2 | 1 | 1 | 1 | 2 | 1 | 1 | 1 | 2 | 1 |
| | ♂ | | SC | 3 | 0 | 2 | 0 | 0 | — | — | — | — | — | — | 0 |
| *B. juncea* | ♀ | | | SC | 0 | 1 | 2 | 3 | 1 | — | — | 1 | — | — | — |
| | ♂ | | | SC | 0 | 0 | 0 | — | 0 | — | — | 0 | — | — | — |
| *B. hirta* | ♀ | | | | SI | 1 | — | 0 | 0 | — | — | — | — | — | — |
| | ♂ | | | | SI | 0 | — | 0 | 0 | — | — | — | — | — | — |

*Source:* Downey et al.[28]

[a] Ratings of 0 to 5 indicate approximately nil, 0.002, 0.02, 0.5, 1, and over 2 hybrids per flower pollinated, respectively; dashes indicate insufficient information to assign a rating.

trolled by maternal nuclear DNA, cytoplasmically controlled by chloroplast or mitochondrial DNA, or an interaction of nuclear and cytoplasmic control. This was carried out using atrazine-resistant and atrazine-susceptible weed biotypes as well as several progenies treated with 3 kg/ha atrazine postemergence. The results supported the hypothesis that atrazine resistance in the wild *B. campestris* biotypes was uniparentally inherited through the female parent, and controlled by cytoplasmic DNA (V. Souza Machado and J. D. Bandeen, unpublished results). Subsequent work on *B. campestris* and *B. napus* rapeseed species by Beversdorf et al.[22] using several backcrosses confirmed this. This was further borne out with Hill reaction studies using isolated chloroplasts of atrazine-susceptible and atrazine-resistant biotypes from wild *B. campestris*. They were carried out to evaluate the effect of atrazine, a photosynthesis inhibitor, on these genotypes. Isolated chloroplasts were extracted from seedlings of the two biotypes as well as from the hybrids of reciprocal crosses. Atrazine at $10^{-5}$ M inhibited the photochemical activity in chloroplasts isolated from the atrazine-susceptible biotypes (S) and $F_1$ cross (SR), where the susceptible biotype was used as the female parent. The atrazine-resistant biotype (R) and $F_1$ cross (RS) where the resistant biotype was used as the female parent were not affected by $10^{-4}$ M atrazine. These results indicated that a differential resistance to atrazine does exist at the chloroplast level and could be involved with a cytoplasmic component in the wild *B. campestris*.[23] This aspect

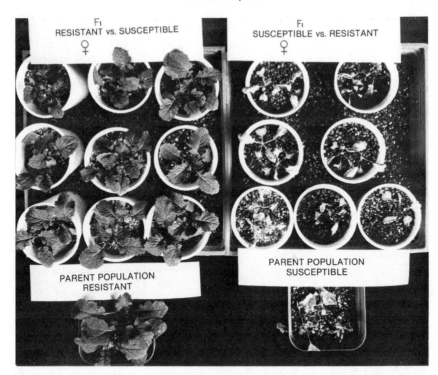

**Figure 13.3** Reciprocal crosses between resistant and susceptible *Brassica* biotypes. Visual phytotoxicity of parents and F₁ reciprocal crosses of *B. campestris* biotypes, after postemergence treatment with atrazine at 3 kg/ha. (V. Souza Machado, previously unpublished.)

has been recently elaborated by Darr et al.[23a] and in Chapter 10 at the *in vitro* level.

Work at Liverpool was carried out on the inheritance of atrazine- and simazine-resistant *Senecio vulgaris* from Washington State and a simazine susceptible population from the Botanic Gardens at Ness, U.K., having a "rayed" marker gene. Scott and Putwain[23b] indicated that uniparental inheritance of triazine resistance was involved and the rayed marker trait was inherited in a semidominant nuclear manner.

Solymosi[23c] studied the inheritance of triazine resistance in *Amaranthus retroflexus* and observed uniparental inheritance through the female parent using cotyledon size as a nuclear marker.

Gasquez et al.[23d] tested reciprocal crosses of triazine-resistant and triazine-susceptible *Solanum nigrum*. Results of F₁ and F₂ progeny showed that triazine resistance was maternally inherited; however, a few hybrids inherited the paternal chloroplast characteristics. Similar results were noted by Darmency[23e] with *Poa annua*.

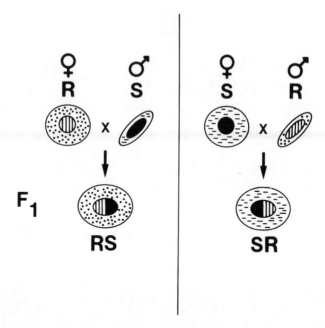

**Figure 13.4** Diagrammatic scheme of reciprocal crosses between resistant (R) and susceptible (S) biotypes of *B. campestris*. The two reciprocal $F_1$ hybrids (RS and SR) have a similar nuclei but different cytoplasms. The cytoplasm of the $F_1$ is similar to the female parent. (V. Souza Machado, previously unpublished.)

Current investigations by the author on the inheritance of atrazine resistance in *Ambrosia artemisiifolia,* a monoecious species, indicated female uniparental inheritance. It would appear from these observations that besides maternal inheritance effects, nuclear/cytoplasmic interaction may be involved.

From the data above it may be summarized that the plastid resistance to triazines can be wholly inherited through cytoplasmic DNA as in the case of *B. campestris,* but the nuclear DNA involvement may also have to be considered in other instances.

## 3  CROP-BREEDING POTENTIAL FOR RESISTANCE

Just as wild relatives of crops[24] have been used to introduce genes for disease resistance, similarly related weeds could be used to transfer herbicide resistance to related crops. The genetic structure of many cultivated plants is not entirely self-contained. It often extends into the weeds that infest the crop. Often the weedy companion is from the same species as the crop. Thus the weed germ pool can be a storehouse of genetic diversity and serve as a bridge to the germ pools of less closely related wild relatives.[25]

**Figure 13.5** Introduction of triazine resistance from a weed to a crop. Rapeseed hybrids in the background (*A*) show resistance to atrazine compared with their male parent 'Candle' *B. campestris* in the foreground (*B*). The female parent of the F₁ was triazine-resistant wild *B. campestris*. Atrazine was sprayed at 3 kg/ha postemergence. (V. Souza Machado, previously unpublished.)

The weed wild turnip rape is the same species *B. campestris* as Polish rape grown commercially in western Canada. Transferring triazine resistance into Polish rape (Fig. 13.5) would permit the use of triazines to conrol broadleaf weeds and improve the present weed control programs. In particular, the triazines would control *Thlaspi arvense* and *Brassica kaber,* which are major problems weeds in some production areas. Trifluralin is now used as a preplant incorporated herbicide, but there is a need for more effective herbicides to control broadleaf weeds at later stages in this crop. The use of triazine herbicides with rapeseed production would not only lower weed control costs but also improve seed grading by reducing *B. kaber* adulteration. Disease and pest incidence would be lowered by eliminating related triazine-susceptible alternate hosts. Combine operations at harvest could be carried out earlier, without the problems encountered with the late growth of weeds. Earlier harvesting would also decrease pod shattering, reducing harvest losses.

In western Canada, Polish rapeseed is grown in rotation with wheat, and the use of triazines may decrease broadleaf weeds in the subsequent wheat crop. The problem of appearance of triazine-resistant volunteer rapeseed seedlings from shattered seed would be remedied by continuing the current use of phenoxy herbicides in wheat. Atrazine residues could be a problem on the heavier soils and where there is a shorter growing season. The use of the less persistent chlorotriazines as cyanazine could prevent these problems where triazine-sensitive crops are planned for the following season. Triazine-resistant *Brassica* weeds have been reported so far only in a limited area in Quebec, with no

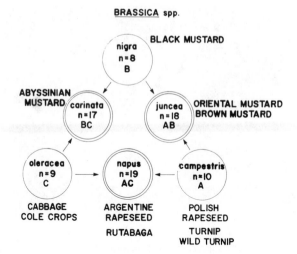

**Figure 13.6** Interspecific relationships of the *Brassica* species: the scheme of U.[27] The haploid chromosome number and haploid genome composition is denoted for each species. (Adapted from U.[27])

reports of spread within the province or of its occurrence in western Canada. Because of the cytoplasmic inheritance, the spread of triazine resistance from a cultivated resistant rapeseed cultivar to closely related weeds would not be a major threat; it is not transmitted through the pollen.[26]

There are three *Brassica* species with different basic chromosome numbers, which have combined to evolve many economic crop species through hybridization and allopolyploidy. This suggested the possibility of transferring triazine resistance from *B. campestris* to some of these closely related species. The three diploid species, *B. oleracea, B. nigra,* and *B. campestris,* have haploid chromosome numbers of 9, 8, and 10, respectively, and are thought to represent a phylogenetically ascending series[27] (Fig. 13.6). Natural hybridization between *B. campestris* (genome AA) and *B. nigra* (BB) have given the allopolyploid leaf mustard or brown/oriental mustard, *B. juncea* (AABB). Similarly, the basic chromosome sets of *B. nigra* (BB) and *B. oleracea* (CC) have given the allopolyploid, Abyssinian mustard *B. carinata* (BBCC). *B. campestris* (AA) and *B. oleracea* (CC) crosses have given the Argentine rape and rutabaga species *B. napus* (AACC). These *Brassica* oil seed, condiment, and vegetable crops have evolved within closely related species which form part of a larger interspecific, intergeneric gene pool. However, with all these crops there is a need for developing more effective pre- and postemergence chemical weed control methods to eliminate broadleaf weeds. The problem could be solved by breeding resistance to triazine herbicides into these crops.

The possibility of obtaining crosses within and between *Brassica* species does vary with the species and cultivars chosen, as well as the condition of the plants and their environment. A recent review by Downey, et al.[28] indicated that although some interspecific crosses have been made fairly easily, others have produced seed only after thousands of pollinations, while some species have never been successfully intercrossed. These problems are all the more compounded when cytoplasmic inheritance is involved and only one-way crosses must be made rather than reciprocal crosses. Generally, the interspecific crosses are more successful if an allopolyploid species such as *B. napus* or *B. juncea* is used as the female parent, particularly if the allopolyploid has one genome in common with the pollen parent. Hybrids between monogenomic species are more difficult to obtain, even with bud pollinations. Most investigators have tended to make large numbers of crosses in the expectation that some pollen grains would effect fertilization and by chance some embryos would survive natural barriers. The high rate of failure of the crosses has been directly attributed to an imbalance between the developing embryo and the endosperm as well as the incompatibility systems prevalent in the *Brassicas*. Another approach to transferring genetic information between the *Brassica* is protoplast fusion (see Chapter 14).[29] The results of interspecific *Brassica* fusions have been reported by Glebe and Hoffmann.[30]

Beversdorf et al.[22] reported the transfer of triazine resistance from the resistant weed biotype of wild *B. campestris* to rapeseed cultivars of Polish rape *B. campestris* and Argentine rape *B. napus*. The program was initiated to incorporate the cytoplasm of the triazine-resistant wild *B. campestris* ($2n = 20$) into *B. campestris* cultivars 'Candle' and 'Torch' and *B. napus* cultivars 'Tower', 'Altex', and 'Regent' ($2n = 38$) through hybridization and backcrossing. Only three fertile 38-chromosome progenies were identified among backcross ($BC_1$) progeny of wild *B. campestris* X 'Tower' (recurrent pollen parent), despite hundreds of pollinations having been attempted. Many of these pollinations produced deceptive results; pods developed without seed, a type of "false pregnancy." All $F_1$'s and $BC_1$'s of all crosses between wild *B. campestris* (female) and cultivated rapeseed (recurrent pollen parents) were resistant to 3 kg/ha atrazine and cyanazine in growth room tests. In field trials, $BC_3$'s and $BC_4$'s survived the same postemergence applications of cyanazine, atrazine, and metribuzin (0.6 kg/ha), while their pollen parents 'Torch', 'Candle', 'Tower', 'Altex', and 'Regent' were killed by all the treatments. Hill reaction analysis showed that there had been no loss of triazine resistance through the five generations of backcrossing, as reported by Weiss-Lerman et al.[31] Selection for low erucic acid and glucosinolates in rapeseed, important economic traits are being monitored among the backcross progenies (W. D. Beversdorf, personal communication).[32] No obvious deleterious effects of the resistant cytoplasm on agronomic characters have been observed among backcross progeny, although careful analysis will not be completed until adequate seed stocks for extensive field trials are available. Triazine resistant from 'Tower $BC_1$' is being successfully transferred to the cultivars 'Altex' and 'Regent'. Currently, progenies of 'Tower $BC_5$' are being multiplied in the field for subsequent testing in herbicide and variety trials, as a prerequisite stage for future registration as new cultivars (W. D. Beversdorf, personal communication).[32]

The initial intraspecific transfer of triazine resistance into rutabaga cultivar genotypes from the triazine-resistance 'Tower BC$_1$' and rapeseed genotype ('AtraTower', $2n = 38$) has been reported.[31] F$_1$ seedlings were obtained from crosses between 'AtraTower' (female) and the rutabaga cultivar 'Laurentian' (male) using bud pollinations. The F$_1$ seedlings were screened for atrazine resistance using *in vivo* and *in vitro* methods. Two-week-old F$_1$ seedlings and 'Laurentian' seedlings were sprayed with 3 kg/ha atrazine. All the 'Laurentian' seedlings developed phytotoxic symptoms and died, but none of the F$_1$ were killed. Hill reaction experiments using isolated chloroplasts were carried out. No photochemical inhibition was evident with the F$_1$ at concentrations higher than those inhibiting 'Laurentian' plastids. Further backcrosses using 'Laurentian' pollen produced a biennial rutabaga genotype with swollen root traits and a triazine-resistant cytoplasm which was confirmed in four backcross cycles using leaf chlorophyll fluorescence techniques (V. Souza Machado and A. Ali, unpublished results). A similar program has been initiated with the rutabaga cultivar 'York'. This could lead to the development of pre- and postemergence control of broadleaf weeds in rutabaga using triazine herbicides. However, progress in the development of rutabaga cultivars is slower than in rapeseed, because of its biennial pattern of flowering. Cold treatment vernalization is being used to hasten flowering.

The potential exists to transfer triazine resistance to European winter rape cultivars and other *Brassica* species such as *B. oleracea* ($2n = 18$), which includes such crops as cabbage, cauliflower, kohlrabi, broccoli, brussels sprouts, and kale. Current research has produced intraspecific hybrids ($2n = 28$) between *B. napus* and *B. oleracea* (V. Souza Machado and A. Ali, unpublished results). *Raphanus sativus* (radish) and synthetic forage crops such as *Raphanobrassica* (radicole) could also be considered. With horticultural crops that normally do not "bolt," the shedding of seed at harvest, which normally results in a volunteer weed in the subsequent season, is not a major problem. The principal problem, which has yet to be evaluated, is the possibility of undesirable agronomic traits associated with the triazine-resistant cytoplasm of the wild *B. campestris*. Another broadleaf crop meriting research work on the transfer of triazine resistance is sugar beet. As triazine resistance in the related *Chenopodium album* is cytoplasmically inherited,[19a] it might be feasible to transfer it interspecifically into sugar beet cultivars. Triazine resistance from *Solanum nigrum* might be utilized in potatoes.

Gressel et al.[29] drew attention to the possibility of triazine resistance being transferred to various *Brassica* crops using the technique of protoplast fusion. This is a research field that could lead to the successful production of cultivars in the future. On a note of caution, the extensive use of the cytoplasm of wild *B. campestris* to produce triazine-resistant cultivars of rapeseed, rutabaga, and other horticultural crops could, however, lead to a greater uniformity of the genetic base of these *Brassica* crops. This genetic uniformity could be interpreted as a general vulnerability to future disease and pest epidemics. A lesson has been illustrated in the past with the 1970 southern corn leaf blight epidemic on hybrid corn, in which the extensively used Texas cytoplasm proved vulnerable. However, this criticism could be levied at most of our major crops, which are uniform genetically and therefore impressively vulnerable.[34]

## 4 CONCLUDING REMARKS

The breeding potential of transferring triazine resistance into economic crops is a goal that has yet to be fully achieved. The encouraging point to note so far with the rapeseed and rutabaga programs has been the stability of resistance maintained over five generations.[31,33] Provided that plant vigor and yield are comparable with current recommended varieties, the use of triazine herbicides in the rapeseed crop would be a significant step in controlling the major problem broadleaf weeds which currently plague this crop in Canada.

It may be argued that breeding for herbicide resistance is not a widespread application, because of relatively few cases of simple inheritance and the narrow variation for tolerance which occurs in some economic crop species. All the same, the search for herbicide resistance and tolerance in weeds, or in wild species related to economic crops, should continue, to transform some of these crops into herbicide-resistant and herbicide-tolerant cultivars. Utilizing these genetic and plant breeding potentials together with a program of intergrated weed control involving crop rotation, herbicide rotation, and mechanical cultivation could significantly widen the basis for crop-weed management in the future.

## REFERENCES

1  Anonymous, *Selected References to the Selection and Breeding of Cereal and Other Crop Varieties for Increased Tolerance of Herbicides, 1953–1972,* Weed Research Organization, Annotated Bibliography No. 46, Oxford, 1972.

2  J. D. Hayes, Varietal resistance to spray damage in barley, *Nature (Lond.),* **183,** 551 (1959).

3  J. D. Hayes, R. K. Pfeiffer, and M. S. Rana, The genetic response of barley to DDT and barban and its significance in crop protection, *Weed Res.,* **5,** 191 (1965).

4  A. F. Wiese and J. R. Quinby, Inheritance of 2,4-D and propazine resistance in grain sorghum, *Abstr. Meet. Weed Sci. Soc. Am.* (Bushland, Tex.), 29, (1969).

5  C. O. Grogan, E. F. Eastin, and R. D. Palmer, Inheritance of susceptibility of a line of maize to simazine and atrazine, *Crop Sci.,* **3,** 451 (1963).

6  G. E. Scott and C. O. Grogan, Location of a gene in maize conditioning susceptibility to atrazine, *Crop Sci.,* **9,** 669 (1969).

7  J. L. Geadelmann and R. N. Andersen, Inheritance of tolerance to HOE 23408 in corn, *Crop Sci.,* **17,** 601 (1977).

8  V. E. Comstock and R. N. Andersen, An inheritance study of tolerance to atrazine in a cross of flax, *Crop Sci.,* **8,** 508 (1968).

**9**   R. E. Stafford, V. E. Comstock, and J. H. Ford, Inheritance of tolerance in flax (*Linum usitatissimum* L.) treated with MCPA, *Crop Sci.,* **8**, 423 (1968).

**10**  D. J. Edwards, Jr., W. L. Barrentine, and T. C. Kilen, Inheritance of sensitivity to metribuzin in soybeans, *Crop Sci.,* **16**, 119 (1976).

**11**  V. Souza Machado, Inheritance of tolerance to metribuzin herbicide in the tomato, Ph.D. dissertation, University of Guelph, Guelph, Ontario, 1976.

**12**  A. E. Smith and R. E. Wilkinson, Differential absorption, translocation and metabolism of metribuzin by soybean cultivars, *Physiol. Plant,* **32**, 253 (1974).

**13**  C. L. Foy and P. Castelfranco, Distribution and metabolic rate of $C^{14}$ labelled simazine and four related alkylamino triazines in relation to phytotoxicity, *Plant Physiol. Suppl.,* **35**, 28 (1960).

**14**  R. H. Shimabukuro, G. L. Lamoureaux, D. S. Frear, and J. E. Bakke, Metabolism of *s*-triazines and its significance in biological systems, in *Pesticide Terminal Residues,* Butterworth, Toronto, 1971, pp. 323–342.

**15**  G. R. Stephenson, J. E. McLeod, and S. C. Phatak, Differential tolerance of tomato cultivars to metribuzin, *Weed Sci.,* **24**, 161 (1976).

**16**  I. Gogerty, The monster pests, *The Furrow,* July/Aug., 2, 1979.

**17**  A. B. Schooler, A. R. Bell, and J. D. Nalewaja, Inheritance of siduron tolerance in foxtail barley, *Weed Sci.,* **20**, 167 (1972).

**18**  R. Jacobsohn and R. N. Andersen, Differential response of wild oat lines to diallate, triallate and barban, *Weed Sci.,* **16**, 491 (1968).

**19**  V. Souza Machado and J. D. Bandeen, Cross-pollination and $F_1$ segregation of atrazine tolerant and susceptible biotypes of lamb's-quarters, *Res. Rep. Can. Weed Comm., East. Sect.* (Fredericton, New Brunswick), 305, (1977).

**19a** S. I. Warwick and L. Black, Uniparental inheritance of atrazine resistance in *Chenopodium album* L., *Can. J. Plant Sci.,* **60**, 751 (1980).

**20**  V. Souza Machado, J. D. Bandeen, G. R. Stephenson, and D. Lavigne, Uniparental inheritance of chloroplast atrazine tolerance in *Brassica campestris, Can. J. Plant Sci.,* **58**, 977 (1978).

**21**  G. R. Stringam, Inheritance of chlorotic cotyledon in *Brassica campestris* L. *Can. J. Genet. Cytol.,* **11**, 924 (1969).

**22**  W. D. Beversdorf, J. Weiss-Lerman, L. R. Erickson, and V. Souza Machado, Transfer of cytoplasmically-inherited triazine resistance from bird's rape to cultivated rapeseed (*Brassica campestris* L. and *B. napus* L.), *Can. J. Genet. Cytol.,* **22**, 167 (1980).

**23**  V. Souza Machado, J. D. Bandeen, and G. R. Stephenson, Hill reaction studies with atrazine susceptible and tolerant biotypes of bird's rape, *Res. Rep., Expert Comm. Weeds. East. Sect.* (Windsor, Ontario), 339, (1978).

**23a** S. Darr, V. Souza Machado, and C. J. Arntzen, Uniparental inheritance of a chloroplast photosystem: II. Polypeptide controlling herbicide binding, *Biochim. Biophys. Acta,* **634**, 219 (1981).

**23b** K. R. Scott and P. D. Putwain, Maternal inheritance of simazine resistance in a population of *Senecio vulgaris, Weed Res.,* **21**, 137 (1981).

**23c** P. Solymosi, Inheritance of herbicide resistance in *Amaranthus retroflexus, Novenytermeles,* **30**, 57 (1981).

**23d** J. Gasquez, H. Darmency and C. P. Compoint, Etude de la transmission de la resistance chloroplastique aux triazines chez *Solanum nigrum,* C. R. Acad. Sci. (Paris) *Sec. D,* **292**, 847 (1981).

**23e** H. Darmency, I.N.R.A., Dijon, France, (personal communication) 1980.

**24**  J. R. Harlan, Evolutionary dynamics of plant domestication, *Jap. J. Genet.,* **44**, 337 (1969).

**25** P. Busey, Genetics and the weed problem, *World Crops,* 128 (May–June 1976).

**26** R. A. E. Tilney-Bassett, The genetics of plastid variegation, in *Genetics and Biogenesis of Mitochondria and Chloroplasts,* C. W. Birky, Jr., P. S. Perlman, and R. J. Byers (Eds.), Ohio State University Press, Columbus, Ohio, 1975, pp. 268–308.

**27** N. U, Genome analysis in *Brassica* with special reference to the experimental formation of *B. napus* and peculiar mode of fertilization, *Jap. J. Bot.,* **7**, 389 (1935).

**28** R. K. Downey, A. J. Klassen, and G. R. Stringam, Rapeseed and mustard, in *Hybridizations of Crop Plants,* Monogr. Am. Soc. Agron. 1980, pp. 495–509.

**29** J. Gressel, S. Zilkah, and G. Ezra, Herbicide action, resistance and screening in cultures vs. plants, in *Frontiers of Plant Tissue Culture,* T. A. Thorpe (Ed.), International Association for Plant Tissue Culture, Calgary, Alberta, 1978, pp. 427–436.

**30** Y. Y. Glebe and F. Hoffmann, "Arabidobrassica": plant genome engineering by protoplast fusion, *Naturwissenschaften,* **66**, 547 (1979).

**31** J. Weiss-Lerman, W. D. Beversdorf, and V. Souza Machado, Transfer of triazine resistance to cultivated rapeseed, *Abstr. Meet. Weed Sci. Soc. Am.* (Toronto), **37**, (1980).

**32** W. D. Beversdorf, Univ. of Guelph, Canada (personal communication) 1980.

**33** V. Souza Machado, J. D. Bandeen, and T. Adlington, Transfer of triazine resistance to rutabaga, *Res. Rep. Expert Comm. Weeds, East. Sect.* (Ste. Foy, Quebec), **410**, (1979).

**34** Anonymous, *Genetic Vulnerability of Major Crops,* National Academy of Science, Washington, D.C., 1972, pp. 1–2, 257–262.

# Herbicide Resistance in Plant Cell Cultures

CAROLE P. MEREDITH

Department of Viticulture and Enology
University of California
Davis, California

PETER S. CARLSON

Department of Crop and Soil Sciences
Michigan State University
East Lansing, Michigan

## 1  INTRODUCTION

Cultured plant cells can be manipulated in a variety of ways under carefully controlled conditions and as such are routinely used for biochemical, physiological, developmental, and genetic studies. Callus cultures can be established from most tissues of virtually any plant, as can liquid suspension cultures. Typically, a small sterile piece of plant tissue (e.g., a hypocotyl section or leaf disk) is placed on an agar-solidified nutrient medium containing mineral salts, vitamins, phytohormones, and sucrose. The tissue proliferates and forms an unorganized mass of cells—a callus. A liquid suspension culture of small clumps of cells is obtained by dispersing the callus in a liquid medium. By appropriate enzymatic treatment of a cell suspension or intact plant organs, cell walls can be removed, releasing protoplasts (isolated cells bounded only by their cell membranes). Modifying the composition of the nutrient medium will, in many cases, induce the development of whole plants from the cultured cells.[1] Whole-plant regeneration can be routinely accomplished from initial explants or newly established callus cultures of most plants. It is still difficult, however, to induce plant development from many established callus and suspension cultures and from many protoplasts.[2]

Because cultured plant cells are in many ways unicellular organisms and each cell has the potential to develop into an entire plant, a number of the techniques of microbial genetics can be applied to higher plants. These techniques can supplement the traditional methods of plant breeding to yield new variants of agronomic and/or academic value.

Populations of cultured plant cells generally contain occasional spontaneous variants which differ from the rest of the cell population in any of a number of characteristics and which may or may not be the result of mutation.[3] It is also possible, by several methods, to obtain haploid cell cultures,[4,5] thereby facilitating the selection of recessive mutations which might be masked in diploid cells. There is evidence, however, that recessive mutations appear in diploid organisms at a much higher frequency than might ordinarily be expected, possibly as homozygotes resulting from somatic recombination after a single mutation event.[6] The frequency of mutants may be increased by the use of chemical or physical mutagens.[7,8]

By means of classical microbial mutant selection procedures, these variants can be identified and isolated. For example, drug-tolerant cell lines have been selected in tobacco, carrot, *Datura innoxia,* and other species by subjecting cell populations to drug-containing culture medium.[9] Whereas most cells are inhibited by a drug, rare tolerant variants survive and proliferate. By similar methods, numerous variants that can tolerate other phytotoxic substances (e.g., salt,[10] aluminum,[11] pathotoxins[12,13]) have also been obtained.

There is evidence that differential tolerance among genotypes of a species can be maintained in cell cultures derived from these genotypes (e.g., copper and zinc tolerance in *Agrostis stolonifera,*[14] disease resistance in tobacco[15] and soybean[16] and triazine resistance in photoheterotrophic cell cultures of resistance and susceptible biotypes of *Chenopodium album*[16a]). However, tolerance or resistance that is based on characteristics specific to differentiated structures (e.g., cuticle thickness[17]) rather than cellular metabolism may not be expressed in cell cultures. Similarly, tolerance selected in cultured cells may not necessarily carry over to regenerated plants.

A number of cases have now been reported in which a trait selected in cultured cells has been retained in regenerated plants and transmitted genetically to progeny. Toxin-tolerant cell lines were obtained from corn callus cultures subjected to *Helminthosporium maydis* Race T pathotoxin. Plants regenerated from the selected lines were resistant to the toxin and the trait was transmitted maternally to the progeny.[13] Several drug-resistant variants have also been shown to carry over to regenerated plants and to be transmitted sexually, as have several variants resistant to amino acid analogs.[9] Selection for resistant variants in plant cell cultures has been recently reviewed by Siegemund.[9a]

The mutant selection procedures described above can be used to select for *novel* genetic variation. If a desired trait can be identified in a close relative of the species in question, protoplast fusion techniques may be useful to incorporate the trait. Under the appropriate conditions plant protoplasts readily fuse with each other, even those from unrelated species. In several cases, somatic hybrid plants have been regenerated from fused protoplasts of closely related species.[1] Although in most cases the parental species are members of the same genus, several intergeneric somatic hybrid plants have now been reported, including potato/tomato[18] and carrot/*Aegopodium podagraria*.[19]

Somatic hybridization, although in its infancy at present, may one day be a routine means of transferring genetic information from one species to another when sexual hybridization is impossible. Gressel[20] has proposed that protoplast fusion might be used to transfer triazine resistance from *Brassica campestris* to cultivated *Brassica* species (see also Chapter 13).

## 2  HERBICIDE-RESISTANT CROPS

Although the emphasis of this volume is on the unwelcome appearance of herbicide resistance in weeds, there is much to be said for deliberately generating herbicide resistance in crop species (see Chapter 12).

Plant cell culture techniques can be employed to obtain such herbicide tolerance, as will be described below. Even without plant regeneration, herbicide-tolerant cell cultures can be valuable in studies of herbicide metabolism and mode of action. Comparisons between resistant and sensitive cell lines can provide clues as to the basis for herbicide toxicity and selectivity.

## 3  TOLERANCE IN CELL CULTURES

Herbicide tolerance in plant cell cultures can take several forms, ranging from a temporary physiological modulation in a cell population to a permanent heritable change which is expressed and genetically transmitted by plants regenerated from the cell culture (Table 14.1). Reports to date fall into the following classes:

1  Tolerance is expressed by cultured cells, but is lost if the cells are grown away from the herbicide for one or more culture passages.
2  Tolerance is retained by the cell culture even after one or more culture passages without herbicide.
3  Tolerance is stable in the absence of the inhibitor and is also expressed in plants regenerated from the cultured cells or in cell cultures derived from the regenerated plants.
4  Tolerance is stable, is retained through the regeneration process, and is also transmitted to the progeny of regenerated plants.

Cell lines of the first type have probably not undergone any genetic change but, rather, a biochemical adaptation to the presence of the inhibitor. For example, the addition of a herbicide to a cell culture may activate or induce the synthesis of enzyme(s) which can metabolize or detoxify the herbicide. The presence or activity of such enzymes may well be dependent on the continued

**Table 14.1 Herbicide-Tolerant Variants Selected in Plant Cell Cultures**

| Herbicide | Species | Selection Method | Stability[a] | Regeneration | Phenotype Expressed in Plants | Inheritance | References |
|---|---|---|---|---|---|---|---|
| 2,4-D | *Nicotiana sylvestris* | Stepwise[b] in suspension culture | ? | No | — | — | 21 |
| | Tobacco | Stepwise in callus culture | No | Yes | ? | ? | 24 |
| | Carrot | Stepwise in suspension culture | No | No | — | — | 22 |
| | Carrot | Treatment of suspension followed by plating (repeated several times) | Yes | No | — | — | 36 |
| | *Trifolium repens* | Single treatment in suspension culture | ?[c] | No | — | — | 29, 30 |
| 2,4,5-T | *T. repens* | Single treatment in suspension culture | ?[c] | No | — | — | 29, 30 |
| 2,4-DB | *T. repens* | Single treatment in suspension culture | ?[c] | No | — | — | 29, 30 |

| Propham | Tobacco | Mutagenesis of protoplasts and plating on selective medium | ? | Yes | Yes | ? | 44 |
| Phenmedipham | Tobacco | Mutagenesis of leaves, herbicide treatment, then *in situ* selection of green islands | Yes | Yes | Yes | Yes | 45 |
| Asulam | Rapeseed | Sustained treatment of white callus | Yes | No | — | — | 37 |
| Paraquat | Soybean | ? | Yes | No | — | — | 38 |
| | Tobacco | 1. Callus on selective medium 2. Mutagenized callus on selective medium 3. Stepwise selection in callus 4. Plating of suspensions on selective medium | Yes | Yes | Yes | ? | 40, 41 |
| Picloram | Tobacco | Plating of suspension on selective medium | Yes | Yes | Yes[d] | Yes | 46 |

**Table 14.1** (Continued)

| Herbicide | Species | Selection Method | Stability[a] | Regeneration | Phenotype Expressed in Plants | Inheritance | References |
|---|---|---|---|---|---|---|---|
| Bentazon | Tobacco | Mutagenesis of leaves, herbicide treatment, then *in situ* selection of green islands | Yes | Yes | Yes | Yes | 45 |
| Atrazine | Alfalfa | Mutagenized callus and suspensions screened with high herbicide concentration | ? | No | — | — | 31 |
| | Soybean | Stepwise in suspension culture | ? | — | — | — | 21 |
| Amitrole | Tobacco | Sustained treatment of green callus | Yes | Yes | ? | ? | 32, 33 |
| | | Treatment of suspension followed by selective plating | Yes | Yes | ? | ? | 34 |

[a] Retention of tolerance in the absence of the herbicide.

[b] Gradually increasing herbicide concentrations.

[c] Tolerance retained when grown in low 2,4-D concentrations (not tested in the absence of 2,4-D).

[d] Phenotype expressed in callus derived from regenerated plants.

presence of the herbicide, such that growth of the cell culture for one or more culture passages without the herbicide may result in loss of the enzyme activity and thus loss of the herbicide tolerance. Such inducible tolerance may appear more frequently in cell cultures than in intact plants. Although the same enzyme may be induced in the plant, the response may be so slow that vital processes (e.g., translocation, ion uptake) may be irreversibly interrupted before the herbicide can be detoxified. Plant cells in culture, however, are surrounded by nutrients and do not depend on the maintenance of a complex body of interacting organs and tissues for survival. Cultured cells may thus be better able to survive until sufficient herbicide has been detoxified.

Loss of tolerance may also be observed in the case of a mixed population of sensitive and tolerant cells. Even if a population originally contains only tolerant cells, it may become mixed as a result of reversions (back mutations) to the sensitive state. Although the tolerant cells will be more fit in the presence of the herbicide and will increase in frequency in the population, in the absence of the herbicide the original cell type will often (but not always) be more fit than the variant (see Chapters 9, 15, and 17). If, then, the cell culture is grown in the absence of the herbicide, sensitive cells will gradually outnumber the tolerant cells. When the culture is again challenged with the herbicide, it may appear that tolerance has been lost, as the cell population will now consist largely of sensitive cells.

In cases of the second type, the stability of the tolerance is a preliminary indication of genetic change. However, although stability is a criterion for mutation, it is in itself an insufficient basis for attributing the tolerance to mutation. Nor is expression of the tolerance in regenerated plants or cell cultures derived from them a definitive proof of mutation, although here again this is a requirement. A true mutation, even when not expressed, will be unaffected by the developmental state of the plant cells and will not be lost during the differentiation process.

It is entirely possible, however, that herbicide tolerance based on a true mutation may not be expressed in plants regenerated from a tolerant cell line but only in cell cultures derived from the plants. Cell cultures and whole plants represent different developmental states with different patterns of gene expression. If herbicide tolerance in a cell culture is based on a mutation in a gene that is not expressed in the appropiate tissue in the whole plant, one would not expect tolerance to be expressed in the plant. One would expect, however, that a cell culture derived from the regenerated plant and maintained under the same cultural conditions as the original cell culture would exhibit the tolerance.

If the tolerance can be detected in the progeny of regenerated plants or in cell cultures derived from progeny plants (Class 4), a true genetic change is indicated. The pattern by which the trait is inherited will provide information about the nature of the mutation: whether it is monogenic or polygenic, nuclear or cytoplasmic, dominant, codominant, or recessive.

Herbicide tolerance of this fourth type, whose genetic basis is confirmed and whose inheritance pattern is established, can then be incorporated into a breeding program to develop a herbicide-tolerant crop variety. Tolerance of the third type (retained by the plant but not transmitted genetically) may also be of agronomic use in a vegetatively propagated crop. In this situation, genetic transmission of the trait may be irrelevant, but stability, of course, is essential.

## 3.1 Tolerant Cell Lines

There have been several examples of herbicide-tolerant cell lines for which *stability* of the tolerance was not demonstrated. In some cases it was clearly shown that the cell line lost its tolerance when grown in the absence of the herbicide and subsequently retested, whereas in other cases stability was apparently not examined.

Using a suspension culture of *Nicotiana sylvestris* (obtained from a haploid plant), Zenk[21] obtained a cell line that was completely resistant to 3 X 10$^{-4}$ M 2,4-D and inhibited only slightly by 10$^{-3}$ M, whereas the original cell suspension was completely inhibited by 3 X 10$^{-4}$ M. The resistant line was obtained by subjecting the original suspension to gradually increasing 2,4-D levels over a 1½ year period. No information was presented regarding the stability of the resistance, but it was apparently based on the increased ability of the selected strain to metabolize 2,4-D. An atrazine-tolerant diploid soybean cell suspension was selected by similar methods.

An analogous selection experiment was conducted with carrot cells.[22] After being grown for one culture passage in the absence of 2,4-D, the selected cell line lost its tolerance. Widholm suggests that in both the *N. sylvestris* and the carrot lines, 2,4-D-metabolizing enzymes were induced by the herbicide and that when the inducer was removed the enzyme levels dropped rapidly, resulting in loss of the herbicide tolerance.[23]

A 2,4-D-tolerant tobacco line was recently described.[24] This callus line was also obtained by gradually increasing the herbicide concentration. The selected strain was able to grow in the presence of 120 mg/l 2,4-D (5.4 X 10$^{-4}$ M) while control callus was completely inhibited by 10 mg/l (4.5 X 10$^{-5}$ M). This variant could also grow in the presence of 100 mg/l IAA (5.7 X 10$^{-4}$ M) but only when a small amount of 2,4-D was also included. 2,4-D tolerance was lost when the callus was grown in the absence of 2,4-D. These results suggest that in this case also, 2,4-D-metabolizing enzymes may have been induced by 2,4-D. These enzymes can apparently also metabolize IAA, but only as long as the inducer (2,4-D) is present. The metabolism of 2,4-D in plant cell cultures has been investigated for a number of species.[25-28]

With suspension cultures of perennial white clover increased tolerance could be demonstrated after a 5 day pretreatment with herbicide. Three phenoxy herbicides (2,4-D, 2,4,5-T, and 2,4-DB) were used, and pretreatment with any one resulted in increased tolerance to all three.[29] Herbicide concentrations in the 10$^{-5}$ M range were used for both the pretreatment and subsequent testing. The increased tolerance is attributed to enrichment of the cell population for cells that are physiologically adapted to the herbicide. Sensitive cells are presumably eliminated from the population early in the pretreatment period. Tolerance in this case is not due to metabolism of the herbicides, as most of the 2,4-D and 2,4,5-T remains intact.[30]

Atrazine tolerance has been selected in both callus and suspension cultures of alfalfa.[31] Cells were first mutagenized with ethyl methanesulfonate and then exposed to high concentrations of the herbicide. A number of tolerant variants were obtained from both callus and suspensions. Plants could not be regenerated, however, and the stability of the tolerance was apparently not tested.

## 3.2  Stable Tolerant Cell Lines

Herbicide-tolerant cell lines in this category differ from those described above in that these cell lines have been shown to retain their tolerance *in the absence of the herbicide.* Typically, variants that were initially tolerant to a herbicide are grown for one or  more culture passages without the herbicide and subsequently retested for tolerance. If the tolerance is due to a genetic change, it will generally be stable with or without the selective pressure imposed by the herbicide. Tolerance based on physiological or biochemical adaptation may or may not be stable.

Barg and Umiel[32] investigated the response of green tobacco callus to amitrole. At the end of the second culture passage several amitrole-tolerant callus sectors were isolated on the basis of exceptional growth, reduced bleaching, or increased differentiation in the presence of normally inhibitory amitrole concentrations. These variants were grown for 1 year without amitrole, retested with $10^{-4}$ M amitrole, then grown for another 8 months and again tested. All the lines retained a significant degree of amitrole tolerance after this time. Shoots were regenerated from all the variants,[33] but the level of tolerance in these plants and inheritance data have not yet been reported.

Amitrole tolerance in tobacco cell cultures has also been examined by Richter and McDaniel.[34] Tobacco suspension cultures were incubated for several weeks with inhibitory amitrole concentrations and then plated on agar containing amitrole. A number of variants with various degrees of amitrole tolerance were isolated. Several lines retained substantial tolerance when retested with $3.8 \times 10^{-4}$ M amitrole after 5 months on medium with no herbicide. Plants have now been regenerated from several of the tolerant lines and progeny of some of the plants have been obtained. The amitrole tolerance of these regenerated plants has not yet been determined.[35]

A stable 2,4-D-tolerant carrot line was obtained by Gressel.[36] Carrot suspensions were treated with $4 \times 10^{-3}$ M 2,4-D for 3 hr, rinsed, and plated. Surviving calli were resuspended and subjected to the same treatment and selection. After four such cycles a line was obtained which retained 20% viability in $4 \times 10^{-3}$ M 2,4-D. This tolerance was retained by cells grown for one year without 2,4-D.

Asulam tolerance has been reported in achlorophyllous callus cultures of rapeseed.[37] Cells were maintained for three culture passages on medium containing inhibitory but not lethal concentrations of asulam. When the callus was returned to normal medium and subsequently retested for asulam tolerance, it grew better than the controls. As the tolerance apparently appeared uniformly throughout the callus tissue, rather than in small isolated sectors, it is probably due to some stable physiological adaptation rather than a genetic change.

Hughes[38] isolated a soybean cell line which was completely resistant to paraquat. The variant line was selected from cells treated with $10^{-3}$ M paraquat and was maintained at $10^{-4}$ M. This line grew normally at $10^{-3}$ M paraquat and retained its resistance after being grown for several months without the herbicide. This line was also tolerant to the related herbicide, diquat.[39] Whereas paraquat sensitive cells were killed by 5 X $10^{-6}$ M diquat, the paraquat-resistant cells were only slightly inhibited by this concentration and could still grow at 5 X $10^{-4}$ M diquat. $10^{-3}$ M diquat was completely inhibitory.

## 3.3 Tolerance Retained in Regenerated Plants

Even if a stable, tolerant cell line can be selected, obtaining tolerant plants from the cultured cells is no easy task. For many species, regenerating plants is extremely difficult with normal cell cultures, let alone variants. And if plants can be regenerated they may lose the tolerance expressed in culture. Nonetheless, there are examples in the literature, albeit all with tobacco, of regenerated plants that retain the herbicide tolerance selected *in vitro*.

Paraquat tolerance has been obtained in achlorophyllous tobacco by several selection methods.[40] Stepwise callus selection (with 5 X $10^{-6}$ M paraquat followed by $10^{-5}$ M), direct callus selection with $10^{-5}$ M, callus X-ray mutagenesis followed by selection with $10^{-5}$ M, and plating of suspension cultures on $10^{-5}$ M all yielded tolerant variants.[41] All variants were allowed to grow for at least 4 weeks without herbicide and then rechallenged with $10^{-4}$ M paraquat. All four selection methods yielded stable tolerance. Plants were regenerated from most of the stable cell lines from each of the selection experiments. Of 40 plants, 13 retained tolerance to either $10^{-4}$ or $10^{-5}$ M paraquat. Callus cultures derived from six tolerant and 10 sensitive plants were also tested and all were tolerant of either $10^{-4}$ or $10^{-5}$ M paraquat, with the exception of one sensitive callus line derived from a tolerant plant. Experiments are now in progress to determine whether the paraquat tolerance is transmitted genetically, the pattern of its inheritance, and the mechanisms of the resistance.[41]

## 3.4 Tolerance Transmitted Genetically

To date, tobacco is the only species in which herbicide tolerance selected for by *in vitro* methods has been shown to be expressed in regenerated plants and transmitted to the progeny of these plants. Much of the pioneering work in plant tissue culture was done with tobacco. Consequently, a vast store of information has been amassed and the *in vitro* growth of tobacco cells has been characterized in detail. The requirements for the induction of plant regeneration are well defined and plants can routinely be obtained from established callus and suspension cultures as well as from protoplasts. Not surprisingly, then, tobacco is a very popular subject for mutant selection experiments, as efforts can be concentrated on the selection and characterization of the variant and need not be expended on developing or improving basic *in vitro* procedures.

A tobacco callus line was selected with increased tolerance to the fungicide carboxin (5,6-dihydro-2-methyl-*N*-phenyl-1,4-oxathiin-3-carboxamide).[42] The tolerance was retained after 18 cell divisions in the absence of carboxin. Several fertile plants were regenerated and callus derived from these plants retained the tolerance. (The plants themselves were not tested, as intact tobacco plants are naturally carboxin resistant.) The trait was transmitted to the progeny and appeared to segregate as a dominant nuclear mutation.[43]

When mutagenized tobacco protoplasts were plated on medium containing 5 $\mu$g/ml propham (2.8 X $10^{-5}$ M) a number of growing colonies were obtained.[44] Of the 29 plants regenerated from these colonies, 10 were fertile. Only the progeny of these 10 plants were subsequently tested for propham tolerance. Although the progeny plants themselves showed no increase in tolerance (as measured by seedling root growth), protoplasts derived from progeny of one of the 10 fertile regenerated plants exhibited a very slight but consistent increase in propham tolerance. It is not clear what proportion of the progeny of this plant exhibited the increased tolerance. Although these results suggest a possible genetic basis for the observed tolerance, without inheritance data they remain inconclusive.

Two cases stand as clear evidence that plant cell culture techniques can indeed be utilized to obtain herbicide-tolerant mutants. In one case, an interesting combination of *in vivo* and *in vitro* manipulations was employed; in the other a classical microbial genetic approach proved successful.

Whereas bentazon and phenmedipham are both toxic to tobacco plants, achlorophyllous tobacco callus is unaffected by these herbicides. To circumvent this complication, Radin and Carlson[45] devised a novel selection scheme (Fig. 14.1). They sprayed the immature leaves of mutagenized haploid tobacco plants with either bentazon or phenmedipham. As the leaves expanded, small green islands appeared on otherwise yellow leaves. These green sectors were excised and placed on a culture medium known to induce shoot regeneration. Plants were regenerated from almost all of the excised sectors, and 21% of the bentazon-selected plants and 13% of the phenmedipham-selected plants retained resistance to their respective herbicides.

When the tolerant plants (diploidized with colchicine) were crossed to wild-type plants, the $F_1$ progeny were all herbicide sensitive, indicating that either the tolerance was not genetic or was based on recessive mutation(s). In the $F_2$ progeny, herbicide tolerance reappeared in eight of 15 bentazon selections and two of seven phenmedipham lines, confirming a genetic basis for the tolerance

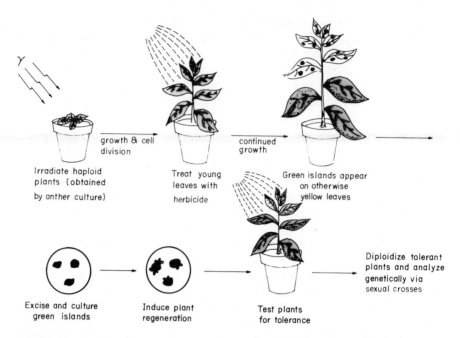

growth & cell
division

continued
growth

Irradiate haploid
plants (obtained
by anther culture)

Treat young
leaves with
herbicide

Green islands appear
on otherwise
yellow leaves

Excise and culture
green islands

Induce plant
regeneration

Test plants
for tolerance

Diploidize tolerant
plants and analyze
genetically via
sexual crosses

**Figure 14.1**    Procedure used by Radin and Carlson[45] for *in situ* selection of bentazon- and phenmedipham-tolerant tobacco mutants.

in these isolates. The $F_2$ segregation ratios implied single gene mutations in most cases, with the exception of three bentazon lines, in which two genes were indicated. Complementation tests to show whether the lines were due to mutations at the same gene indicated that the eight bentazon-tolerant mutants fell into four groups, with four, two, one and one allele in each, respectively. The two phenmedipham mutants were not at the same gene. It is very interesting that so many genes can confer resistance.

Chaleff and Parsons[46] used classical microbial mutant selection methods to isolate picloram-tolerant mutants in tobacco (Fig. 14.2). Tobacco cell suspensions were plated on agar medium containing $5 \times 10^{-4}$ M picloram and after 1 to 2 months of incubation, growing colonies appeared. These colonies were isolated and any that continued to grow after a second picloram passage were induced to regenerate shoots. Of seven variant lines isolated from one experiment, plants could be regenerated from only six. Callus derived from the regenerated plants was picloram tolerant for five of the six tolerant cell lines. In four of these five lines, tolerance was transmitted to the progeny and was expressed in both plants and callus derived therefrom. In all four cases, segregation ratios are consistent with those expected for dominant single-gene mutations. Two of these four mutations, plus an additional mutation obtained in a subsequent experiment, have now been shown to be linked.[47]

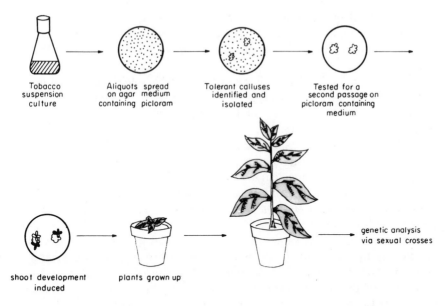

**Figure 14.2**  Procedure used by Chaleff and Parsons[46] for *in vitro* selection of picloram-tolerant tobacco mutants.

## 4  CONSIDERATIONS AND CONCERNS

Selecting herbicide-tolerant variants in plant cell cultures is complicated by the absence of many differentiated functions in cultured cells. For example, most cell cultures do not photosynthesize, nor do they transpire or translocate. If the mode of action of a herbicide is based on the inhibition of a differentiated function such as photosynthesis, it may not be toxic to a heterotrophic cell culture. This difficulty may be overcome by developing novel selection schemes such as the one devised by Radin and Carlson,[45] or by using freshly isolated leaf cells.[48] Alternatively, it may be possible to induce the expression of the differentiated function in the cultured cells. This is clearly possible with photosynthesis[49] and can probably be accomplished for other processes.

Conversely, a herbicide may inhibit cell cultures in spite of the absence of the primary target function. Again taking photosynthesis as an example, herbicides known to be photosynthesis inhibitors have been shown to inhibit dark-grown heterotrophic cells.[50,51] Presumably, secondary modes of action are involved in these cases.

It becomes obvious then that programs for selecting herbicide tolerance in cell cultures must be very carefully conceived. A cell culture appropriate to the particular herbicide must be obtained; that is, one should be certain that the target process is indeed operational in the cell culture. Herbicide concentration is also critical. The lowest inhibitory concentration should be used so as to maximize the likelihood that observed inhibition is due to the primary effect of the herbicide. Higher concentrations may well bring in to play other inhibitory mechanisms which bear little relation to the mode of action of the herbicide in the whole plant.

Metribuzin serves to illustrate the problems of herbicide selectivity in plant cell cultures. In dark-grown soybean suspension cultures, varietal differences in metribuzin sensitivity were retained.[52] The sensitive variety was severely inhibited by $10^{-6}$ M metribuzin, whereas the tolerant variety was unaffected. In dark-grown tomato suspensions, on the other hand, four varieties with differing metribuzin sensitivity were all unaffected by 9 X $10^{-5}$ M metribuzin (20 ppm), a concentration lethal to seedlings.[53] Inhibition could be demonstrated at 7 X $10^{-4}$ M, but all four varieties were uniformly affected.

In both the soybean and tomato cultures, a secondary nonphotosynthetic mechanism is obviously involved beyond the photosynthetic inhibition thought to be the primary mode of action. The differential sensitivity of the soybean varieties is the result of differential metabolism of the herbicide, and this detoxification mechanism is maintained in the cultured cells.[52] The differential sensitivity of the tomato varieties is presumably based on some mechanism that is not functioning in the achlorophyllous cell cultures.

Another major difficulty is not restricted to the selection of herbicide-tolerant variants but is common to most efforts to select variants in plant cell cultures. Cultured cells and whole plants represent different developmental states which are characterized by distinct patterns of gene expression. Although there are certainly many genes that are expressed in both cell cultures and plants, there are undoubtedly many others whose expression is restricted to one state or the other. Additionally, different tissues in the whole plant also differ in patterns of gene expression, as will different types of cell cultures (i.e., callus vs. suspension, compact vs. friable, heterotrophic vs. autotrophic). To select for a mutation conferring herbicide tolerance in a cell culture system, the mutation must, of course, be in a gene that is expressed in the cultured cells. This gene must also be expressed in the appropriate tissue of the whole plant if the tolerance is to be carried through to regenerated plants. The loss of a variant trait in regenerated plants and its subsequent recovery in cell cultures derived from such plants may well reflect these differences in gene expression.[54] The range of possible genetic mechanisms by which tolerance may be achieved is thus limited at the outset by the use of cultured cells. It seems advisable then to conduct selection experiments on a rather large scale, so as to generate a considerable number of variants. After the elimination of unstable and nongenetic variants and those which are not expressed in regenerated plants, there may yet remain a few potentially useful mutants.

An additional stumbling block is the difficulty involved in regenerating plants from cell cultures. Establishing a cell culture from virtually any plant is a straightforward process, but inducing that culture to shift developmental gears and regenerate a plant is presently possible for only a limited number of species. It is not by chance that most examples of herbicide tolerance in cultured plant cells (and *all* examples in which plants were regenerated) have employed tobacco, one of the easiest of all plants to regenerate. Although many crops still resist regeneration efforts, progress is steady and we can expect that the regeneration problem will eventually be transcended.

These obstacles might seem virtually insurmountable, but the recent accomplishments by Chaleff and Parsons[46] and Radin and Carlson[45] are most encouraging. They serve as convincing examples of the potential of plant cell culture methods for generating herbicide-tolerant variants for both the laboratory and the field.

## REFERENCES

1   I. K. Vasil, M. R. Ahuja, and V. Vasil, Plant tissue cultures in genetics and plant breeding, *Adv. Genet.,* **20,** 127 (1979).

2   E. Thomas, P. J. King, and I. Potrykus, Improvement of crop plants via single cells *in vitro—* an assessment, *Z. Pflanzenzücht.,* **82,** 1 (1979).

3   C. P. Meredith and P. S. Carlson, Genetic variation in cultured plant cells, in *Propagation of Higher Plants through Tissue Culture,* K. W. Hughes, R. Henke, and M. Constantin (Eds.), U.S. Dept. of Energy, Knoxville, Tenn., 1978, p. 166.

4   K. J. Kasha (Ed.), *Haploids in Higher Plants,* University of Guelph, Guelph, Ontario, 1974.

5   W. A. Keller and G. R. Stringam, Production and utilization of microspore-derived haploid plants, in *Frontiers of Plant Tissue Culture 1978,* T. A. Thorpe (Ed.), International Association for Plant Tissue Culture, Calgary, Alberta, 1978, p. 113.

6   K. L. Williams, Mutation frequency at a recessive locus in haploid and diploid strains of a slime mould, *Nature (Lond.),* **260,** 785 (1976).

7   Z. R. Sung, Mutagenesis of cultured plant cells, *Genetics,* **84,** 51 (1976).

8   M. L. Christianson and M. O. Chiscon, Use of haploid plants as bioassays for mutagens, *Environ. Health Perspect.,* **27,** 77 (1978).

9   P. Maliga, Resistance mutants and their use in genetic manipulation, in *Frontiers of Plant Tissue Culture 1978,* T. A. Thorpe (Ed.), International Association for Plant Tissue Culture, Calgary, Alberta, 1978, p. 381.

9a  F. Siegemund, Selektion von Resistenzmutanten in pflanzlichen Zellkulturen—eine Übersicht, *Biol. Zbl.,* **100,** 155 (1981).

10  P. J. Dix and H. E. Street, Sodium chloride-resistant cultured cell lines from *Nicotiana sylvestris* and *Capsicum annuum, Plant Sci. Lett.,* **5,** 231 (1975).

11  C. P. Meredith, Selection and characterization of aluminum-resistant variants from tomato cell cultures, *Plant Sci. Lett.,* **12,** 25 (1978).

12  P. S. Carlson, Methionine sulfoximine-resistant mutants of tobacco, *Science,* **180,** 1366 (1973).

13  B. G. Gengenbach, C. E. Green, and C. M. Donovan, Inheritance of selected pathotoxin resistance in maize plants regenerated from cell cultures, *Proc. Natl. Acad. Sci. USA,* **74,** 5113 (1977).

14  L. Wu and J. Antonovics, Zinc and copper tolerance of *Agrostis stolonifera* L. in tissue culture, *Am. J. Bot.,* **65,** 268 (1978).

15  J. P. Helgeson, G. T. Haberlach, and C. D. Upper, A dominant gene conferring disease resistance to tobacco plants is expressed in tissue cultures, *Phytopathology,* **66,** 91 (1976).

16 M. J. Holliday and W. L. Klarman, Expression of disease reaction types in soybean callus from resistant and susceptible plants, *Phytopathology,* **69,** 576 (1979).

16a J. P. Blein, Mise en culture de cellules de jeunes plantes de *Chenopodium album* sensibles ou résistantes à l'atrazine, *Physiol. Vege.,* **18,** 703 (1980).

17 T. Uchiyama and N. Ogasawara, Disappearance of the cuticle and wax in outermost layer of callus cultures and decrease of protective ability against microorganisms, *Agric. Biol. Chem.,* **41,** 1401 (1977).

18 G. Melchers, M. D. Sacristan, and A. A. Holder, Somatic hybrid plants of potato and tomato regenerated from fused protoplasts, *Carlsberg Res. Commun.,* **43,** 203 (1978).

19 D. Dudits, Gy. Hadlaczky, G. Y. Bajszar, Cs. Koncz, G. Lazar, and G. Horvath, Plant regeneration from intergeneric cell hybrids, *Plant Sci. Lett.,* **15,** 101 (1979).

20 J. Gressel, A review of the place of *in vitro* cell culture systems in studies of action, metabolism and resistance of biocides affecting photosynthesis, *Z. Naturforsch.,* **34c,** 905 (1979).

21 M. H. Zenk, Haploids in physiological and biochemical research, in *Haploids in Higher Plants—Advances and Potential,* K. J. Kasha (Ed.), University of Guelph, Guelph, Ontario, 1974, p. 339.

22 J. M. Widholm, Selection and characterization of biochemical mutants, in *Plant Tissue Culture and Its Bio-technological Application,* W. Barz, E. Reinhard, and M. H. Zenk (Eds.), Springer-Verlag, Berlin, 1977, p. 112.

23 J. M. Widholm, The selection of agriculturally desirable traits with cultured plant cells, in *Propagation of Higher Plants through Tissue Culture,* K. W. Hughes, R. Henke, and M. Constantin (Eds.), U.S. Dept. of Energy, Knoxville, Tenn., 1978, p. 189.

24 H. Ono, Genetical and physiological investigations of a 2,4-D resistant cell line isolated from the tissue cultures in tobacco: I. Growth responses to 2,4-D and IAA, *Sci. Rep. Fac. Agric. Kobe Univ.,* **13,** 273 (1979).

25 C.-S. Feung, R. H. Hamilton, and R. O. Mumma, Metabolism of 2,4-dichlorophenoxyacetic acid: V. Identification of metabolites in soybean callus tissue cultures, *J. Agric. Food Chem.,* **21,** 637 (1973).

26 C.-S. Feung, R. H. Hamilton, and R. O. Mumma, Metabolism of 2,4-dichlorophenoxyacetic acid: VII. Comparison of metabolites from five species of plant callus tissue cultures, *J. Agric. Food Chem.,* **23,** 373 (1975).

27 C.-S. Feung, R. H. Hamilton, and R. O. Mumma, Metabolism of 2,4-dichlorophenoxyacetic acid: 10. Identification of metabolites in rice root callus tissue cultures, *J. Agric. Food Chem.,* **24,** 1013 (1976).

28 G. H. Davidonis, R. H. Hamilton, and R. O. Mumma, Metabolism of 2,4-dichlorophenoxyacetic acid in soybean root callus and differentiated soybean root cultures as a function of concentration and tissue age, *Plant Physiol.,* **62,** 80 (1978).

29 T. H. Oswald, A. E. Smith, and D. V. Phillips, Herbicide tolerance developed in cell suspension cultures of perennial white clover, *Can. J. Bot.,* **55,** 1351 (1977).

30 T. H. Oswald, personal communication cited in ref. 23, 1978.

31 D. T. Hansen, Utah State University, Logan, Utah, personal communication, 1979.

32 R. Barg, and N. Umiel, Development of tobacco seedlings and callus cultures in the presence of amitrole, *Z. Pfanzenphysiol.,* **83,** 437 (1977).

33 R. Barg and N. Umiel, Selection for herbicide resistance in tissue culture and phenotypic variation among the resistant mutants, in *Production of Natural Compounds by Cell Culture Methods,* A. W. Alfermann and E. Reinhard (Eds.), Gesselschaft fur Strahlen- und Umweltorschung mbH, Munich, 1978, p. 337.

34 P. L. Richter and C. N. McDaniel, Isolation of herbicide (amitrole) tolerant cell lines, *Plant Physiol. Suppl.,* **63,** 147 (1979) (abstr.).

35 C. N. McDaniel, Rensselaer Polytechnic Institute, Troy N.Y., personal communication, 1979.

**36**  J. Gressel, Genetic herbicide resistance: projections on appearance in weeds and breeding for it in crops, in *Plant Regulation and World Agriculture,* T. K. Scott (Ed.), Plenum, New York, 1979, p. 85.

**37**  J. Flack and H. A. Collin, Selection of resistance to asulam in oil seed rape, *Abstr. 4th Int. Congr. Plant Tissue Cell Cult.,* Calgary, Alberta, 171 (1978).

**38**  K. W. Hughes, Isolation of a herbicide-resistant line of soybean cells, in *Plant Cell and Tissue Culture,* W. R. Sharp, P. O. Larsen, E. F. Paddock, and V. Raghavan (Eds.), Ohio State University Press, Columbus, Ohio, 1977 (abstr.).

**39**  K. W. Hughes, Diquat resistance in a paraquat resistant soybean cell line, *Abstr. 4th Int. Congr. Plant Tissue Cell Cult.,* Calgary, Alberta, 170 (1978).

**40**  O. K. Miller and K. W. Hughes, Selection of paraquat-resistant variants of tobacco from cell cultures, *In Vitro,* **16**, 1085 (1980).

**41**  K. W. Hughes, University of Tennessee, Knoxville, Tenn., personal communication, 1979.

**42**  J. C. Polacco and M. L. Polacco, Inducing and selecting valuable mutation in plant cell culture: a tobacco mutant resistant to carboxin, *Ann. N. Y. Acad. Sci.,* **287**, 385 (1977).

**43**  J. C. Polacco, University of Missouri, Columbia, Mo., personal communication, 1979.

**44**  D. Aviv and E. Galun, Isolation of tobacco protoplasts in the presence of isopropyl *N*-phenylcarbamate and their culture and regeneration into plants, *Z. Pflanzenphysiol.,* **83**, 267 (1977).

**45**  D. N. Radin and P. S. Carlson, Herbicide-tolerant tobacco mutants selected *in situ* recovered via regeneration from cell culture, *Genet. Res., Camb.,* **32**, 85 (1978).

**46**  R. S. Chaleff and M. F. Parsons, Direct selection *in vitro* for herbicide-resistant mutants of *Nicotiana tabacum, Proc. Natl. Acad. Sci. USA,* **75**, 5104 (1978).

**47**  R. S. Chaleff, Further characterization of picloram-tolerant mutants of *Nicotiana tabacum, Theor. Appl. Genet.,* **58**, 91 (1980).

**48**  J. Gressel, Weizmann Institute of Science, Rehovot, Israel, personal communication, 1979.

**49**  Y. Yamada, F. Sato, and M. Hagimori, Photoautotrophism in green cultured cells, in *Frontiers of Plant Tissue Culture 1978,* T. A. Thorpe (Ed.), International Association for Plant Tissue Culture, Calgary, Alberta, 1978, p. 453.

**50**  L. S. Jordan, T. Murashige, J. D. Mann, and B. E. Day, Effect of photosynthesis-inhibiting herbicides on non-photosynthetic tobacco callus tissue, *Weeds,* **14**, 134 (1966).

**51**  E. C. Metcalf and H. A. Collin, The effect of simazine on the growth and respiration of a cell suspension culture of celery, *New Phytol.,* **81**, 243 (1978).

**52**  T. H. Oswald, A. E. Smith, and D. V. Phillips, Phytotoxicity and detoxification of metribuzin in dark-grown suspension cultures of soybean, *Pestic. Biochem. Physiol.,* **8**, 73 (1978).

**53**  B. E. Ellis, Non-differential sensitivity to the herbicide metribuzin in tomato cell suspension cultures, *Can. J. Plant Sci.,* **58**, 775 (1978).

**54**  J. M. Widholm, Regeneration of plants from 5-methyltryptophan-resistant tobacco cell cultures, *Abstr. 4th Int. Congr. Plant Tissue Cell Cult.,* Calgary, Alberta, 138 (1978).

# CHAPTER FIFTEEN

# Evolution of Heavy Metal Resistance–An Analogy for Herbicide Resistance?

A. D. BRADSHAW

Department of Botany
University of Liverpool
Liverpool, United Kingdom

## 1 INTRODUCTION

The toxicity produced by the heavy metals left by recent and ancient seekers of industrial wealth has produced some remarkable examples of the evolution of specific resistances. These have always been termed examples of tolerance because the plants do not exclude the metal—they therefore tolerate it—but they may equally well be termed examples of resistance, as the levels are totally toxic to other members of the same species. In many ways these appear to be an excellent analogy for the evolution of herbicide resistance which we now are following, and expecting, wherever herbicides are being used. There are remarkable numbers of species in which metal tolerance has been reported. In 1971, Antonovics et al.[1] counted 37; the number now is probably double. Is this what we are to expect with herbicide resistance, or are there fundamental, ecological, and evolutionary differences that could have mitigating effects?

## 2 HEAVY METAL TOXICITY

The salient characteristic of heavy metals in the environment is that they are present in the soil more or less permanently. Certainly, metals such as lead, zinc, and copper cannot be catabolized or evaporated, and they are extremely resistant to leaching. Thus, heavy metals can present a continuous selection pressure operating at all stages of the life history of a plant species, from the earliest seedling stage through to the established adult. There is little opportunity for the direction of selection to be reversed.

There are many substances in the environment, however, such as organic matter and phosphate, which can complex metals and render them unavailable. At the same time some metals are at least slightly soluble, so that their concentrations can be affected by weather conditions. Zinc salts accumulate on soil surfaces in arid conditions and are leached away in wet conditions. As a result, symptoms of zinc toxicity and plant death on lead and zinc wastes vary with season and are most common in the summer.

Mine wastes are notoriously variable. The exact metal content of any pile of waste will depend on the particular part of the ore body from which it came. The metal content of waste materials has decreased as extraction techniques improved. At the same time the metals have spread into neighboring habitats in different degrees as a result of air and water-mediated pollution. As a result, despite the permanence of metal contamination, a long-established mine site is a remarkable complex of toxic and less toxic environments with fluctuation in selection pressures, in time as well as in space.[2,3]

The spatial characteristics of metal-contaminated habitats are very interesting, especially their area and the distance separating them from normal habitats. The first investigators of evolutionary differentiation looked for it within species in populations occupying very large areas of habitat and widely separated from each other. Yet metal-contaminated sites, particularly those on which evolutionary processes have been studied, are usually small, rarely more than 5 ha and usually less than 1 ha. Sometimes they can be only a few meters across, for instance where they have been caused by localized dumping of waste material or by the dissolution of zinc coatings on galvanized materials. They have sharp, well-defined edges, so that normal and contaminated habitats are separated by only a few meters.

The characteristics of metal-contaminated sites are therefore similar in many ways to those of herbicide-contaminated habitats. The major difference is in the much greater persistence of heavy metal toxicity, acting from the pre-emergence to the adult stage continuously from one year to the next. This constancy applies even where the level of contamination is low. This is approached only by situations in which persistent herbicides such as triazines are applied annually. But even with this group of herbicides, toxicity falls progressively after application and is affected by weather conditions.

There is little in metal-contaminated habitats equivalent to the selection produced by occasional (or even regular) use of nonpersistent herbicides from which a weed population can "escape" assisted by its phenotypic plasticity[4] (Chapter 17). There are no corollaries with cases where because of the sensitivity of the crop, the herbicide has to be applied at a concentration which only reduces weed growth e.g., in the control of wild oats by barban.[5]

It is therefore where persistent herbicides are applied that the rate of appearance of herbicide resistance most resembles that of metal resistance[6,7] (Chapters 2 and 3). This rate of appearance is helped by the enormous areas over which such herbicides are being applied, far greater than areas of metal contamination, a point to be returned to later.

## 3 OCCURRENCE OF RESISTANT POPULATIONS

Turesson[8] carried out his pioneering studies on evolutionary differentiation in plants in the 1920s. Despite this, the first clear-cut indication of a distinct metal-resistant population able to be separated by its behavior from other populations of the same species did not appear until a decade later. Then, Prat[9] showed that copper mine populations of *Silene vulgaris* would survive on copper mine waste and normal populations would not. Further discoveries were not made until lead and zinc tolerance was found in Wales with the grass *Agrostis tenuis*.[10]

Heavy-metal-contaminated wastes have been produced by mining since the Bronze Age, and even in areas such as Britain which were settled late, metal contamination due to human activities has existed in many sites for 2000 years. There are also naturally occurring areas with heavy metals which must predate these.

It is not surprising, therefore, that, once populations resistant to heavy metals were sought, they have been found in innumerable sites. The recognition of metal resistance was immensely assisted by the discovery that it could be assessed by a very simple rooting test.[11,12]

Sufficient examples of the evolution of metal resistance now exist that we can make a number of generalizations.[1,13]

1   Resistance occurs in young (less than 5 years old) as well as old areas of contamination. There is no sign of the evolution of resistance being a slow process.

2   Resistance occurs in populations growing on very small contaminated areas (not more than 5 m across) as readily as in larger ones. Zinc resistance has even been found in a population growing underneath a galvanized iron fence.[14] Certainly there is no indication that there is a critical minimal area or population size.

3   Species with resistant populations can be found in all geographic regions of the world. There are no signs that they are absent from certain continents or certain floras. There are no signs that resistance is absent from certain climatic regions; it has been found in species of subarctic, temperate, and tropical floras. It is also found in species adapted to very different soil conditions, extreme in pH or nutrients, as well as in species on neutral fertile soils.

4   Resistance does not appear to be restricted to a few angiosperm families, although it does appear more commonly in certain families such as the

Caryophyllaceae and Gramineae. However, it seems possible that resistance to heavy metals is not found in certain families, and that certain species do not have the ability to evolve resistance even when placed in a metal-contaminated environment that would be expected to select for it.

Thus there is no evidence that we are dealing with a rare character occurring in a few rather odd species. Yet it is a specialized character; possession of metal resistance has never yet been shown to adapt the individuals that possess it to anything other than high levels of particular metals which are not normal parts of the environment of plants.

One of the most remarkable characteristics of heavy metal resistance is its specificity. Populations occuring on copper-contaminated soils are not resistant to lead, and vice versa: resistance to lead, copper, zinc, and nickel in nearly all cases appear to be specific.[12,15] However, some cases of cross resistance have appeared: zinc and nickel in *Agrostis tenuis*;[15] copper and lead in *Deschampsia caespitosa*;[16] and copper, nickel, and zinc in *Mimulus guttatus*.[17] Why these cross resistances occur is not yet clear and must await an understanding of the physiological mechanisms involved. It seems possible that there are two types of resistance: (*a*) one that has major effects and is specific, and (*b*) one that has minor effects and is nonspecific. This is suggested by the data of Jowett:[12] populations resistant to one metal tend to be slightly more resistant to other metals. Cross resistance seems limited, and resistance usually very specific. There is no sign for instance that populations or species possessing resistance to the aluminum that is available in toxic amounts in acid soils are preadapted by this resistance to any heavy metals.

It is too early to predict what is to be the end point of evolution of herbicide resistance. Although the present situation in the evolution of herbicide resistance approaches that presented by heavy metal resistance, much less time has elapsed and selection has been much more intermittent. It is therefore highly unlikely that evolution could have achieved its full potential of results. Nevertheless, if plant species can show a remarkable capacity to evolve one complex character (heavy metal resistance) which is of no adaptive value in normal environments, it suggests that in the end they will have the same ability for a second (herbicide resistance), especially when we find examples of evolution of the second scattered widely and apparently randomly through higher plants. But how distinctive will herbicide resistant populations be? In how many species will they evolve? We cannot yet tell.

The lack of cross resistance for even closely related metals is very surprising. The presence of cross resistance to herbicides within a related group which often differ only in a side chain, for instance in the chloro-*s*-triazines[18,19] (which can be carried through to the asymmetric triazines, the triazinones) (Chapters 2 and 3), is understandable and to be expected. But are we being wise after the event? We can see the potential of evolution and suggest what we expect will happen. But we can only predict what it can do when it has done it.

For this reason then it is worthwhile examining in some detail what evolution has achieved in relation to heavy metals.

## 4  CHARACTERISTICS OF RESISTANCE

The genetic control of heavy metal resistance seems straightforward, if not necessarily simple. Parent/offspring regressions show that copper resistance has high heritability in *Agrostis tenuis*.[20] More detailed diallel analyses show that for zinc resistance in *Agrostis tenuis* and *Anthoxanthum odoratum*[21,22] and lead resistance in *Festuca ovina*,[23] a number of genes are involved with mainly additive effects, but with dominance. In these species there is little evidence of a single gene with major effects. But in *Mimulus guttatus* it seems likely that only two loci are involved.[24,25]

Generally, metal resistance shows some dominance, but this is variable and there may also be more complex interactions.[25,26] Part of this could be because dominance is in the process of evolution, determined by genes not yet fixed by natural selection, as suggested by Antonovics.[27] But it could also be produced by a threshold effect where resistance does not appear until a certain number of genes are present. Because of this, dominance will be changed as the level of metal at which resistance is tested is increased.[17]

What are the mechanisms that permit resistant plants to grow in relatively high concentrations of heavy metals? In all the chemical analyses made so far there have not been signs of exclusion mechanisms. Metal-resistant plants have always been found to contain as much or more metals in their roots as normal plants, hence the use of the term "tolerance." As metals cannot be detoxified by being broken down, some other systems must operate. The balance of evidence suggests that the metals are usually complexed in a form that renders them unavailable and innocuous.[28] It is not clear whether this occurs in the cell wall or in the vacuole. Earlier work with *Agrostis tenuis* using cell macerates suggests the former,[29] recent work with *Deschampsia caespitosa* using intact roots gives clear evidence for the latter.[30] It is possible that there is more than one mechanism. Various investigators have looked for tolerant enzyme systems but without producing convincing evidence so far.

It seems unlikely that present investigations on the mechanism of metal resistance will throw much light on mechanisms of herbicide resistance, where resistance can readily be produced by mechanisms causing breakdown of the active molecule (e.g., atrazine/corn[31] or by subtle differences in plant morphology causing variation in uptake). However, resistance to atrazine has recently been shown in a few species to be related to lowered chloroplast binding[32] (Chapter 10). Any evidence, therefore, obtained from whatever source on the mechanisms that can evolve to negate the effects of new or abnormal toxic materials will always be of interest and potential value. These mechanisms, in metal-tolerant plants, occur in material that in all other respects has the same biochemical and physiological background as normal nontolerant material. This makes investigation of tolerance and resistance mechanisms much easier, because all differences can be attributed to the tolerance or resistance. Nevertheless, insufficient work has been carried out to clarify exactly the biochemical nature of the mechanisms.

## 5 THE EVOLUTIONARY PROCESS

The evolution of a character depends on the occurrence of appropriate variability as well as selection. With metal resistance, the existence of suitable strong selective forces is not in doubt. The complete absence of plants from many metal-contaminated habitats, and damaging or lethal effects that metals can be shown to have in experimental conditions, indicate the potentially powerful selection pressures that can be exerted by heavy metals. Herbicides can clearly have similar effects.

### 5.1 Variability and Selection

Variability on which selection by heavy metals can act was originally suggested by the variation in resistance found between individuals in a normal population as revealed by a rooting test.[14] But only a few individuals can be tested at any one time by this method. The startling amount of variability that can be found in natural populations is best revealed by growing many thousands of seeds together on a toxic material such as sterilized metal-contaminated mine waste or a metal-containing solution. This shows that in normal populations of *Agrostis tenuis* and other species there is a very low frequency of individuals at least partially resistant to any one particular metal (about 2 in 1000). These are not usually as resistant in tests as are naturally occurring resistant plants, but they survive and grow well, and all the other individuals die; the numbers surviving depends on the level of metal contamination[33] (Fig. 15.1).

These produce at least partially resistant offspring, and there is therefore the potential for complete metal resistance to be evolved in a few generations.[34] There is some variation in the resistance of the survivors; and because *Agrostis tenuis* is outbreeding, there is some further segregation even when only resistant individuals are intercrossed. As the character is controlled by several genes it seems possible that the polygenes which contribute to the resistance are at such low frequency that it is not possible in a single cycle for selection to pick out individuals with the maximum number of genes for resistance. Several cycles of crossing and selection may be necessary before maximum resistance is achieved, just as in other situations where selection is acting on polygene systems.[35,36]

Such variability is, however, not to be found in all species, and there seems to be a correlation between the occurrence of resistant individuals in normal populations of a species and its ability to evolve resistance.[34] *Dactylis glomerata*

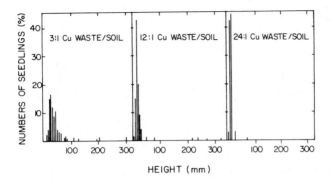

**Figure 15.1** Height of seedlings of a normal population of *Agrostis tenuis* on soil with three different levels of copper contamination. Those seedlings that achieved a height of more than 50 mm were those which had survived and continued growth. The copper waste from Parys Mountain mine in northern Wales contained 1200 ppm $Cu^{2+}$. (Redrawn from Walley et al.[33])

appeared to be an anomaly; it possesses copper-resistant individuals but had not evolved resistant populations. Recently, some copper-resistant *Dactylis* populations have been found.[37]

The principle that evolution of resistance depends on the presence of the appropriate variation in the populations being selected is not remarkable; it is a crucial tenet of the Darwinian theory of evolution. But the fact that so far it has usually been possible to find this variability quite easily in populations is somewhat surprising, because in other evolutionary situations, such as warfarin resistance in rats,[38] the appropriate variability has often been very rare indeed, apparently produced only by a mutation of very low frequency. Recently, in *Mimulus guttatus,* which evolves copper resistance, it has not yet been possible to find genes for resistance in normal populations.[25]

It is not possible to generalize about sources of variability and their significance, except to emphasize that evolution cannot take place without appropriate genetic variability being available. But if the presence of variability that would confer resistance to a particular herbicide is being looked for, it would be unsafe to envisage solely the *Agrostis tenuis* situation, of recognizably resistant individuals occurring within a population at a frequency of $10^{-3}$, which could be picked up by a simple screening test on one or two populations each consisting of only a few thousand individuals: a rat/warfarin or *Mimulus* situation might occur. This would lead to there being no signs of resistance evolving for many years; but as the rare gene was selected for and spread by gene flow, herbicide resistance would eventually be just as certain to occur as if the gene had been common in the first instance. As we shall see later, gene flow can be very important in the spread of metal resistance. A rat or *Mimulus* situation would be very difficult to pick up by a screening test since this would in all likelihood draw a blank, because of limitations on the size of the populations that could be tested. Who knows whether atrazine resistance does occur for instance in flax and soybeans, despite the work already put in?[39,40]

Variants might be picked up in a weed species if a sufficient number of individuals were screened. It would, however, be difficult to simulate the "screening" occurring in the field, where for a single herbicide/weed situation in a small country such as Great Britain probably at least 2 billion individuals are being "screened" annually in herbicide treated fields. It is therefore worthwhile to look for resistance in populations where the herbicide has already been used at least a few times, as this should increase the frequency of any alleles for resistance present. All this is well exemplified by the pattern of occurrence of simazine resistance in *Senecio vulgaris*[41] (Chapter 7).

## 5.2 Rates of Evolution

It might have been thought that the evolution of metal resistance would take a century or more. But quite apart from the fact that early investigations showed metal resistance on recent as well as old mine workings,[42] would this actually be logical? We have already seen that selection for metal resistance is continuous and extremely powerful. As a result, if the variability is present, the evolution of resistance should depend only on (a) the heritability of resistance and (b) the reproductive ability of resistant individuals, both of which are high. The heritability of resistance is at least $0.5$[20–23] and individual plants of species such as *Agrostis tenuis* and *Anthoxanthum odoratum* produce copious amounts of seed in their second year and continue in this manner until they die.

Seedling experiments illustrate that a significant level of metal resistance can evolve in a population in one or two generations. Observations on field populations fit with this: appreciable levels of resistance can be found in populations that have only been exposed to metal contamination for 4 years.[28,43] Such resistance can continue to develop to higher levels by a combination of selection at the adult stage and by interbreeding of the survivors and further selection. It appears that less resistant individuals are eliminated and more resistant individuals appear and increase in frequency. The result is that almost full resistance (compared with long-standing mine populations) can be found in new copper-contaminated populations of *Agrostis stolonifera* at Prescot in Lancashire after 15 years[43] (Fig. 15.2). It appears that adequate variability was present in the original populations so that the only major factors restricting evolution were the absence of seeding because the populations were mown, and because there were edaphic limitations to the growth of individual plants. The restriction on seeding may also have limited intercrossing and the production of segregants with higher resistance.[44]

In herbicide resistance all these factors are obviously important in determining the rate of evolution. However, it is only with the persistent herbicides that the same incessant selection pressure as afforded by metals can occur. When many genes are involved or selection pressure is intermittent, the rate of

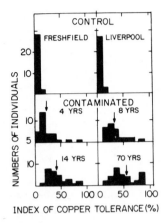

**Figure 15.2** Copper resistance in populations of *Agrostis stolonifera* which have been subjected to copper contamination for different durations near a copper refinery at Prescot, Lancashire. The means are indicated by arrows. The index of copper tolerance is the root elongation in 0.13 mg/l Cu solution divided by the root elongation in a control solution without Cu. (Redrawn from Wu et al.[43])

evolution of resistance will be depressed (Chapter 17). Yet the reproductive ability of weeds, at least annuals, is very high compared with the perennial grass species that have mainly been investigated for metal resistance. There seems little restriction from this source on the rate of evolution of herbicide resistance, as witnessed by the rate of development of triazine resistance within populations of *Senecio vulgaris, Chenopodium album,* and *Amaranthus* sp. once it initially appeared (see Chapters 2 and 3).

## 5.3 Fitness

In the presence of metal contamination it is clear that the fitness of metal-resistant individuals is very high and that there are very strong selection pressures against nonresistant individuals, given selection coefficients of 0.95 or more in favor of resistance.

But what happens in the absence of metal contamination, in normal habitats? It appears that metal-resistant plants are somewhat less fit in these conditions than are normal plants. In the absence of competition the differences are not very great. But when there is competition from other plants, as would occur in normal habitats, the fitness of resistant plants is reduced considerably. The reduction in fitness varies considerably from species to species. In one competition experiment the fitness of zinc-resistant individuals (compared with normal individuals) of *Agrostis tenuis* was 0.16, of *Anthoxanthum odoratum* 0.001, and of *Plantago laneolata* 0.28, measured from dry weight after 1 year. Different results were obtained when the competitive conditions were altered.[45] Thus there can be strong selection pressures operating against metal resistance in populations in normal habitats, although this may not be as strong as the reverse selection on metal-contaminated habitats.

When studying fitness, it is important to remember that the effects of competition are likely to be cumulative. This means that short-term observations do not give a proper estimate of fitness. Observations must be made over the

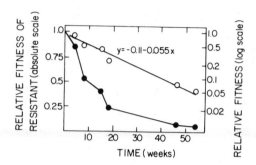

**Figure 15.3** Changes in mean relative fitness of zinc-resistant populations of *Agrostis tenuis, Anthoxanthum odoratum, Plantago lanceolata,* and *Rumex acetosa* (compared with normal populations) when in competition with *Lolium perenne.* Absolute scale, ●; log scale, ○. Fitness was measured as dry matter production. (Redrawn from Hickey and McNeilly.[45])

whole life cycle of annuals and at least over a full year in the case of a perennial[45] (Fig. 15.3). The low fitnesses of metal resistance that are indicated raise awkward questions about the maintenance of low frequencies of metal resistance in normal populations. This adverse selection pressure would tend to eliminate metal resistance. It is possible that resistance is being maintained either by recurrent mutation or by some sort of balanced polymorphism. No hard evidence is yet available.

It is difficult to be certain why metal-resistant plants are less fit than normal plants in normal environments. It is probably because the presence of the metal-resistance mechanism, which complexes the metal or in some other way renders it unavailable, causes an elevated requirement for the metal when it is at low concentration. A pointer to this was found in rooting tests. Root growth of metal-resistant plants has been found to be stimulated by low levels of metal on a number of occasions.[1]

In almost every step that evolution takes there is a cost. It is therefore likely that herbicide resistance will have a cost, affecting fitness in the absence of herbicide. This will mean that during periods when the herbicide is not used, there will be selection against resistance. There is already evidence for this[19,46] (Chapter 17). To quantify this cost precisely enough, to predict the fate of resistance in a population being treated intermittently with herbicide (Chapter 17), may require careful work extending over the whole life cycle of the species being examined.

## 5.4 Geographical Patterns

If selection is very strong, it can exert a dominating effect on the spatial patterns of evolution within species. Localized populations can become differentiated from each other by localized selection. It was once considered that if

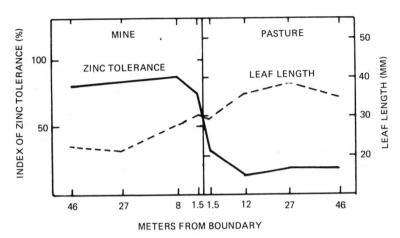

**Figure 15.4** Pattern of zinc resistance in populations of *Anthoxanthum odoratum* across the boundary of the Trelogan mine in Clwyd, Wales, compared with the pattern of length of flag leaf. The selection pressure on zinc resistance is strong, on leaf length weak. The index of zinc tolerance is the root elongation in 12 mg/l Zn solution divided by the root elongation in a control solution without Zn. (Redrawn from Antonovics and Bradshaw.[50])

populations were not separated by considerable distances, gene flow, by pollen and seed, would prevent the development of differences in gene frequencies between them. But evolutionary differentiation can be found occurring in populations only a few meters apart. Theoretical analyses based on known selection pressures and rates of gene flow show that this is logically acceptable.[47,48] We now even know of a case of population differentiation in a perennial grass, *Anthoxanthum odoratum,* occurring over 10 cm.[49]

Such very sharp patterns of population differentiation are commonplace for metal resistance at the boundaries of metal-contaminated situations (e.g., in *Anthoxanthum*)[50] (Fig. 15.4). It would appear that it is the sharpness in the gradient of contamination which determines the sharpness in the gradient of population differentiation. However, there are situations where gene flow can overcome the effects of selection and lead to a blurring of the differences between populations being subject to different selection.[51,52]

Exactly what pattern develops depends not only on the amount of gene flow but also on the strength of the selection operating on different characters or in different habitats. We can obtain evidence of the effects on different characters by examining, in mine boundary populations, a character other than metal resistance. Leaf length, for instance in *Anthoxanthum odoratum,* is likely to be subject to some selection by edaphic factors; but it will not be subject to as severe pressures as metal resistance. As a result, leaf length is more influenced by gene flow[50] (Fig. 15.4).

The effect of habitat can be seen if we compare what happens in contaminated and normal habitats. In the metal-contaminated habitats that have been investigated, the contamination has been severe, the fitness of nonresistant individuals is zero, and as a result selection for metal resistance has overcome any gene flow.[20] But in the adjacent uncontaminated habitats, selection against metal resistance is not so severe, as we have seen, and therefore can be overcome by gene flow. As a result, metal resistance spreads into surrounding populations, particularly those downwind.[51,52] But the degree of spread will depend on the balance between gene flow and selection, which can vary between species. Different patterns of differentiation may, therefore occur in different species.[45]

All these considerations affect only the details of the geographical pattern of evolution. There is no evidence that the evolution of metal resistance has ever been excluded by gene flow from neighboring nonresistant populations. From similar arguments it seems unlikely that the evolution of herbicide resistance will be prevented by gene flow, although it should be delayed. A highly localized pattern of evolution is to be expected. This will occur not only if herbicide-resistant individuals have reduced fitness but also because so many weeds are normally self-fertilizing. This will prevent gene flow, as it does in metal-contaminated situations.[53] We already have evidence of selection against herbicide resistance in normal situations in *Senecio vulgaris*,[46] which is about 90% self-fertilized.

We have, however, been looking at gene flow from one point of view only. It is true that gene flow can be overcome by selection when its products are disadvantageous. What if its products are advantageous? All observed patterns of gene flow whether of pollen or of seed, by wind or animals are strongly leptokurtic; a large proportion of propagules move only a short distance, but a higher number move over long distances than would be expected from normal dispersal processes.[47] The effects of gene flow are also cumulative over generations. When normal populations of *Agrostis tenuis* surrounding a large copper mine, Parys Mountain, were examined for the occurrence of resistant individuals, significantly elevated numbers were found over 5 km away from the mine.[51]

Gene flow can therefore carry genes for resistance out of one population into another over large distances. If selection favors such genes, they can then migrate from one population to another very easily. This will of course be important only in outbreeding species. However, gene flow can also occur by seed in all species, whether inbreeding or outbreeding, in a leptokurtic manner. Miners moving from one mine working to another may well have helped in this.

Gene flow of this sort will be extemely important in the spread of herbicide resistance. It will be facilitated particularly by uniformity of cropping and herbicide treatment over vast areas of land, providing contiguous areas suitable for the spread of resistant genotypes, and also by the development of contract harvesting, which inevitably means movement of machinery transporting, with great precision, resistant individuals of weed species from one area appropriate to them to another.

## 6 CONCLUSIONS

Understanding the way in which metal resistance has evolved has meant delving into evolutionary mechanisms. This has revealed that evolution in plants has the potential to be rapid and localized, even in the very restricted environments of metal-contaminated sites. As such, the evolution of metal resistance is a model for what can happen with the use of herbicides on plant populations when those herbicides exert the same sort of selection pressure, which is when they are powerful and persistent. For other situations involving herbicides, the evolution of metal resistance shows us that to understand evolution we need to know, and be able to define precisely, all the operative factors, especially the variability available, the intensity of selection, the population biology, and the power of reproduction of the species.[13] Without knowledge of these parameters we will be able only to treat the evolution of herbicide resistance as a series of historic events, when its importance requires us to be able to understand how and why it occurs, and to predict when and under what circumstances it will occur in the future.

## REFERENCES

1  J. Antonovics, A. D. Bradshaw, and R. G. Turner, Heavy metal tolerance in plants, *Adv. Ecol. Res.,* **7,** 1 (1971).

2  M. S. Johnson, T. McNeilly, and P. D. Putwain, Revegetation of metalliferous mine spoil contaminated by lead and zinc, *Environ. Pollut.,* **12,** 261 (1977).

3  R. A. H. Smith and A. D. Bradshaw, The use of metal tolerant plant populations for the reclamation of metalliferous wastes, *J. Appl. Ecol.,* **16,** 595 (1979).

4  A. D. Bradshaw, Evolutionary significance of phenotypic plasticity in plants, *Adv. Genet.,* **13,** 115 (1965).

5  J. Smith and R. J. Finch, Chemical control of *Avena fatua* in spring barley, *Proc. Br. Crop Prot. Conf. Weeds,* **3,** 841 (1978).

6  J. D. Bandeen, J. V. Parochetti, G. F. Ryan, B. Maltais, and D. V. Peabody, Discovery and distribution of triazine resistant weeds in North America, *Abstr. Meet. Weed Sci. Soc. Am.,* San Francisco, 108 (1979).

7  J. Gressel and L. A. Segel, The paucity of plants evolving genetic resistance to herbicides: possible reasons and implications, *J. Theor. Biol.* **75,** 349 (1978).

8  G. Turesson, The genotypical response of the plant species to the habitat, *Hereditas,* **3,** 211 (1922).

9  S. Prat, Die Erblichkeit der Resistenz gegen Kupfer, *Ber. Dtsch. Bot. Ges.,* **52,** 65 (1934).

10  A. D. Bradshaw, Populations of *Agrostis tenuis* resistant to lead and zinc poisoning, *Nature (Lond.),* **169,** 1089 (1952).

11  D. A. Wilkins, A technique for the measurement of lead tolerance in plants, *Nature (Lond.),* **180,** 37 (1957).

12  D. Jowett, Populations of *Agrostis* spp. tolerant of heavy metals, *Nature (Lond.)*, **182**, 816 (1958).

13  A. D. Bradshaw and T. McNeilly, *Evolution and Pollution*, Arnold, London, 1981.

14  A. D. Bradshaw, T. S. McNeilly, and R. P. Gregory, Industrialization, evolution and the development of heavy metal tolerance in plants, in *Ecology and the Industrial Society*, Brit. Ecol. Soc. Symp. 5, G. T. Goodman et al. (Eds.), Blackwell, Oxford, 1965, pp. 327–343.

15  R. P. G. Gregory and A. D. Bradshaw, Heavy metal tolerance in populations of *Agrostis tenuis* Sibth. and other grasses, *New Phytol.*, **64**, 131 (1965).

16  R. M. Cox and T. C. Hutchinson, Metal co-tolerances in the grass *Deschampsia caespitosa, Nature (Lond.)*, **279**, 231 (1979).

17  W. R. Allen and P. M. Sheppard, Copper tolerance in some California populations of the monkey flower, *Mimulus guttatus, Proc. R. Soc. B*, **177**, 177 (1971).

18  J. D. Bandeen and R. D. McLaren, Resistance of *Chenopodium album* to triazine herbicides, *Can. J. Plant Sci.*, **56**, 411 (1976).

19  S. R. Radosevich and A. P. Appleby, Relative susceptibility of two common groundsel (*Senecio vulgaris* L.) biotypes to six *s*-triazines, *Agron. J.*, **65**, 553 (1973).

20  T. McNeilly and A. D. Bradshaw, Evolutionary processes in copper tolerant populations of *Agrostis tenuis, Evolution*, **22**, 108 (1968).

21  D. W. Gartside and T. McNeilly, Genetic studies in heavy metal tolerant plants: I. Genetics of zinc tolerance in *Anthoxanthum odoratum, Heredity*, **32**, 287 (1974).

22  D. W. Gartside and T. McNeilly, Genetic studies in heavy metal tolerant plants: III. Zinc tolerance in *Agrostis tenuis, Heredity*, **33**, 303 (1974).

23  C. Urquhart, Genetics of lead tolerance in *Festuca ovina, Heredity*, **26**, 19 (1971).

24  M. R. Macnair, Major genes for copper tolerance in *Mimulus guttatus, Nature (Lond.)*, **268**, 428 (1977).

25  M. R. Macnair, The genetics of copper tolerance in the yellow monkey flower, *Mimulus guttatus*: I. Crosses to nontolerants, *Genetics*, **91**, 553 (1979).

26  D. A. Wilkins, The measurement and genetical analysis of lead tolerance in *Festuca ovina, Rep. Scott. Plant Breed. St.*, 85 (1960).

27  J. Antonovics, Evolution in closely adjacent plant populations. VI. Manifold effects of gene flow, *Heredity*, **23**, 507 (1968).

28  W. Ernest, Physiological and biochemical aspects of metal tolerance, in *Effects of Air Pollutants on Plants*, T. A. Mansfield (Ed.), Cambridge University Press, London, 1975, pp. 115–133.

29  R. G. Turner and C. Marshall, The accumulation of zinc by subcellular fractions of roots of *Agrostis tenuis* Sibth. in relation to zinc tolerance, *New Phytol.*, **71**, 671 (1972).

30  A. Brookes, J. C. Collins, and D. A. Thurman, The mechanism of zinc tolerance in grasses, *J. Plant Nutr.*, 3, 695 (1981).

31  R. H. Shimabukuro, Atrazine metabolism and herbicidal activity, *Plant Physiol.*, **42**, 1269 (1967).

32  K. Pfister, S. R. Radosevich, and C. J. Arntzen, Modification of herbicide binding to photosystem II biotypes of *Senecio vulgaris, Plant Physiol.*, **64**, 995 (1979).

33  K. A. Walley, M. S. I. Khan, and A. D. Bradshaw, The potential for evolution of heavy metal tolerance in plants: I. Copper and zinc tolerance in *Agrostis tenuis, Heredity*, **32**, 309 (1974).

34  D. W. Gartside and T. McNeilly, The potential for evolution of heavy metal tolerance in plants: II. Copper tolerance in normal populations of different plant species, *Heredity*, **32**, 335 (1974).

35  F. W. Robertson, Selection response and the properties of genetic variation, *Cold Spring Harbor Symp. Quant. Biol.*, **20**, 166 (1955).

36 C. M. Woodworth, E. R. Leng, and R. W. Jugenheimer, Fifty generations of selection for protein and oil in corn, *Agron. J.,* **44,** 60 (1952).

37 C. Ingram, University of Liverpool, personal communication, 1981.

38 R. J. Berry, *Inheritance and Natural Selection,* Collins, London, 1977, pp.

39 R. N. Andersen, A search for atrazine resistance in soybeans, *Abstr. Meet. Weed Sci. Soc. Am.,* 157, 1969.

40 R. N. Andersen and R. Behrens, A search for atrazine resistance in flax (*Linum usitatissimum* L.), *Weeds,* **15,** 85 (1967).

41 R. J. Holliday and P. D. Putwain, Evolution of herbicide resistance in *Senecio vulgaris* L.: I. Variation in susceptibility to simazine between and within populations, *J. Appl. Ecol.,* **17,** 779 (1980).

42 D. Jowett, Population studies on lead tolerant *Agrostis tenuis, Evolution,* **18,** 70 (1964).

43 L. Wu, A. D. Bradshaw, and D. A. Thurman, The potential for evolution of heavy metal tolerance in plants: III. The rapid evolution of copper tolerance in *Agrostis stolonifera, Heredity,* **34,** 165 (1975).

44 A. D. Bradshaw, The evolution of heavy metal tolerance and its significance for vegetation establishment on metal contamination sites, in *Heavy Metals in the Environment,* T. C. Hutchinson, (Ed.), Toronto University Press, Toronto, 1975, pp. 599–622.

45 D. A. Hickey and T. McNeilly, Competition between metal-tolerant and normal plant populations: a field experiment on normal soil, *Evolution,* **29,** 458 (1976).

46 S. G. Conard and S. R. Radosevich, Ecological fitness of *Senecio vulgaris* and *Amaranthus retroflexus* biotypes susceptible or resistant to atrazine, *J. Appl. Ecol.,* **16,** 171 (1979).

47 S. K. Jain and A. D. Bradshaw, Evolutionary divergence among adjacent plant populations: I. The evidence and its theoretical analysis, *Heredity,* **21,** 407 (1966).

48 R. M. May, J. A. Endler, and R. E. McMutrie, Gene frequency clines in the presence of selection opposed by gene flow, *Am. Nat.,* **109,** 659 (1975).

49 R. W. Snaydon and M. S. Davies, Rapid population differentiation in a mosaic environment: IV. Populations of *Anthoxanthum odoratum* at sharp boundaries, *Heredity,* **37,** 9 (1976).

50 J. Antonovics and A. D. Bradshaw, Evolution in closely adjacent plant populations: VIII. Clinal patterns in *Anthoxanthum odoratum* across a mine boundary, *Heredity,* **25,** 349 (1970).

51 A. D. Bradshaw, Pollution and evolution, in *Effects of Air Pollutants on Plants,* T. A. Mansfield (Ed.), Cambridge University Press, London, 1976, pp. 135–159.

52 T. McNeilly, Evolution in closely adjacent plant populations: III. *Agrostis tenuis* on a small copper mine, *Heredity,* **23,** 99 (1968).

53 J. Antonovics, Evolution in closely adjacent plant populations: V. Evolution of self fertility, Heredity, **23,** 219 (1968).

# CHAPTER SIXTEEN

# Practical Significance and Means of Control of Herbicide-Resistant Weeds

**J. V. PAROCHETTI**

U.S. Department of Agriculture, SEA-Extension
Washington, D.C.

**M. G. SCHNAPPINGER**

CIBA-GEIGY Corporation
Greensboro, North Carolina

**G. F. RYAN**

Western Washington Research and Extension Center
Washington State University
Puyallup, Washington

**H. A. COLLINS**

CIBA-GEIGY Corporation
Greensboro, North Carolina

## 1 INTRODUCTION

Some specific examples of alternative control methods for the more widely distributed weeds that have become herbicide resistant are discussed in this chapter. The principles that must be considered before selecting various alternative methods of control are introduced first, followed by a discussion of several specific resistant weed species.

If a weed escapes a herbicide application, one should not be too quick to assume that the weed has become resistant. The first principle is to determine if the failure to control the particular weed species was because of certain environmental factors, faulty herbicide application, or if the species was al-

309

ways inherently tolerant to the herbicide(s) used. As an example, *Panicum dichotomiflorum* is somewhat tolerant to atrazine. If atrazine is applied as the sole herbicide for several years in continuous corn where *P. dichotomiflorum* or other tolerant grasses are present, they will tend to increase in population. This phenomenon is a shift in the specific population of weeds of the type discussed in Chapter 4. For this reason, atrazine is now used mostly in combination with another herbicide that is more effective on these annual grasses.

Environmental factors affect herbicide performance. Rainfall or irrigation is critical soon after application to activate many soil applied herbicides. Soil pH, organic matter, and clay content are important factors that can affect the performance of soil applied herbicides. With the triazine herbicides, low soil pH (acidic conditions) accelerates dissipation, whereas a pH above 8 (basic or calcareous conditions) retards dissipation. If triazine herbicides dissipate inordinately rapidly, the resulting weeds that escape may be misinterpreted as resistant biotypes. Similarly, the higher the organic matter and clay content of the soil, the greater the amount of herbicide that is bound. This may also result in ineffective weed control.

Most postemergence herbicides require at least a short period of time without rainfall immediately following treatment for maximum effect. For maximum postemergence herbicide activity, the weeds must be actively growing, not under drought stress. Optimum light may also be necessary for maximum effect, especially with herbicides that inhibit photosynthesis, such as triazine herbicides.

Some herbicides do not perform well when applied after the weed seeds germinate. This often contributes to poor herbicide performance if the herbicide application is delayed on soil that has been tilled for several days prior to application. As an example, alachlor is not as effective in controlling germinated *P. dichotomiflorum,* even when it is applied before grass emergence.

Finally, one must consider the proper application of herbicides, including the correct rate based on the the texture and organic matter of the soil, the anticipated or observed weed species, the type of application equipment that will provide uniformity of application, and the incorporation method, timing, and equipment for those herbicides that should be soil incorporated.

These factors must be examined thoroughly to determine if they contribute to the lack of control of a weed population. If the weed species present has evolved because of herbicide resistance, a further examination of the field will reveal that only one dominant weed species is usually present. This can be verified by comparison with the unintentionally untreated areas that are occasionally missed during herbicide applications. These untreated areas will probably be populated by several to many weed species. In addition to the example of *P. dichotomiflorum* or other tolerant grasses, one may observe perennial weeds, such as *Sorghum halepense, Agropyron repens,* or *Cyperus* sp. to be widespread if the herbicide used is not effective against these weed species.

Once a weed species is suspected to be resistant to an applied herbicide, scientific confirmation is recommended, especially if no other known cases of resistance to that particular weed species have been documented in the immediate vicinity (i.e., within 30 to 50 km). Proper scientific confirmation (see Chapter 7 for appropriate methods) should provide justification and motivation to undertake alternative herbicide, cultural, and crop rotation measures necessary to effect adequate control in these fields in future years.

When such failures are suspected or have been confirmed to be due to herbicide resistance, it is further highly advisable to make every reasonable effort to prevent the resistant biotype from going to seed or spreading. This is especially important if it is the first such observable occurrence of resistance by the species to the herbicide in the area or field. It is even more critical if the seeds produced by this species are characteristically long-lived (i.e., remain dormant but viable for years) in soil and have a competitive or undesirable nature. To prevent the resistant weed from spreading or going to seed, either another type of herbicide having a different mode of action, or cultivation, or both, would likely be the best alternative. The herbicide would almost certainly have to be one having postemergence activity (e.g., 2,4-D type) if the weeds were already growing. If the area of infestation is along a railroad or where soil tillage is impossible, either another herbicide, mowing, burning, hand removal, and so on, could be used.

## 2  SELECTING A CONTROL PROGRAM

### 2.1  Containment

Following confirmation that a weed species is newly resistant, all reasonable efforts should be made to contain it to the farm or to that specific field, and in preventing seed production and distribution. This will involve an immediate voluntary quarantine of the farm. All machinery must be cleaned of adhering seeds before leaving the field or farm. One must also make certain that seeds of the weed species are not transported in silage, manure, or in seed crops. It may also require timely tillage to incorporate weed residues in the soil to minimize wildlife, wind, and water transport of seeds from the site. We are fortunate that triazine resistance in weeds has been confirmed to be maternally inherited in all cases studied to date, so it is not transmitted by pollen. However, this may not be true of resistance to other herbicides.

### 2.2  Use Different Herbicides or Adopt New Technology

Of the three options, substitution of a different or additional herbicide to control the newly resistant weed is probably the first solution the farmer employs. Farmers may be reluctant to change any other practice, especially crops or tillage, unless absolutely necessary or there is a proven need. The decision to select a substitute or additional herbicide is limited by the number and, more important, by the different chemical groups of herbicides that are available or can soon be made available by registration. The number of different types of herbicides available, mostly with different modes of action, is very important. Often, once resistance has developed for one member of the chemical group such as the triazines, other chemicals in that group are also generally ineffective in controlling the resistant weed species (see Chapters 9 and 10).

Another optional technology is a change in application methods. Specific examples would include the wick-applicator and recirculating sprayer for the application of certain currently registered herbicides specifically to the weeds and not to the sensitive crop.

Substitute or additional herbicides and innovative application techniques will add additional costs to the production of the crop. These additional costs must be weighed against other alternative weed control practices, which may include changing the cropping methods.

### 2.3  Alter the Current Cropping System

The appearance of a resistant weed species affords strong motivation to change cropping systems, especially when the resistant weed species becomes more dominant and prolific than the crop. Nevertheless, farmers are often reluctant to grow a different crop or to change their current cropping system, especially if their system has been successful. This is understandable, as they have acquired knowledge and experience in growing, harvesting, and marketing that crop at a profit. Changing the crop that is grown is the last option most farmers will choose, but it is still a viable choice. By selecting an alternative crop, the grower can utilize many practices that will greatly reduce competition from the herbicide-resistant weed. The grower can select a herbicide that is cost effective in controlling the resistant weed, or tillage can be utilized at different times of the year. Also, a cropping system that is more competitive, such as the permanent sward, or where the crops grow and mature in different seasons, will break the cycle of the problem weed. An example of a different crop which has a different cycle would be winter grain as a substitute for spring or summer culture of corn and soybeans.

### 3  CONTROL TECHNIQUES FOR TRIAZINE-RESISTANT WEEDS IN NORTH AMERICA

### 3.1  *Senecio vulgaris*

When resistance of *Senecio vulgaris* to triazine herbicides was first documented, other herbicides were evaluated as possible substitutes for simazine in a nursery weed control program.[1,2] The triazine-resistant biotype was 99 to 100% controlled in pot tests by dichlobenil, norea, and fluometuron at 2.2 kg/ha, and by chloroxuron at 4.5 kg/ha (Table 16.1).

Dichlobenil and norea were both registered and available for nursery use at that time, and could partially replace simazine. However, because dichlobenil is not tolerated by some ornamentals, and not all weed species are controlled, dichlobenil could not completely replace simazine where triazine-resistant *Senecio* was a problem. Dichlobenil could not replace atrazine in most Christmas tree plantations because several conifer species are susceptible to dichlobenil. Norea has not been commercially available since 1971. The two other substituted ureas, chloroxuron and fluometuron, have been under consideration for use in ornamentals but have not been registered for this use in the United States.

Directed sprays of dinoseb or paraquat have been used in nurseries and Christmas tree plantations to control *Senecio* seedlings where preemergence control has not been adequate with available herbicides. Conscientious use of this practice should help to reduce the seed production of this triazine-resistant biotype. However, it is a more expensive and less satisfactory method than the use of preemergence herbicides. It is from such situations that some weeds have become resistant to paraquat (Chapter 3).

Table 16.1  Preemergence Control of *Senecio vulgaris* from Two Locations with Several Herbicides

| Herbicide | Rate (kg/ha) | % Controlled[a] Seed Source[b] | |
|---|---|---|---|
| | | R | S |
| Atrazine | 2.2 | 0 | 100 |
| | 4.5 | 0 | 100 |
| | 6.7 | 0 | 98 |
| Simazine | 2.2 | 14 | 100 |
| | 4.5 | 0 | 100 |
| | 6.7 | 0 | 100 |
| Dichlobenil | 2.2 | 100 | 100 |
| Nitrofen | 1.1 | 0 | 7 |
| | 2.2 | 14 | 44 |
| Chloroxuron | 2.2 | 58 | 75 |
| | 4.5 | 100 | 99 |
| Fluometuron | 1.1 | 95 | 98 |
| | 2.2 | 100 | 100 |
| Norea | 2.2 | 98 | 100 |

[a] Based on number of surviving seedlings related to number in check pots.

[b] Source R, simazine or atrazine in use since 1958; source S, triazine herbicides not in continuous use. (Modified with permission from Ryan[1].)

Napropamide, applied in combination with simazine or with oxadiazon in a highway landscape experiment, completely controlled a population of *Senecio* that was only 44% controlled by simazine and 71% controlled by oxadiazion alone.[3] Napropamide was not applied separately in this experiment. In other studies, control of *Senecio* by napropamide varied from poor to excellent.[2,4] Control of *Senecio* by oxadiazon in field and container studies has varied from fair to good.[2,4-6]

The combination of napropamide with other herbicides, such as simazine or oxadiazon, has been suggested for use in ornamentals[7] or in young orchards or vineyards.[8] In some cases the napropamide plus oxadiazon combination controlled *Senecio* longer than either herbicide separately.[2]

Oxyfluorfen is now available for *Senecio* control in conifer seedbeds and for dormant conifers in container and field nurseries. It will be effective for controlling this weed in situations where it can be used. It is effective both pre- and postemergence on small *Senecio* seedlings.

There are, therefore, several herbicides available to replace or to use with triazines for controlling *Senecio*. Some have limited crop tolerance, and in other cases the cost will be increased because of application techniques, or the need for more than one herbicide to cover the complete weed spectrum.

## 3.2  *Amaranthus* spp.

In the United States, many growers with land in a rolling terrain prefer not to prepare a seedbed, which provides the opportunity for significant soil erosion. Also, limitations in time and the costs of tillage and the conservation of soil moisture encourage the use of no-tillage methods wherever possible. Although the several modifications of no-till, minimum tillage, or conservation tillage provide some definite advantages, there are also some significant disadvantages, including the lack of alternative weed control programs in case of failures due to dry weather, resistant weeds, and so on. Once the decision has been made to plant a crop without tillage, the farmer is more limited in the list of effective herbicides available, cannot incorporate the herbicide into the soil (which can often help in activating the herbicide or in eliminating the early flush of weeds), and cannot cultivate should failures occur later in the season.

In some areas, especially under no-till cropping, some triazine-resistant weeds, such as *Amaranthus* spp., have developed and spread. Past grower practices in most affected areas have centered around no-tillage, using paraquat plus atrazine plus simazine as a standard herbicide treatment. Initial reaction by growers to the presence of resistant *Amaranthus* was to continue no-tillage corn and use postemergence applications of either dicamba or 2,4-D. In many cases, two postemergence treatments were required to obtain a desirable level of efficacy. As the *Amaranthus* began to spread to a greater number of fields, it was realized that postemergence treatments were not only time consuming but costly.

Growers began to change their practices and returned to plowing their fields, which facilitated using preplant-incorporated herbicides such as the thiocarbamates, including butylate and EPTC. The preparation of a seedbed also allowed more effective use of preemergence surface-applied herbicides, such as metolachlor, alachlor, and pendimethalin, which have not been as successful under no-tillage conditions due to their variability in controlling annual grasses, and pendimethalin has sometimes caused stand reductions in no-till corn. The presence of *Amaranthus,* which is not controlled by triazine herbicides (Chapters 2 and 3), has therefore resulted in several changes in grower practices, including a change to herbicides with different chemistry, as well as additional applications of postemergence chemicals and/or alternative cultural practices. Those continuing with the standard triazine mixtures have had to make one or several postemergence treatments, while some growers have changed to conventional tillage that facilitated using either preplant-incorporated or preemergence herbicides, which are often more effective. Those growers electing to continue no-tillage have added either alachlor or metolachlor to their triazine mixtures and monitor their fields for late emergence of escaping *Amaranthus.* Table 16.2 presents data from a typical field study on control of resistant *Amaranthus hybridus* in no-till corn.

Although alachlor and metolachlor have generally been effective in controlling *Amaranthus,* higher rates of application have been required with no-till than with conventional tillage. Table 16.3 presents a similar field trial conducted in the same area and year as Table 16.2, but under conventional tillage. From these data and other trials, metolachlor appears to be the most consistent of these two herbicides.[9] Also, data from field research trials have indicated that metolachlor performed somewhat better or for a longer period than did

Table 16.2  Control of Triazine-Resistant *Amaranthus hybridus* in Field Corn Grown without Tillage (New Windsor, Md., 1978)

| Herbicides[a] | Rate (kg/ha) | % *Amaranthus* Controlled | |
| --- | --- | --- | --- |
| | | 30 Days | 60 Days |
| Control | — | 0 | 0 |
| Atrazine + simazine | 1.7 + 1.7 | 5 | 30 |
| Alachlor + atrazine | 2.0 + 1.7 | 60 | 45 |
| Alachlor + atrazine | 2.5 + 1.7 | 78 | 65 |
| Metolachlor + atrazine | 2.0 + 1.7 | 82 | 67 |
| Metolachlor + atrazine | 2.2 + 1.7 | 78 | 65 |
| Metolachlor + atrazine | 2.5 + 1.7 | 97 | 78 |
| Pendimethalin + atrazine | 1.7 + 1.7 | 70 | 68 |

[a]All treatments except control included paraquat (0.6 kg/ha).

(Unpublished data by Schnappinger.)

Table 16.3  Control of Triazine-Resistant *Amaranthus hybridus* in Field Corn Grown with Conventional Tillage (Hereford, Md., 1978)

| Herbicide | Rate (kg/ha) | Application Method[a] | % *Amaranthus* Controlled |
|---|---|---|---|
| Control | — | — | 0 |
| Butylate + atrazine | 3.5 + 1.1 | PPI | 99 |
| EPTC + atrazine | 3.5 + 1.1 | PPI | 98 |
| Vernolate + atrazine | 3.5 + 1.1 | PPI | 99 |
| Alachlor + atrazine | 1.8 + 1.4 | PRE | 94 |
| Alachlor + atrazine | 2.1 + 1.4 | PRE | 96 |
| Metolachlor + atrazine | 1.8 + 1.4 | PRE | 96 |
| Metolachlor + atrazine | 2.1 + 1.4 | PRE | 97 |
| Pendimethalin + atrazine | 1.8 + 1.4 | PRE | 97 |
| Simazine + atrazine | 1.4 + 1.4 | PRE | 0 |

[a]PPI, preplant-incorporated; PRE, preemergence.
(Unpublished data by Schnappinger.)

alachlor.[10,11] Metolachlor, on the other hand, has been especially ineffective in controlling triazine-resistant *Chenopodium album* and *C. missouriense.*

In contrast to conventional tillage, where resistant *Amaranthus* control is very consistent with several herbicides, the results with no tillage have been variable and in many cases late season weed control has been only moderate to fair. Whereas growers have been able to rely on slowly degrading herbicides for season-long control in conventional tillage, they have had to continue monitoring their no-tillage fields for late emerging *Amaranthus,* and in some cases apply either dicamba or 2,4-D as postemergence treatments.

### 3.3  *Echinochloa crus-galli*

Atrazine and most other *s*-triazine herbicides have generally provided good control of *Echinochloa crus-galli.* Even under no-till methods, control with atrazine and simazine, alone or in combinations, have given satisfactory results. In one area of Maryland, a triazine-resistant *Echinochloa* appeared about 1976. It is still limited to approximately a 25-km radius, and its practical significance to growers is difficult to determine at this time.

In areas where *Echinochloa* was not controlled by either atrazine or simazine, attempts were first made to obtain control using postemergence herbicides.[11a] Of the herbicides tested, the triazines cyanazine and ametryn were ineffective. Linuron treatment resulted in fair control, and the only treatment that resulted in satisfactory control was a directed postemergence application of paraquat (Table 16.4).

Even though no dependable and safe herbicide program is available to control resistant *Echinochloa* after it has emerged, most growers with the problem

Table 16.4  Control of Triazine-Resistant *Echinochloa crus-galli* in No-Till
Field Corn by Postemergence and Post-directed Herbicide Applications
(Winchester, Md., 1979)

| Herbicide | Rate (kg/ha) | Application Method | % *Echinochloa crus-galli* Controlled | |
|---|---|---|---|---|
| | | | 13 July[a] | 20 Sept.[a] |
| Control | — | — | 0 | 0 |
| Paraquat | 0.6 | Post-directed[b] | 80 | 73 |
| Ametryn | 1.7 | Post-directed | 30 | 0 |
| Linuron | 1.1 | Post-directed | 43 | 68 |
| Cyanazine | 2.2 | Postemergence[c] | 0 | 0 |
| Atrazine | 2.2 | Postemergence | 0 | 13 |

[a]Weed control was measured on July 13 and on September 20.

[b]Post-directed treatments applied on July 2.

[c]Postemergence treatments applied on May 23.

(Modified with permission from Schnappinger et al.[11a].)

still grow no-till corn. They usually add either metolachlor or alachlor to their triazine herbicide combinations. As with the control of *Amaranthus* spp. and other weeds, the continuation of no-till where triazine-resistant *Echinochloa* is present limits the grower to fewer herbicide options than with conventional tillage. Results from a field study summarized in Table 16.5 indicate that metolachlor in combination with triazine herbicides provided consistently acceptable control under minimum tillage. Control resulting from pendimethalin plus atrazine was barely acceptable.

Results of tests conducted in conventionally tilled corn indicated that a variety of herbicides are effective in controlling *Echinochloa* (Table 16.6). The grower has the option of using either incorporated herbicides, such as the thiocarbamates, or any one of several preemergence herbicides.

## 4  CONTROL TECHNIQUES FOR TRIAZINE-RESISTANT WEEDS IN EUROPE

As with the triazine-resistant weeds in North America, because of their relatively recent development and the availability of alternative herbicides, cultivation, and other easy means of control, the published reports or references to methods for their control in Europe are limited. Based on the data available and the similarity in the resistant biotypes and conditions, however, we feel that the same principles will apply. While some of the same alternative herbicides have been successfully used, depending on the crop and weed problems, some herbicides or combinations not available in North America have been useful in Europe. Table 3.3 provides some data on the amount of control obtained with

Table 16.5   Control of Triazine-Resistant *Echinochloa crus-galli* in Field Corn without Tillage (Westminster, Md., 1979)

| Herbicide | Rate (kg/ha) | % *Echinochloa crus-galli* Controlled | | |
|---|---|---|---|---|
| | | 30 Days | 60 Days | 130 Days |
| Control | — | 0 | 0 | 0 |
| Atrazine + simazine[a] | 2.2 + 2.2 | 20 | 17 | 15 |
| Metolachlor + atrazine | 2.5 + 2.2 | 80 | 85 | 77 |
| Metolachlor + simazine | 2.5 + 2.2 | 78 | 90 | 77 |
| Metolachlor + atrazine + simazine | 2.5 + 1.7 + 1.7 | 80 | 87 | 79 |
| Pendimethalin + atrazine | 1.7 + 1.7 | 72 | 72 | 73 |
| Alachlor + atrazine | 2.2 + 2.2 | 68 | 65 | 67 |
| Alachlor + atrazine | 2.8 + 2.2 | 68 | 50 | 43 |
| Metolachlor + cyanazine | 2.5 + 2.2 | 83 | 90 | 75 |
| Cyanazine + atrazine | 2.8 + 2.2 | 48 | 57 | 57 |

[a]All treatments except control included paraquat (0.6 kg/ha).
(Modified with permission from Schnappinger et al.[11a].)

a number of alternative herbicides on triazine-resistant *Chenopodium album* collected from Canada and Switzerland. The degree of control was generally satisfactory and similar to that from nontriazine herbicides, although there were some significant differences between the two sources in a few cases.

Under some conditions (e.g., Austria), good control of triazine-resistant *Amaranthus retroflexus* and *C. album* in corn has been obtained with bromofenoxim at rates of 1 to 2 kg/ha applied postemergence to both crop and weeds.[12] Szith and Furlan[13,14] found that atrazine-resistant *Bidens tripartita* was readily controlled with preemergence treatments of linuron and chlorbromuron, although some corn injury occurred. Postemergence applications of pyridate and bromofenoxim gave satisfactory control but were better tolerated by corn than the urea derivatives. They also found the latter treatments to provide best results against triazine-resistant *Polygonum convolvulus*.

Hartmann[15] reported that the atrazine-resistant weeds which had developed in continuous corn in Hungary were very effectively controlled with other herbicides or combinations.

Several extensive field trials have been conducted in France on the control of triazine-resistant *C. album, A. retroflexus,* and *Solanum nigrum.* Darrigrand et al.[16] reported that of several herbicides and combinations, pendimethalin + atrazine (1.2 + 0.8 to 2.0 + 2.0 kg/ha) gave best and most consistent preemergence control of all three biotypes. Effective postemergence treatments included bromofenoxim at 1.0 kg/ha (inconsistent on *A. retroflexus*), bentazon (1.5 kg/ha) + oil, pyridate (1.5 kg/ha), and dinoterb (1.0 kg/ha). Some corn phytotoxicity was observed from all postemergence treat-

Table 16.6  Control of Triazine-Resistant *Echinochloa crus-galli* in Field Corn Grown with Conventional Tillage (Westminster, Md., 1979)

| Herbicide | Rate (kg/ha) | Application Method[a] | % *Echinochloa crus-galli* Controlled | | |
|---|---|---|---|---|---|
| | | | 30 Days | 60 Days | 130 Days |
| Control | — | — | 0 | 0 | 0 |
| EPTC | 4.5 + 1.1 | PPI | 60 | 78 | 70 |
| Butylate + atrazine | 4.5 + 1.1 | PPI | 55 | 73 | 75 |
| Pendimethalin + atrazine | 1.7 + 1.7 | PRE | 75 | 77 | 80 |
| Alachlor + atrazine | 1.7 + 1.7 | PRE | 73 | 72 | 71 |
| Alachlor + atrazine | 2.2 + 1.7 | PRE | 73 | 72 | 70 |
| Alachlor + cyanazine | 2.2 + 1.7 | PRE | 78 | 73 | 69 |
| Metolachlor + atrazine | 1.7 + 1.7 | PRE | 82 | 87 | 83 |
| Metolachlor + atrazine | 2.2 + 1.7 | PRE | 85 | 90 | 89 |
| Metolachlor + simazine | 1.7 + 1.7 | PRE | 72 | 87 | 83 |
| Atrazine + simazine | 1.7 + 1.7 | PRE | 10 | 0 | 13 |

[a] PPI, preplant-incorporated; PRE, preemergence.
(Modified with permission from Schnappinger et al.[11a].)

ments, especially dinoterb. Both pre- and postemergence treatments were needed for adequate weed control. Simonin et al.[17] found that postemergence treatments of dinoterb (1.0 kg/ha), dinoseb (1.0 kg/ha), DNOC (1.5 kg/ha), and bromofenoxim (1.25 kg/ha) and preemergence applications of butralin + atrazine (1.0 + 2.4 kg/ha) were effective against triazine-resistant *C. album*. Except for bromofenoxim, the postemergence treatments were also effective against *S. nigrum*. The postemergence treatments again caused foliar injury but did not reduce corn yield.

In Austria, Neururer[18] concluded that the various strategies which had been or should be used to prevent or contain herbicide-resistant or tolerant weeds included the following:

1  Use of alternative herbicides or those specifically aimed at the problem weed.

2  Use of economic threshold values to avoid supplementary selection pressure.

3  Use of a reasonable degree of nonchemical weed control measures.

## 5  THE SPECIFIC PROBLEMS OF RAILROAD, ROADSIDE, AND INDUSTRIAL (NONSELECTIVE) WEED CONTROL

Railroads and other noncropland sites do not afford the opportunity for cultivation, crop rotation, or other cultural practices in the battle against weeds, including those that may be resistant to triazines or other herbicides. Therefore, the fight continues to be waged with herbicides, often in tank mixtures.

Alternative control methods have been of very limited usefulness. In some instances, contact-type herbicides such as sodium chlorate, diquat, paraquat, ametryn, MSMA, DSMA, or glyphosate are applied postemergence in tank mix combination with a residual-type herbicide. This may include atrazine, bromacil, diuron, prometon, simazine, tebuthiuron, and hexazinone, among others. The contact herbicides typically burn down the existing vegetation, leaving the residual component of the mixture to provide preemergence control of germinating weeds.

In most instances, it is desirable to make herbicide applications preemergence to weeds. The use of two residual herbicides is common in this instance, with a triazine herbicide often one of the tank mix partners.

In addition to triazine-resistant weeds in croplands, there have also been increasing instances of herbicide resistance evolving in noncropland or uncultivated sites, such as railroad rights-of-way, roadsides, and industrial sites. In fact, it is especially under such conditions of repeated annual applications of the same herbicides, at high rates, without tillage or other means to prevent the weeds that escape from surviving and producing seed, that the possibility of resistance is greatest, the spread of the resistant populations most extensive (via railroad cars, automobiles, etc.), and the alternative methods of control most limited once herbicide resistance has become established.

Some herbicide applicators have observed that with repeated usage of atrazine, simazine, and other triazine herbicides, alone or in combinations, there have been some failures to control such grass weeds as *Digitaria sanguinalis, Panicum dichotomiflorum,* and *Panicum capillare.* Two broadleaf weeds exhibiting triazine resistance on railroad lines are *Kochia scoparia* and *Salsola kali.*

Along railroads and railroad rights-of-way, one might surmise that a major factor contributing to the lack of weed control is the leaching of residual herbicide(s) below the effective weed germination zone, due to the physical nature of the railroad bed and right-of-way. It is easy to misconstrue such failures as resistance.

However, this has not always been the cause of failures on railroads in recent years. Seedlings from seeds of *Kochia scoparia* and *Panicum capillare* plants collected on railroad sites where the weeds survived and grew vigorously following applications of atrazine and simazine were indeed resistant to normal levels of these herbicides, as confirmed at the CIBA-GEIGY Research Facility at Vero Beach, Florida. Although the total extent of the spread of triazine resistant weeds has not been confirmed, it can be confidently assumed that resistant *Kochia scoparia* is infesting railroad lines throughout all the northern states to some extent, and it extends into Kansas, Oklahoma, and Texas to the south. Triazine-resistant *Panicum capillare* can be found on railroads in most northeastern states from New York to Michigan. Several other weeds have been assumed to be resistant, but have not yet been confirmed.

Realizing that resistance to triazines exists within certain weed species found in such noncropland sites, there has been a concerted effort to introduce the use of herbicide mixtures in some railroad situations where resistance has been observed. Two examples of tank mixtures suggested are atrazine plus diuron and simazine plus diuron. Bromacil has also been an effective addition in combination with these herbicides.

Where *Digitaria sanguinalis, Panicum dichotomiflorum,* and *Panicum capillare* are the dominant grasses, use of these and other tank mixtures have been found satisfactory. Diuron has also been suggested for use alone. More recently, a tank mix combination of metolachlor with atrazine or simazine has received registration in the United States.

## 6 CONCLUDING REMARKS

It is important for growers who suspect herbicide-resistant weed species to have these confirmed prior to initiating any changes in control practices. It is possible that the weed species present are inherently tolerant to the herbicides used.

Once herbicide-resistant weed species have been identified, it is important to determine the extent that the species are distributed in the immediate geographical region. If a small infestation is localized to one farm or one field, containment of the species within that area may be possible. Quarantine methods are recommended in such cases whenever feasible.

Control options that are available to the grower include substituting herbicides in an existing cropping system, altering tillage operations, the use of herbicides applied at different times during the cropping season, and crop rotations.

A more drastic control technique is to change the type of crop that is grown in the weed-infested areas. This will take advantage of the fact that a grower can alter or break the life cycle of the weed and reduce the population of that weed.

In all likelihood, controlling herbicide-resistant weeds will result in a greater cost to the producer. The extent of this cost will depend on the expense of the additional tillage operations, reduction in value of the alternative crops grown, increased cost of the substitute herbicides, or additional herbicides needed to control the resistant weed. The current cost of 2,4-D or dicamba to control triazine-resistant weeds in no-tillage corn would be minimal. However, if additional preemergence herbicides are used, the increased cost would be substantially higher.

Assessing the economics in the control of a resistant weed is much more complex if growers need to grow different crops. The cost of controlling the herbicide-resistant weed would not be an isolated factor in two different cropping systems.

For growers with perennial crops such as ornamentals, nurseries, and tree plantations, the options for controlling resistant weeds tend to be limited to alternative herbicides. Alternative herbicides may be more expensive and/or require more-time-consuming application techniques. Thus the economic consequences of herbicide-resistant weeds will vary considerably with different cropping situations.

Considering the costs and the alternatives, it may well be worth considering the many means available that will greatly delay the appearance of herbicide resistance (Chapter 17). This includes the normal use of crop rotation and multiple herbicides. To our knowledge, resistance has yet to appear under such conditions.

Control of herbicide-resistant weeds along railroads, roadsides, and industrial sites presents some special problems. Alternative methods of vegetation control are much more limiting, herbicides are more essential, are used at higher rates, and repeatedly. Herbicide resistance could be expected to occur more frequently, spread more rapidly, and would become a source of infestation into croplands. For these and other reasons, herbicide-resistant weeds along railroads justify special attention and efforts to find more satisfactory alternative control measures, which they have not received up to the present time. In fact, this source of resistant weeds has been virtually ignored. If weeds develop cross resistance to other types of residual herbicides, most of which act as photosynthesis inhibitors, we may have little or no economically feasible means for their control with herbicides along railroads and other noncropland sites.

There are several other triazine-resistant weeds, and a few cases of weed resistance to other types of herbicides mentioned in Chapters 2 and 3, which are not included in this chapter. This review is not intended to be a complete record of all data from alternative herbicides or other control procedures that have been studied or used on these weeds. Most triazine-resistant species are not discussed here because either they are recent developments or have not been extensively studied for specific methods of control. In most cases, the other triazine-resistant broadleaf weeds can be controlled by the same procedures as those used for resistant *Amaranthus* spp., and the resistant annual grasses can be managed similar to resistant *Echinochloa crus-galli*. Unfortunately, some of the most useful preemergence herbicides for control of other weeds in many crops (e.g., metolachlor and alachlor) fail to give consistent control of *Chenopodium album*. It should not become a serious problem, however, as it can be quite readily controlled by postemergence applications of 2,4-D and dicamba in corn, and there are also other options available.

## REFERENCES

1  G. F. Ryan, Resistance of common groundsel to simazine and atrazine, *Weed Sci.*, **18**, 614 (1970).

2   G. F. Ryan, Control of bittercress, common groundsel, and barnyardgrass in two nursery container media, *Proc. West. Soc. Weed Sci.,* **29,** 156 (1976).

3   G. F. Ryan, N. Rosenthal, and R. L. Berger, Chemical weed control in roadside vegetation on highway right-of-way, *Wash. State Transport. Comm. Tech. Rep.,* **34,** 1 (1979).

4   C. L. Elmore, W. A. Humphrey, T. W. Mock, and R. G. Snyder, Common groundsel control in *Buxus microphylla, Flower Nursery Rep.,* 2 (May–June 1977).

5   R. E. Bailey and J. A. Simmons, Oxadiazon for weed control in woody ornamentals, *Weed Sci.,* **27,** 396 (1979).

6   C. L. Elmore, W. A. Humphrey, and T. W. Mock, Two preemergence herbicides in container-grown ornamentals, *Flower Nursery Rep.,* 4 (Summer 1979).

7   D. V. Peabody, D. G. Swan, and R. Parker, *Washington State Weed Control Handbook,* Washington State University Cooperative Extension Service, 1980, p. 119.

8   A. Lange, C. Elmore, B. Fischer, H. Kempen, and E. Stevenson, Napropamide and oryzalin, two new selective herbicides for weed control in young orchards and vineyards, *Proc. West. Soc. Weed Sci.,* **29,** 202 (1976).

9   H. J. Strek and J. B. Weber, Alachlor and metolachlor comparisons in conventional and reduced-tillage systems, *Proc. South. Weed Sci. Soc.,* **34,** 33 (1981).

10   R. L. Ritter and T. C. Harris, Control of triazine resistant redroot pigweed in conventional and no-tillage corn, *Proc. Northeast. Weed Sci. Soc.,* **35,** 41 (1981).

11   C. Buchholz, E. R. Higgins, M. G. Schnappinger, and S. W. Pruss. Longevity of control with metolachlor and alachlor, *Proc. Northeast. Weed Sci. Soc.,* **35,** 124 (1981).

11a  M. G. Schnappinger, J. R. Hensley, W. C. Bay, D. L. Greene, and S. W. Pruss, Triazine resistant barnyardgrass control in field corn in Maryland, *Proc. Northeast. Weed Sci. Soc.,* **35,** 30 (1981).

12   H. M. LeBaron, Biochemistry Department, CIBA-GEIGY Corp., Greensboro, N.C., personal communication, 1981.

13   R. Szith and H. Furlan, Der dreiteilige zweizahn (*Bidens tripartita*) ein neues atrazinresistentes unkraut im mais, *Der Pflanzenarzt,* **32,** 6 (1979).

14   R. Szith and H. Furlan, Der windenknöterich (*Polygonum convolvulus* L.) ein "neues" atrazinresistentes unkraut im maisbau, *Der Pflanzenarzt,* **33,** 95 (1980).

15   F. Hartmann, The atrazine resistance of *Amaranthus retroflexus* L. and the expansion of resistant biotype in Hungary, *Növényvédelem,* **15,** 491 (1979).

16   M. Darrigrand, S. Soffietti, H. Haudecoeur, and C. Souchet, Possibilities actuelles de lutte dans les cultures de mais contre les dicotyledones resistantes aux triazines, *Compt. Rend. COLUMA 10 Conf.,* Vol. 1, 119 (1979).

17   A. Simonin, J. P. Guinefoleau, and J. C. Lebosse, Essais de lutte contre les mauvaises herbes dévenues resistantes aux triazines dans le mais, *Compt. Rend. COLUMA 10 Conf.,* Vol. 1, 130 (1979).

18   H. Neururer, Das "resistentwerden" von unkräutern, *Der Pflanzenarzt,* **32,** 8 (1979).

# CHAPTER SEVENTEEN

# Interrelating Factors Controlling the Rate of Appearance of Resistance: The Outlook for the Future

J. GRESSEL

Department of Plant Genetics
The Weizmann Institute of Science
Rehovot, Israel

L. A. SEGEL

Department of Applied Mathematics
The Weizmann Institute of Science
Rehovot, Israel

## 1  INTRODUCTION

Shortly after the introduction of each type of modern pesticide from antibiotics through rodenticides, resistance was quick to appear: that is, all types of pesticides except herbicides, where the appearance was very slow. Resistance to herbicides has now occurred over limited areas and mainly, but not exclusively, in one group, the triazines. This appearance of resistance has been discussed at length in previous chapters. In considering the literature over the past 10 to 20 years, we can ask why resistance has not appeared earlier. Are the triazines the harbingers of things to come with other herbicides? Why are the triazines so different from the phenoxy acid herbicides, which have been used in continuous culture situations for longer periods of time? These questions have been considered in a series of recent reviews[1-7] which we can but reconsider and update.

Many of the factors controlling the rate of appearance of resistance have been delineated and reviewed recently for insecticide resistance[8,9] and the interpretations are similar to ours. There are also some unique features of the herbicide/weed relationship in addition to those considered in the insecticide/

325

insect relationship. The evolution of weeds as a special group of plants is the subject of an extensive review, but the evolution of herbicide resistance was not touched upon.[10] Changing weed patterns as a result of obliteration of the prevalent population by herbicides is beyond the scope of this chapter and was discussed in Chapter 4. A recent symposium dealt with methods of avoiding and reducing development of resistance to all types of pesticides, including herbicides, and these reviews have been published in ref. 11.

Ever since the question of when to expect resistance was posed to one of us some 21 years ago by L. G. Holm,[12] we have been gathering case histories from the areas of pesticides, heavy metal toxicities, and population genetics, and have tried to make a coherent picture of what may be going on. The rate of appearance of resistances seems to be governed by an interaction between special features of the weeds and various features of the herbicides. We first describe each of these features separately, and then put forth an equation that describes the interrelationships in a simplified manner. Finally, we substitute possible numbers into the equation which represent the various scenarios occurring in the field. This should allow us to assess the importance of triazine resistance and the likelihood of similar occurrences. It should also allow us to consider herbicide "designs" and treatment procedures to defer the appearance of weeds resistant to these and other types of herbicides as well.

## 2  FACTORS AFFECTING THE RATE OF RESISTANCE

### 2.1  Genetic Component

For genetic resistance to occur, one or more alleles for resistance must be present at some level in the field population of a weed. It is highly unlikely for a completely new gene to appear *de novo,* although one could be transferred from other species. Thus we have it that streptococci seem not to have a gene or plasmid for penicillin resistance, and fungi which have developed resistance to many a fungicide are still controlled by the venerable Bordeaux mixture, in use for about a century.[13] In a careful analysis of selection for heavy metal tolerance of pasture species on mine tailings, Bradshaw's group found that only about a third of the species previously in the area developed genetic resistance (see Chapter 15).

The modes of inheritance will govern the initial frequency of the resistant gene in the population, or the starting point for enrichment, but will hardly affect the rate of enrichment. It is clear from cases of differential tolerance and resistance that genes for herbicide tolerance appear in nature (Appendix Table A1).

Without knowing the genetics, it is hard to predict the field frequency of resistance to any given herbicide. The initial frequency depends on the number of genes involved, the dominance, and the ploidy. To an extent, a herbicide can be "designed" so that its corresponding genotype has an infinitesimally arbitrarily low frequency of resistant individuals within a population. A herbicide with a multiplicity of differing sites of action because it possesses different reactive groups will have a corresponding frequency of resistance which is the compound frequency of the genes for resistance to each group. This compounded frequency can be obtained by using herbicide mixtures as well. The only alternative ways a plant can evade such a herbicide is by evolving a detoxification mechanism (Chapter 8) or by evolving a change in the weed phenology to germinate after the herbicide is no longer effective (see Chapter 6), which is as effective as biochemical modes of resistance.

As discussed in Chapter 12, basically all classical modes of inheritance of resistance have been found in plants, from monogene dominant through polygenic. The diversity of modes of inheritance are especially striking for atrazine (see Chapter 13). Atrazine tolerance is quantitatively inherited in flax. Corn resistance is controlled by a dominant nuclear gene, whereas in *Brassica campestris* and some other weeds the trait is maternally inherited.

Typically, monogene-dominant phenotypes are found at frequencies of about $10^{-5}$ to $10^{-6}$ (1 in 100,000 to 1 in 1,000,000). Monogene recessive phenotypes are expected at $10^{-9}$ to $10^{-11}$, and multigenes at multiples thereof. The actual frequencies of recessive-mutant frequencies in diploid populations have not been determined in great numbers. There is some reason to believe that they may be present at a frequency that is much closer to dominant allele frequencies than seems logically possible.[14] The triazine resistant strains of *Amaranthus, Chenopodium, Senecio,* and *Brassica* spp. seem to be affected at the same chloroplast site and are thus probably similarly maternally inherited. It might be conjectured that the reason they appeared first is their high frequency in field populations. An outstanding problem in this respect is that geneticists cannot estimate typical frequencies of cytoplasmic mutant alleles within plant populations.

There exists another genetic possibility worthy of consideration. Often herbicide tolerance is a function of the rate of detoxification. A plant is tolerant if it can remove the toxin before it kills. An increase in the rate of detoxification can afford greater protection. One way to achieve this is to duplicate the genes for the detoxifying enzymes. At present it is debated whether such a method of increasing tolerance acts in insecticide/insect relationships.[15]

## 2.2 Selection Pressure of Herbicides: The Unknown Effective Kill

If there *are* genes for resistance, it is empirically evident that the higher the rate of kill, the more rapid the enrichment for resistance. This was especially evident in cases of mine tailings containing heavy metals. Pasture revegetation by resistant strains selected as a result of the metals occurred within a year or two in these cases of near 100% kill (Chapter 15). Most herbicides are applied at rates giving 90 to 95% temporary kill. This is a lower kill rate than usually used for insecticides and fungicides, which suggests that this may be a factor in the slow appearance of herbicide resistance. Higher herbicide rates would still often leave some "escapees" (plants that are not resistant), but would often injure or kill the crops. At 90% kill, without intervening factors, there would be a tenfold increase in resistants per year. If the initial frequency of resistants was $10^{-10}$, resistance would become manifest in 10 years if no other factors were involved. There are at least three intervening biological factors, one related to the herbicide and two related to weed biology. Whereas the farmer and weed scientist ascertain temporary kill after application of a herbicide, these are not sufficient data to determine the selection pressure of a herbicide. For the purpose of measuring selection pressure we must define a new term, "effective kill," which is quite dissimilar to the "knockdown" measured soon after herbicide treatment. Effective kill would be the reduction in weed seed yield over the year or agronomic season brought about by herbicide treatment. The difference between knockdown and effective kill is governed by an interaction between herbicide persistence and late germination of weed seed after the herbicide is gone, as well as the capacity of escapees of any given species to be plastic (i.e., to expand in a "Parkinsonian" manner to "fill the space available" after herbicide thinning). Weed seeds do not germinate simultaneously like crop seed (see ref. 16) and there can be considerable germination after a rapidly degraded herbicide is gone. The "plasticity" of weed stands can be seen by controlling seeding density;[17] heavily thinned stands have a proportionately greater weed seed yield. This has been effectively demonstrated by Isensee et al.[18] They applied massive doses of 13 herbicides and followed the revegetation of 19 species. The herbicide levels were so high that the first effect was a contact killing of all plants, followed later by revegetation. The herbicides had varying selectivities; thus various numbers of plants germinated and grew to maturity in the first 2 years. We plotted their data on resulting plant numbers and area covered (Fig. 17.1). It is apparent that even under these conditions of massive herbicide "overkill," the plants managed to expand (Fig. 17.1). A 50% reduction in plant number resulted in a doubling in area covered.

Broader aspects of the significance of phenotypic plasticity to weed evolution are discussed in depth by Bradshaw.[19]

Because of the paucity of agroecological data on effective kill, one can only estimate the results of applying different types of herbicides (Fig. 17.2). The variation in effective kill among herbicides can be presumed to be quite large, between 40 and 90% based on only a cursory examination of fields in fall, between the time crops are harvested until the time of first frost or tillage. This undergrowth of weeds with seed can be negligible when a highly persistent herbicide is used, or considerable when an ephemeral rapidly degraded herbicide is employed. If crop yields, quality, and ease of harvesting are equal from two such treatments, then as we shall elaborate below, there may be an advantage to the latter treatment. There are additional advantages from general ecological considerations, as well as those resulting from lack of herbicide residues in the soil.

It is clear also that the genetic mechanisms controlling resistance may also govern its spread. If the species is an outbreeder and the trait nuclear, resist-

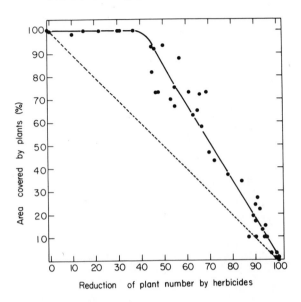

**Figure 17.1** Example of Parkinsonian plasticity: revegetation following massive herbicide treatment. Calculated and plotted from Table 1 in Isensee et al.[18] Plots were treated with 2 to 400 times the recommended agricultural doses of 13 herbicides. Revegetation occurring was measured 1 and 2 years after treatment. Each point represents the effect of one herbicide (in three plots) on plant number and area covered for either the first or second year. The dashed line represents a linear proportionality between plant number and area covered, and the deviation (solid line) represents Parkinsonian plasticity. (From Gressel and Segel[1] by permission.) (Copyright by Academic Press, London.)

ance can be spread more rapidly by pollen than with an inbreeding species. If the trait is maternally inherited, as has been found with triazine resistance (Chapter 13), the pollen transfer should have no effect on the rate of spread.

### 2.3 Competition and Fitness within a Weed Species

When pressure is brought to bear on a wild population to select for a given trait, the selected individuals are less "fit." This has been described by Haldane[20] as the "cost" of selection. Despite any philosophical reservations we may have about the relationship between selection and loss of fitness, it seems to be a general phenomenon. A decrease in fitness is known to occur with bacteria resistant to antibiotics, fungi, insects resistant to their pesticides, and even to rats resistant to warfarin.[21] The genetic explanations for fitness tend to be complex.[22] The "wild-type" weed can be more fit than the selected individual at any of a number of stages in the life cycle because of the following factors: (a) the proportion of seeds germinating at a given time, (b) the rate of germination, (c) success in establishment following self-thinning,[23-25] (d) any of the

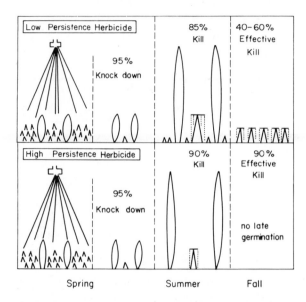

**Figure 17.2**  Effect of herbicide kill and persistence on the effective kill. In this example two selective herbicides kill 95% of the weeds, $\Lambda$, without affecting the crop, $\emptyset$. In both cases the weed grows more than it would have had it not been thinned by herbicides because of a Parkinsonian plasticity. The effective kill, as measured by seed output, is only 90% by midsummer. The nonpersistent herbicide allows late summer germination of weeds as an understory beneath the crops; if viable seed is shed before winter, this further lowers the effective kill and hence the selection pressure. In both scenarios the crop yield may be essentially identical. The necessary actual measurements of effective kill have not, to our knowledge, been reported.

physiological characters resulting in differences in growth rate, (*e*) Parkinsonian plasticity and (*f*) the seed size and yield per flower and per plant. Examples of differences in fitness at various stages of growth can be found among weeds resistant to the triazines. Triazine-resistant *Chenopodium* seed germinates later than the susceptible, whereas resistant *Amaranthus* sp. and *Senecio vulgaris* are less fit later in growth, possibly because of less effective photosynthesis (see Chapters 9 and 10). All of these factors affecting fitness become compounded into the ''reproductive fitness'' of each strain. In the classical case, when the wild type and the selected type are grown separately, there may be no difference between them in yield per unit area (see ref. 26). The differences in fitness may become apparent only when the selected and wild type are competing in nonselective conditions.

The ''mutual interference'' between strains can be measured by interplanting, using methods delineated by De Wit and Vanden Bergh.[27] Whether strain fitness differences were tested between wild types and strains tolerant to heavy metals (see ref. 26) or to triazine herbicides (see Chapter 9), the tolerant types were always less fit. The magnitude of this fitness differential is astounding to the uninitiated. Both with heavy metals[26] and with herbicide-tolerant strains,[28] the resistants were only about half as fit as the wild type, sometimes less. In the cases of *S. vulgaris* and *Amaranthus* sp., the wild type had a slightly higher yield than the triazine-resistant. When triazine-resistant and triazine-susceptible *S. vulgaris* were grown together without herbicide, the susceptible had twice as high a yield of dry matter and a three times greater yield of seed per plant[28] (see Figs. 9.5 and 9.6).

The differential in fitness has many interesting implications. The type selected can continue to exist only where the selection pressure is strong enough to keep the wild type down. This is why antibiotic-resistant bacteria can exist only in hospitals; they rapidly disappear elsewhere. It also means that the herbicide-resistant plant, when it does appear, will have a lower yield than the susceptible plant when the herbicide is not present. If we slightly enrich the population for resistant individuals by treating with herbicide this year, the seed yield from their plants coming up next year will typically only be a fraction of the seed produced by the wild type. Even during the current year there may be differences in yield between the resistant and wild-type strains which are a function of the type of herbicide used. If we use a herbicide that is rapidly degraded, the difference between the resistant biotype and the more fit susceptible wild type will still hold at later stages of growth when the herbicide is no longer present. When a persistent herbicide is used, this fitness differential may not take hold.

We can see from the above that fitness is an important factor in delaying the appearance of resistance. We can also see that it is likely to be more important with the less persistent herbicides.

In some cases the fitness of the resistant biotype can be so low that it poses no danger of competition at all. A corn strain selected for resistance to paraquat is albino (Appendix Table A1). The fitness of the fungus *Cladosporium* resistant to the fungicide triforine [$N,N'$-(1,4-piperazinediylbis(2,2,2-trichloroethylidene))bis-formamide)] is so low that the phytopathologists consider the resistance problem to be nonexistent.[29] Such examples of fungicide-resistant fungi are becoming well recognized.

## 2.4 The Spaced-Out Germination of Weed Seeds

Special attributes of weed seeds make them a unique factor in pesticide resistance. By contrast, most of the factors described in previous sections are common to the interactions of bacteria, fungi, insects, and rodents with their pesticides.

One aspect has been mentioned in the Section 2.2; this is the spaced-out germination *within* a given season. Another very important aspect is the spaced-out germination of seeds from the immense pool of weed seed in the soil (see ref. 30) over a period of years. The population dynamics of entering and leaving the seed bank for different species has been extensively covered.[31] There are strong genetic implications to this slow germination over a period of years. The general case has been looked at mathematically[32] and we have analyzed the particular case of how this reservoir of susceptible seed in the soil affects enrichment for resistance.[1]

Every time we apply selection pressure by treating with a herbicide, we end up by enriching the soil population of weed seed by a certain proportion of resistant weed seed, assuming that there is already some low frequency of resistant individuals in natural populations and these are capable of growth and reproduction (as described above). In the following year only a portion of the resistant seed will germinate, but susceptible seed from the same year and from previous years will also germinate. Thus the proportion of resistant seed germinating will be less than the proportion of resistant seed introduced to the soil in the preceeding year. This smaller proportion of resistant seed leaving the seed bank than that coming in will occur every year a herbicide is used. This means that the soil seed bank will exert a strong "buffering" influence that will help delay the rate of enrichment for resistance. This will be compounded with the delaying effect of the lesser fitness of the resistant strains.

## 3  QUANTITATIVE INTERACTIONS BETWEEN THE FACTORS

### 3.1  The Mathematical Interrelations

If we know which of the factors affecting herbicide resistance are more important quantitatively, we may be able to consider specific modifications of agricultural practices to delay the appearance of resistance. To do this we must somehow quantify the effects of selection pressure, the herbicide persistence, and the seed bank on the rates of enrichment for resistance. This should also allow us to see how modifying any of these parameters will affect the rate at which resistant seedlings will appear. Such an analysis, coupled with guesses at initial frequencies of resistant biotypes, would help estimate the number of years it would take for resistant strains to become noticeable. Such modeling has been done for the evolution of insecticide resistance by Georghiou and Taylor.[8,9] Their insect system lacks an equivalent for the large soil seed reservoir. We have independently integrated the factors governing the rates of evolution of herbicide-resistant weeds, including the effects of the seed bank, with similar conclusions and suggestions. The series of mathematical considerations, simplifications, and estimations (16 equations) from the latter[1] analysis culminated rather simply as

$$N_n = N_0 \left(1 + \frac{f\alpha}{\bar{n}}\right)^n \qquad (17.1)$$

The proportion of resistants of a given species in the $n^{th}$ year of continued treatment of a given herbicide ($N_n$) equals the proportion in the field prior to herbicide treatment ($N_0$), times the factor in parentheses to the power of $n$, the number of years of treatment. $N_0$ itself is a function of the frequency of natural mutation to the resistant biotype and the fitness of such a biotype. Actually, in a consideration of the rate of enrichment for resistance, $N_0$ is immaterial. In this simplified model the mutation frequency need only be considered for the initial frequency. Even in those cases where the herbicide may increase mutation frequency[33] the rate at which resistance appears will not be seriously affected. This is because the selection effect of the herbicide will be many times greater than any slight mutational effect the herbicide may have.

The factor in parentheses governs the rate of increase of resistance. It contains the overall fitness ($f$) of the resistant compared to the susceptible biotype, which in the known cases of heavy metal resistance[26] is about 0.5 and for herbicides is between 0.3 and 0.5.[28] The selection pressure ($\alpha$) is defined as the proportion of the remaining resistants divided by the proportion of remaining susceptibles. Thus, in calculating, if no resistants are killed and 95% of the susceptibles are killed, $\alpha = 1/0.05 = 20$. In the case of heavy metal resistance (Chapter 15), $\alpha$ will approach infinity because of the extremely high rate of kill. If we need a seed set to measure selection pressure and guess that effective kill is 40% then $\alpha$ will be lowered to $1/0.6 = 1.7$. All this is divided by $\bar{n}$, the only complex factor in the equation, the average life span of the species in the soil seed bank. If we were dealing with crops that germinate immediately, then $\bar{n} = 1$ year. With most weed species $\bar{n}$ would be between 2 and 5 years (see ref. 16). Because $\bar{n}$ is in the denominator, it depresses the rate at which resistance will increase. Upon looking at the possible numbers that can be inserted into the equation, we can see empirically that in most cases, the low effective selection pressure $\alpha$ has the greatest effect on reducing the rate of appearance of herbicide resistance. The fitness differential would be a less important modifier and the seed bank would be a major modifier.

## 3.2 Plotting the Model

The interrelations are even more apparent when we use equation 17.1 to generate lines from different scenarios of kill and average seed-bank life span, with the fitness fixed at 0.5 (Fig. 17.3). We have arbitrarily started in year zero from a field frequency of $10^{-10}$ (e.g., where resistance might be inherited by a single recessive gene in a diploid weed). If we have any other initial field frequency, it is possible to move the frequency scale in Fig. 17.3 to fit the case, or to use the right-hand scale. From the *slopes* it is clear that we should be enriching yearly for herbicide resistance. If we follow the slopes, we see that it will take many years until we reach a frequency of resistants that will be noticeable (i.e., more than the 5 to 10% that remain anyway after a herbicide treatment) (Fig. 17.3).

Thus, we will not realize that we are enriching for herbicide resistance until it is upon us. This rate of enrichment is not easy to measure, even in laboratory experiments.

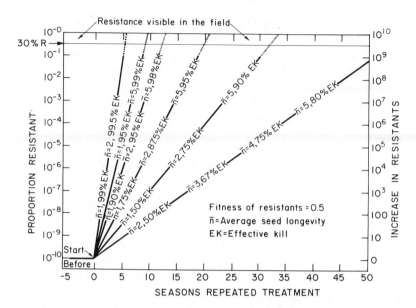

**Figure 17.3** Effects of various combinations of selection pressure ( α ) and average soil seedbank longevity (n̄) on the rates of enrichment of herbicide resistance. The values are plotted for a fitness of 0.5, which seems to be typical of strains resistant to a wide variety of phytotoxic compounds. Resistance would become apparent in the field only when there are more than 30% resistant plants. The scale on the right indicates the increase in resistance from any unknown initial frequency of resistants in the population, whereas the scale on the left starts from a theoretically expected frequency of a recessive monogene. Plotted from equations in ref. 1. (Reprinted with permission from Proceedings of Symposia –IX International Congress of Plant Protection. ©1981 Entomological Society of America.)

To our knowledge, only one group has attempted to make such measurements in the field. Nosticzius et al.[34] started making weed counts in 1974 in cornfields after the fields had been in corn/triazine monoculture for over 5 years. In 1974 the level of unchecked *Amaranthus retroflexus* was about the same as it had been in previous years. Note the rapid fivefold increase *A. retroflexus* at the expense of *Echinochloa crus-galli* and *Digitaria sanguinalis* (Fig. 17.4), which more or less follows the predictions of the model.

When the selection pressure is lower, or when the average seed bank longevity is long, the lesser fitness of resistants will have a greater effect. This is best seen in Fig. 17.5. At the low effective selection pressures of the less persistent herbicides, fitness exerts it greatest effect. Under these circumstances, differential fitness will also have its greatest effect in depressing the rate of appearance of resistance.

Various researchers have tried isolating herbicide-resistant biotypes in experimental field trials, with little success until very recently.[35] It has been possible to isolate partial tolerants by repeated selection (Chapter 12). A typical case

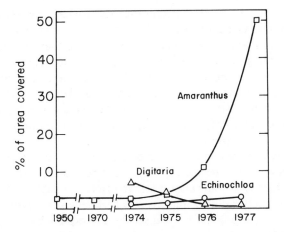

**Figure 17.4** Changes in weed populations in a monoculture cornfield treated annually with atrazine. *Amaranthus retroflexus, Echinochloa crus-galli,* and *Digitaria sanguinalis,* the foremost weeds, were counted. The field in the Agricultural Combine of Babolna, Hungary, received atrazine in corn from 1970. (Data are plotted from Table 1 in Nosticzius et al.[34])

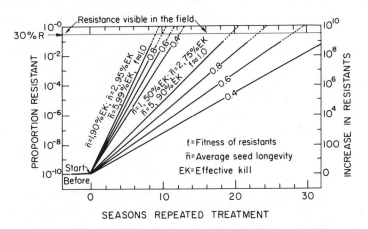

**Figure 17.5** Effect of the lesser fitness of the resistant strains on the appearance of herbicide resistance. Different fitnesses are plotted at various selection pressures ($\alpha$) and seed longevities ($\bar{n}$). Plotted from equations in ref. 1. (Reprinted with permission from Proceedings of Symposia–IX International Congress of Plant Protection. © 1981 Entomological Society of America.)

would be 90% kill of the normal strain and 50% kill of the partially tolerant strain. The selection pressure for such a situation is lower than with resistance (i.e., $\alpha = 0.5/0.1 = 5$). Thus, the rate of increase of partial tolerance to herbicides should be rather slow, in seeming contradiction to the large number of reports on finding so many tolerant biotypes (Appendix Table A1). In any genetic framework of resistance, it is expected that there would be many more individuals which are partially tolerant than those which are resistant, especially if quantitative polygenic inheritance is involved. It is obvious that the $\alpha$ value will vary for each level of tolerance. It is probable that the initial frequency will also vary; the greater the tolerance, the lower the frequency. This is likely because of the usually polygenic mode of inheritance of tolerance (see Chapter 12). In terms of the model equation, this can be summarized as follows: Although $\alpha$ is lower with tolerants, $N_0$, the frequency of a typical strain with low levels of tolerance before spraying, is probably a few orders of magnitude higher than $N_0$ for resistants.

In cases where there are a number of not too specific detoxification pathways interacting with a herbicide, there may be a large number of genes involved, each having a somewhat additive effect. In such a case there would not be the seemingly sudden jump from a herbicide-susceptible field of weeds to a highly resistant population, caused by the rapid exponential increase in resistant populations (Figs. 17.3 and 17.5). Instead, we might expect a short delay of a few years until the individuals with a small partial tolerance are selected for, along the lines of the model. When these partially tolerant individuals became predominant, the chances of their interbreeding will increase considerably. As there is continual selection pressure by the herbicide, one expects a further gradual increase in tolerance when the various genes conferring tolerance begin to interact quantitatively. We plan a further theoretical study of how tolerance is selected for and enriched. Such a gradual enrichment, after a short lag, seems to be the case with *E. crus-galli, Setaria viridis, D. sanguinalis, Veronica persica,* and *Sorghum-halapens* that have built up greater tolerance to atrazine in Italy and France (Table 3.6). There is far more evidence in a more extensive study by Holliday and Putwain.[36] They tested simazine tolerance in 46 populations of *S. vulgaris* from various locales in England, with various simazine treatment histories. There was a sporadic increase in simazine tolerance after the fifth year of treatment (Fig. 17.6). The variability is great and may be in part due to other cultural practices, including the use of other herbicides, yet the regression is statistically significant. It would be of great interest to know if the differences in populations *are* due to detoxification. The "broad-sense heritability" of one population (10 simazine treatments and 52% tolerance at 0.7 kg/ha) was very low: 0.2. This is far lower than described by Faulkner (Chapter 12) for other tolerant systems, and suggests a strong interaction with yet unknown environmental factors.

If we know the initial frequency of resistant individuals before we start spraying, in theory we should be able to estimate the number of years until resistance will appear using a rearrangment of equation 17.1. When various scenarios are plotted out we see that the slopes are not parallel when different combinations of selection pressure and fitness are used (Fig. 17.7). This is done at only one average seed longevity in Fig. 17.7. Thus the initial frequency of resistants has a relatively smaller effect when high-selection-pressure herbicides are used. With low-selection-pressure herbicides, the slopes in Fig. 17.7 are steeper, indicating a greater effect of the initial frequency of resistance on the length of time it will take for 30% resistance to be apparent.

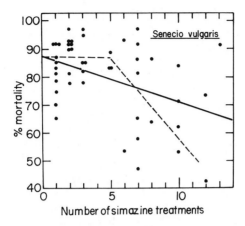

**Figure 17.6** Increased *Senecio vulgaris* tolerance to simazine as a result of repeated treatments. *Senecio* seed was collected at 46 locations in England where the previous treatment history was known. Almost all sites with more than three simazine treatments were also treated at various times with other herbicides. The last treatment of each population was with 0.7 kg/ha simazine under standardized conditions, yielding the results in this figure. The variance due to regression was highly significant ($p < 0.01$). This figure was drawn from data in Figs. 3 and 4 of Holliday and Putwain.[36] The solid line is the calculated regression[36] and the dashed line fits the explanation in the text.

## 4 AGRICULTURAL IMPLICATIONS FROM THE MODELING: EFFECTS OF VARIOUS AGRONOMIC PROCEDURES

We can see from the preceding section that resistance is expected first to the heavily used, repeatedly used, highly persistent, high-kill herbicides. The model "predicts" what we already have learned from earlier chapters; the resistance to triazine herbicides would be rapid and expected. One can have a similar effect by repeatedly using the same nonpersistent herbicide, especially if it has a high rate of kill. This is what happened with the paraquat cases (Chapter 3). What can we learn from the model to prevent a rapid rate of further spread of triazine resistance by concurrent evolution stimulated by selection pressure of this herbicide group? What can we learn to preclude a repeat of the triazine story?

Selection pressure (effective kill) is probably the most important factor affecting the rate of enrichment for herbicide resistance (Fig. 17.3). High effective kill is usually a function of two properties of a herbicide: toxicity and persistence. The latter property will not only govern the effective kill but also the effective fitness. The fitness of the susceptible wild type (as shown in Fig. 17.5) has beneficial relevance only after the herbicide is inactive. Thus herbicides with high effective kill through initial toxicity alone will stimulate a slower evolution to resistance than will herbicides with high kill because of

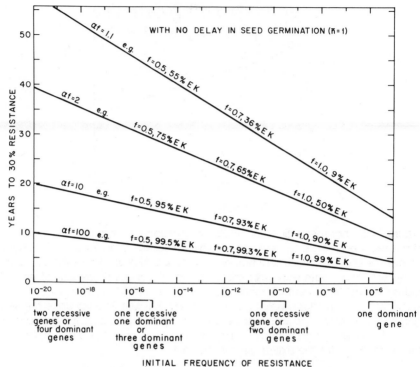

**Figure 17.7** Years to apparent resistance as a function of the initial frequency of resistant mutants in the population and the product of the fitness *f* and the selection pressure ($\alpha$) with $\bar{n} = 1$ (no seed bank). Initial frequencies for various types of presumed mutation frequencies are given, although there is some debate as to whether frequencies of diploid recessives are actually as low as has been calculated.[14] Plotted from equations in ref. 1. (Reprinted with permission from Proceedings of Symposia–IX International Congress of Plant Protection. © 1981 Entomological Society of America.)

toxicity and persistence. We can influence persistence through the choice of herbicides and the choice of herbicide formulations.

The question may be posed as to why triazine resistance appeared in European corn-growing areas (Chapter 3) and in more northern areas in America (Chapter 2) but only to a negligible extent in the much larger U.S. corn belt. Part of the reason may be in crop rotation with concurrent herbicide rotation (Section 4.6) and in a greater propensity to use herbicide mixtures (Section 4.3) in the U.S. corn belt. Assuming the same rotations and the use of only triazine herbicides in all corn-growing areas, another possibility emerges. The U.S. corn belt is at a lower latitude (i.e., has a warmer climate and longer growing season) than the areas where resistance first appeared. Both climate and growing season will decrease the effective kill by triazines, decreasing the rate of enrichment for resistance. In warmer soils the triazines are degraded more rapidly, allowing for earlier mid- to late summer germination of weed seed. Later frosts will allow these weeds to have a higher seed yield. From Fig. 17.3 we see that small decreases in effective kill will substantially delay the appearance of resistance. As yet, there are no comparative ecological studies of effective weed kill in corn at different latitudes.

## 4.1 Germination Stimulants

The soil seed reservoir is an aspect over which we have less control. Over the years, various attempts have been made to stimulate weed seed germination[37] and this approach needs to be assessed in the light of the mathematical models. Many have suggested and looked for ways to force uniform, synchronous germination of weed seeds. Most of these scientists are looking for stimulants and propose what might best be called the "shaving cream" strategy: "Get them up so we can mow them down." If the mowing is done mechanically or by frost and a fixed proportion of seed of all ages germinates, this strategy should have little effect on the rate of appearance of herbicide-resistant weeds. If resistants and susceptibles are equally effected, there would be no change in fitness and thus no effect on the population ratios. If fitness is decreased because resistants suffer relatively more from frost, $\alpha f$ is smaller and resistance is delayed. If the stimulant strongly induces the germination of older weed seed in the soil, the proportion of herbicide-resistant seed germinating could even be greater than that introduced giving a situation akin to $\bar{n} < 1$, which is certainly not desirable.

If the mowing is to be done with a herbicide used as part of the normal herbicide rotation, it may also be counterproductive. The germination induced will strongly increase the effective selection pressure of rapidly degraded herbicides on the species, by having more seed germinate when this herbicide is still active. This will only advance the time when resistance appears. In the terms of the model, a totally effective shaving cream strategy means that the average span of existence in the seed bank will become $\bar{n} = 1$, with all the implications of achieving $\bar{n} = 1$ seen in Figs. 17.3, 17.5 and 17.6.

## 4.2 The Use of Strong Herbicides and Slow-Release Formulations

It is necessary to consider another of the tenets from some weed control texts. The "positive" importance of high selection pressure has been heavily emphasized. It has been stated that: "Hence herbicide applications aimed at control but not elimination are especially dangerous."[38]

There has never been total elimination of a pest with pesticides except in a minuscule number of isolated or localized infestations. It must thus be decided which level of selection pressure (i.e., control) is desirable. More and more practitioners and theoreticians of pesticide use, whether medical use of antibiotics or agricultural use of pesticides, have come to the converse conclusion:

to prevent resistance and help the patient or crop it is best to control the pest just to the level where the desired individuals can compete naturally. We should be wary of the persistent herbicides which are meant to bring about "total kill." Close to total kill may give us totally diminished returns in the not too long run, as resistance will appear more rapidly. Only when complete elimination of the pest species, at least on a localized level, is a feasible and practical objective, will this approach have a positive effect on reduction of resistance to herbicides. The same wariness should be turned to the efforts to achieve slow-release formulations of less persistent herbicides: they only decrease the duration that the fitness differential can be helpful in delaying resistance.

There is now ample experimental evidence that herbicides need not be as persistent as thought. A recent illustrative example has been reported for tomatoes.[39] If tomatoes are transplanted in a weed-free field, there will be no reduction in yield caused by any weeds appearing after 36 days, and those in the field in the first 21 days are also not a problem. Thus tomatoes only need a postemergence herbicide with a 2-week effective persistence to be applied 3 weeks after transplanting. On the other hand, it should be borne in mind that there are some crop situations where high persistence makes sense; those with an open-type vegetation which cannot compete with weeds long into the season and where permanent and total removal of all plant growth is desirable (e.g., railroads, etc.). The use of high-persistence herbicides has caused other problems. A prime example was described recently.[40] In a drought year, corn and sorghum crops were lost in a large geographic area. The presence of residual atrazine in the soil precluded planting winter wheat in the same year.

### 4.3 Herbicide Mixtures

The use of herbicide mixtures is becoming more prevalent, especially with the now more fashionable "no-till" techniques. When this is done, there can be simultaneous selection for resistance to each of the herbicides in each of the susceptible weed species. This happens because each herbicide in the mixture is usually added to kill different weed species. These mixtures may be divided into two types: those where a full rate of each herbicide is used and the weed spectrum killed by each is mutually exclusive, and those where the weed spectrum killed by each is overlapping. The first case is easier to analyze within the model; the model merely has to be applied separately to each weed species. The use of mixtures may broaden the usage of a given herbicide, increasing the number of years over which that herbicide is used. This will bring about a greater enrichment for resistance over any period in the weeds that each herbicide controls. Care must be taken in considering the long run necessity to do this versus the added control one gets from the additional herbicide.

The second situation with overlapping spectra is harder to analyze within the models. Some complex mathematical models have already been suggested to help ascertain if components of such mixtures have only additive effects or have synergic effects.[41,42] If two herbicides are used at their *normal* rates, the frequency of individuals resistant to both should be the compounded frequency of each used separately. If the frequency of resistant individuals to one herbicide was $10^{-10}$ and to the other $10^{-6}$, then when used together the frequency should become $10^{-16}$ when each herbicide has a different site of action. This should considerably lengthen the time to reach resistance; but it may not. If each herbicide has a low effective kill and together they strongly increase the rate of kill, they may well cancel out the effect of lower frequency. This can best be seen in Fig. 17.7. If the two herbicides in this example when used separately are on the line $\alpha f = 10$, it would take 5½ years of use of the one herbicide and 9 years of use of the other to develop resistance. If together they exert a pressure of $\alpha f = 100$, then 30% resistance would be apparent in 7½ years because of the strongly increased selection pressures and not in $5½ + 9 = 14½$ years.

Often, when mixtures are used to kill the same spectrum of weeds, they are used at lower dosages. The mixture gives the same kill as each herbicide used separately. In this case it is not yet clear whether the lower dosages will effect a much lower selection pressure for each herbicide, and thus considerably delay appearance of resistance. This possibly could compensate for the effect of using the herbicide more often in a rotation. If there is a true synergy, there exists the possibility that only plants genetically resistant to both herbicides will survive, which would further delay resistance.

## 4.4  Herbicide Protectants

There is an increasing interest in compounds that protect crops from the action of a herbicide that would otherwise be toxic. These compounds have been termed "safeners," "protectants," or, with semantic inaccuracy, "antidotes."[43] They are used to treat crops before, or along with, the herbicide application. Two such compounds, a diallyldichloroacetamide,* used in combination with a thiocarbamate herbicide; cyometrinil,** applied to sorghum seed to protect it from chloroacetamide herbicide injury, are already marketed, and others are at various stages of development. These compounds will broaden the utilization of some herbicides with expected implications; there will be enrichment for resistance to those particular herbicides if they are used more frequently or in more crops. If the herbicides are highly persistent and already used with other crops in the rotation, it can be detrimental to their long-term usefulness. If the herbicides to be used have a low effective selection pressure, they will pose a far less serious problem, especially if they are to replace highly persistent herbicides.

## 4.5  Herbicide Cross Resistance

The cross resistance between herbicides might be an added complication. In many respects we are more fortunate with herbicides than we have been with

*N,N-diallyl-2,2-dichloroacetamide (R-25788).

** $\alpha$ -[(cyanomethoxy)imino] benzeneacetonitrile (CGA-43089 or Concep).

insecticides. There is a larger variety of herbicides, acting at a wider variety of sites within the organism. Thus less cross resistance is expected. On the other hand, herbicides of one group, the triazines, are used frequently, extensively, and often for different crops. It is to this group that most extensive resistance has appeared (Chapters 2 and 3) and there is cross resistance within the triazines (see Chapters 9 and 10). As the various s-triazines seem to have similar modes of action, the cross resistance between them would be expected. The cross resistance between triazines and ureas that has been reported at the plastid level (Chapter 10) does not seem to act at the level of the plant. Thus its significance here is questionable.

In insects, cases of selected cross resistances to structurally and metabolically *unrelated* insecticides have occurred; when one case appeared, many followed. The weeds developing resistance to triazines remained susceptible to most other herbicides. But continued extensive and exclusive use of the highly persistent triazines, which probably brought about the more rapid evolution of triazine resistance, could reduce the farmers arsenal of this large group of potent herbicides.

## 4.6  Herbicide Rotation

It has been advised that farmers rotate herbicides as well as crops.[44] Will this prevent or delay the appearance of resistance? A related question is: What happens when we stop spraying with a herbicide? This situation is depicted in Fig. 17.8, where the rate of increase was calculated from the model equation but the rate of loss of resistance was calculated on a yearly basis because of the complications imposed by the soil seed reservoir. We have seen that the rate of increase in resistance is affected primarily by the selection pressure, but also depends to some extent on fitness. When herbicide usage is stopped [i.e., there is no longer selection pressure ( $\alpha = 1$ )], the rate of disappearance of resistance is due only to fitness. This rate will generally be much less than the rate of increase obtained with the herbicide.

When we resume herbicide treatment with the same herbicide, we initially induce a more rapid increase, which later parallels the original rate of increase (Fig. 17.8). If we rotate with herbicides of differing groups to which the weeds are not cross resistant, we will delay resistance considerably. A conservative rule of thumb is that one year's cessation of the use of a particular herbicide will delay the appearance of resistance to that herbicide by one year. For example, if it would take 20 years of repeated use to attain field resistance to a given herbicide, it will take 60 years to achieve resistance if the herbicide is used every third year in rotation.

When complete resistance is reached and the particular herbicide is no longer useful, the loss of resistance should be the same as depicted in Fig. 17.8. It must be remembered that the scale on Fig. 17.8 is logarithmic. An example is the data of Conard and Radosevich.[28] It took 6 to 10 years for triazine resistance to appear from a very low (unknown) frequency. In a situation where

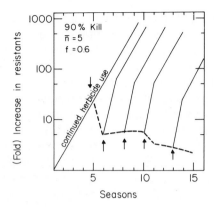

**Figure 17.8** Effect of stopping and restarting herbicide treatment. The ratio of resistant seeds deposited in year $n$ to those deposited in reference year 1 is plotted, giving slopes similar to those in Fig. 17.3. Continual herbicide application is shown with a fitness of 0.6. The lines show the decline in resistance expected when treatment is stopped after year 5 (↓) and the reappearance of resistance when herbicide treatment is restarted after years 1, 3, and 5 (↑). This graph is drawn for the case where 90% of the susceptibles are destroyed by the herbicide, but none of the resistants, and where the average seed longevity is 5 years. (Plotted from equations in ref. 1 and reproduced with permission from Gressel.[2])

$\bar{n} = 1$, Conard and Radosevich[28] have shown that it took six generations for resistance to decrease to a level of 1%, which is many orders of magnitude higher than the initial frequency, and was done in a situation with $\bar{n} = 1$. As $\bar{n} > 1$ in the field, this decrease should be even slower in nature.

After resistance does decay to a hardly noticeable 1%, there will be a rapid resurgence to resistance if the particular herbicide or any of its metabolic relatives are used (see Fig. 17.3 and follow the time from 1% resistance to complete resistance). This may be what has happened with insects when DDT was reintroduced, although there it has been postulated that fitness of the resistant strains had increased due to interbreeding with the wild type.[45] Thus, if we return to what has been said in previous chapters about the triazines, we see that if resistance does become widespread, the loss of this whole group of compounds may essentially become irrevocable, at least as a single herbicide for the control of a specific, newly resistant species of weeds.

## 4.7 No-Till Practices

The use of no-till agricultural techniques requires about a two- to threefold increase in the amounts of herbicides that will be used. This could mean that there will be an overall two- to threefold increase in the rate of enrichment for herbicide resistance. It is then obligatory to choose judiciously herbicides that will not bring about rapid evolution of resistance, especially if the number or effectiveness of available herbicides are reduced because of tillage reduction. Unfortunately, there is insufficient information to make such judicious choices. Whereas we do know much about persistence and kill, we lack the ecological knowledge on effective kill. It may be an insurmountable task to find out the initial frequency of resistance with field or laboratory techniques using whole plants. The initial frequency for picloram resistance was obtained using cell culture techniques (Chapter 14), and these techniques have many uses in herbicide research.[46,47] Another way is to estimate initial frequencies indirectly from information on the modes of action of the herbicide. These data are also too often lacking. If there are multiple sites of action, it might require a simultaneous multigene mutation to resistance, an occurrence of extremely low frequency. Such herbicides have distinct advantages.

## 5 CONCLUDING REMARKS

In conclusion, it is hoped that this analysis of the relationships between herbicides and weed evolution will teach us about what we must start considering in depth, as well as the kinds of data that are needed to ascertain which group of herbicides will be the next to bring about widespread resistance. Within the two groups of herbicides that are probably the most heavily used, we see the immense difference in properties, with the predictable difference in results. Resistance has not appeared to the phenoxy herbicides despite monoculture with continuous use of 2,4-D or its phenoxyacid relatives for more than 30 years. This rapidly degraded group has a much lower effective kill than the triazines. The phenoxy herbicides also have a completely different and more complex, probably multisite, mechanism(s) of action. Resistance to triazines has evolved in 6 to 10 years in a wide variety of locations, climates, and soil types (see Chapters 2 and 3). On the other hand, triazine resistance or tolerance has not yet appeared everywhere and with all weed species where it has been used repeatedly 10 or more times. Does this mean that it will not appear? This is hard to answer, as the increase in resistance is exponential (Fig. 17.3) and the year or two before a field is completely resistant, the level is far below anything detectable or normally observable in the field or laboratory.

We feel that our simple model contains the major parameters affecting evolution to resistance and correctly suggests the interrelationships. We must now obtain the more meaningful, more quantitative data to insert into these equations. We hope that we have demonstrated the potential importance of obtaining such data and the predictive value they would have about the useful agricultural lifetime of different herbicides.

It is obvious that as we use pesticides we enrich for resistance to these pesticides. The logarithmic rate for enrichment means both that we will not see resistance the season before it is a serious problem and, also, that once resistance has appeared to a certain herbicide in one location, its appearance elsewhere will be rapid due to concurrent evolution alone. We can no longer hide our heads in the sand saying "it will not happen," as resistance has occurred and it will continue occurring. Anything we can agronomically and economically do to keep up yields and delay herbicide resistance is worthwhile.

There are some crops and uses that really need persistent herbicides because of an open growth pattern or a requirement for bare ground. The persistent herbicides should be *saved* or used more exclusively for these special needs and not for crops that thoroughly choke out weeds after 6 weeks. Let the crops participate in holding down the weeds and thereby delay the loss of good herbicides. By developing better and more selective postemergence herbicides this can also help in decreasing the need for persistence, as they can be applied much later, and then only when necessary.

In medicine, prophylactic treatments with antibiotics have been reduced because of resistance. Where prophylaxis is necessary, antibiotic rotations are used. Similar approaches are being taken to avoid or reduce resistance to other pesticides[11] and we should seriously consider the same with herbicides.

We must try to rethink other agronomic practices that help to hold down weeds. Even irrigation practices can be of help. In orchard and tree crops, if water is applied only beneath the dense foliage of trees, the weeds between the rows will die of thirst. The same practice applies for fertilization. Unsprayed fence rows could be a good supply of susceptible weed seed, which may lower the effective selection pressure without affecting yield.

Judicious use of herbicides as an adjunct to, rather than a total replacement for other good practices will keep yields up, weeds down, and will permit the use of a given herbicide for a longer period.

## REFERENCES

1  J. Gressel and L. A. Segel, The paucity of plants evolving genetic resistance to herbicides: possible reasons and implications, *J. Theor. Biol.,* **75**, 349 (1978).

2  J. Gressel, Factors influencing the selection of herbicide-resistant biotypes of weeds, *Outlook Agric.,* **9**, 283 (1978).

3  J. Gressel, Genetic herbicide resistance: projections on appearance in weeds and breeding for it in crops, in *Plant Regulation and World Agriculture,* T. K. Scott (Ed.), Plenum, New York, 1979, p. 85.

4  P. Grignac, The evolution of resistance to herbicides in weedy species, *Agro-Ecosystems,* **4**, 377 (1978).

5  K. Ueki and Y. Yamasue, Intraspecific variation of weeds in herbicide susceptibility and resistance development (in Japanese), *J. Pestic. Sci.,* **3**, 445 (1978).

6  R. J. Holliday, P. D. Putwain, and A. Dafni, The evolution of herbicide resistance in weeds and its implications for the farmer, in *Proc. Br. Crop Prot. Conf. Weeds,* 937 (1976).

7  J. Gressel, The avoidance of herbicide resistance, in *Proc. Symp. 9th Int. Congr. Plant Prot.,* Vol. 1, T. Kommendahl (Ed.), Burgess, Minneapolis, Minn., 1981, p. 262.

8  G. P. Georghiou and C. E. Taylor, Genetic and biological influences in the evolution of insecticide resistance, *J. Econ. Entomol.,* **70**, 319 (1977).

9  G. P. Georghiou and C. E. Taylor, Operational influences in the evolution of insecticide resistance, *J. Econ. Entomol.,* **70**, 653 (1977).

10   H. G. Baker, The evolution of weeds, *Annu. Rev. Ecol. Syst.* **5**, 1 (1974).

11   T. Kommendahl (Ed.), *Proc. Symp. 9th Int. Congr. Plant Prot.* Burgess, Minneapolis, Minn., 1981.

12   L. G. Holm, University of Wisconsin, Madison, personal communication, 1960.

13   B. J. M. Ogawa, J. D. Gilpatrick, and L. Chiarappa, Review of plant pathogens resistant to fungicides and bactericides, *FAO Plant Prot. Bull.,* **25**, 97 (1977).

14   K. L. Williams, Mutation frequency at a recessive locus in haploid and diploid strains of a slime mould, *Nature (Lond.),* **260**, 785 (1976).

15   J. Baker, Evolution by gene duplication in insecticide-resistance *Myzus persicae, Nature (Lond.),* **284**, 577 (1980).

16   H. A. Roberts and P. M. Feast, Seasonal distribution of emergence in some annual weeds, *Exp. Hort.,* **21**, 36 (1970).

17   J. L. Harper and D. Gajic, Experimental studies of the mortality and plasticity of a weed, *Weed Res.,* **1**, 91 (1961).

18   A. R. Isensee, W. C. Shaw, W. A. Gentner, C. R. Swanson, B. C. Turner, and E. A. Woolson, Revegetation following massive application of selected herbicides, *Weed Sci.,* **21**, 409 (1973).

19   A. D. Bradshaw, Evolutionary significance of phenotypic plasticity in plants, *Adv. Gen.,* **13**, 115 (1965).

20   J. B. S. Haldane, More precise expressions for the cost of natural selection, *J. Genet.,* **57**, 351 (1960).

21   G. G. Partridge, Relative fitness of genotypes in a population of *Rattus norvegicus* polymorphic for warfarin resistance, *Heredity,* **43**, 239 (1979).

22   L. Van Valen, Selection in natural populations: III. Measurement and estimation, *Evolution,* **19**, 514 (1965).

23   M. Westoby, Self-thinning driven by leaf area not by weight, *Nature (Lond.),* **265**, 330 (1977).

24   R. Dewsberry, Plant self-thinning dynamics, *Planta,* **136**, 249 (1977).

25   A. R. Watkinson, Density-dependence in single species populations of plants, *J. Theor. Biol.,* **83**, 345 (1980).

26   D. A. Hickey and T. McNeilly, Competition between metal tolerant and normal plant populations: a field experiment on normal soil, *Evolution,* **29**, 458 (1975).

27   C. T. De Wit and J. P. Van den Bergh, Competition between herbage plants, *Neth. J. Agric. Sci.,* **13**, 212 (1965).

28   S. C. Conard and S. R. Radosevich, Ecological fitness of *Senecio vulgaris* and *Amaranthus retroflexus* biotypes susceptible or resistant to atrazine, *J. Appl. Ecol.,* **16**, 171 (1979).

29   A. Fuchs, S. P. de Ruig, J. M. van Tuyl, and F. W. de Vries, Resistance to triforine: a nonexistent problem?, *Neth. J. Plant Pathol.,* **83**, 189 (1977).

30   H. A. Roberts and P. A. Dawkins, Effect of cultivation on the numbers of viable weed seeds in soil, *Weed Res.,* **7**, 290 (1967).

31   G. R. Sagar and A. M. Mortimer, An approach to the study of the population dynamics of plants with special reference to weeds, *Appl. Biol.,* **1**, 1 (1977).

32   D. A. Levin and J. B. Wilson, The genetic implications of ecological adaptations in plants, in *Structure and Functioning of Plant Populations,* Verhandelingen der Koninklijke Nederlandse Akademic van Wetenschappen, Afdeling Natuurkunde, Tweede Reeks, deel 70, 1978, p.75.

33   J. Plewa, University of Illinois, Urbana, personal communication, 1979.

34   A. Nosticzius, T. Muller, and G. Czimber, The distribution of *Amaranthus retroflexus* in monoculture cornfields and its herbicide resistance (in Hungarian), *Bot. Kozl.,* **66**, 299 (1979).

35  R. J. Holliday and P. D. Putwain, Evolution of resistance to simazine in *Senecio vulgaris* L., *Weed Res.,* **17**, 291 (1977).

36  R. J. Holliday and P. D. Putwain, Evolution of herbicide resistance in *Senecio vulgaris;* variation in susceptibility to simazine between and within populations, *J. Appl. Ecol.,* **17**, 799 (1980).

37  P. K. Fay and R. S. Gorecki, Stimulating germination of dormant wild oat (*Avena fatua*) seed with sodium azide, *Weed Sci.,* **26**, 323 (1978).

38  J. L. Harper, Ecological aspects of weed control, *Outlook Agric.,* **1**, 197 (1957).

39  G. H. Friesen, Weed interference in transplanted tomatoes (*Lycopersicon esculentum*), *Weed Sci.,* **27**, 11 (1979).

40  O. C. Burnside and M. E. Schultz, Soil persistence of herbicides for corn, sorghum and soybeans during the year of application, *Weed Sci.,* **26**, 108 (1978).

41  P. M. Morse, Some comments on the assessment of joint action in herbicide mixtures, *Weed Sci.,* **26**, 58 (1978).

42  J. C. Streibig, A method for determining the biological effect of herbicide mixtures, *Weed Sci.,* **29**, 469 (1981).

43  F. M. Pallos and J. E. Casida (Eds.), *Chemistry and Action of Herbicide Antidotes,* Academic Press, New York, 1978, p. 171.

44  A. L. Abel, The rotation of weedkilling, *Proc. 1st Br. Weed Control. Conf.,* 249 (1954).

45  J. Kelding, Persistence of resistant populations after the relaxation of the selection pressure, *World Rev. Pest Control,* **6**, 115 (1967).

46  J. Gressel, A review of the place of *in vitro* cell culture systems in studies of action, metabolism and resistance of biocides affecting photosynthesis, *Z. Naturforsch.,* **34c**, 905 (1979).

47  J. Gressel, Uses and drawbacks of cell cultures in pesticide research, in *Plant Cell Cultures,* F. Sala et al. (Eds.), Elsevier/N. Holland, Amsterdam, 1980, p. 379.

# Summary of Accomplishments Conclusions, and Future Needs

## H. LeBARON

Biochemistry Department
CIBA-GEIGY Corporation
Greensboro, North Carolina

## J. GRESSEL

Department of Plant Genetics
The Weizmann Institute of Science
Rehovot, Israel

## 1 THE NATURE AND CHARACTERISTICS OF HERBICIDE RESISTANCE

From the research conducted to date on herbicide resistance in plants as reviewed and summarized in this book, we have learned much about where, when, why, and how it has occurred, what to do to prevent it from occurring elsewhere, and how to reduce its consequences once it occurs. Perhaps of even greater importance, we have learned some good reasons and ways that humankind may benefit materially and scientifically from the practical applications or use of herbicide resistance.

Although insect and disease resistance to chemicals is possibly the greatest problem in the continuous use of commercial insecticides, acaricides, and fungicides, and a serious deterrent to the development of new pesticides, there are important reasons why resistance of weeds to herbicides can avoid becoming major problems, and can even be more of a blessing than a curse. Over human history and our continuous fight with weeds, they have been called many names or defined by many terms, such as, "a weed is a plant that is growing where it is not wanted." Perhaps more today than ever before, it is appropriate for us to be optimistic and generous by stating that "a weed is a plant whose virtues have not yet been discovered." Some plant breeders, geneticists, plant physiologists, and others have recognized weeds as a useful reservoir of genetic material or as plants to study. Not until the concepts of modern genetic engineering are further developed, including the possible transfer of traits from one plant to another (even one that is genetically unrelated) via recombinant DNA, which we expect will come to fruition within the next few years, will it be possible for us to use weeds to their true potential. Certainly, much technology must yet be developed before this vision becomes a reality, but great scientific effort and progress are being made; the necessary breakthroughs should just be a matter of time.

There are no easy answers to the problems we face, scientifically or in other phases of life. In a world of finite resources, where hunger is so common, and with a growing population, we must make some difficult decisions. This has never been more urgent than it is today. Technology has made many things possible that are not economically feasible.

In his recent book *Instant Evolution: We'd Better Get Good at It,* Thomas P. Carney[1] points out that evolution is a matter of changing genetic makeup. "Instant evolution" is, therefore, a thing of the present. We are, indeed, becoming as gods. While this status implies unprecedented opportunities to "create" for the benefit of humankind, it also brings with it uneasiness and fear based on our past abuses of such technology and power. Carney appropriately stresses our need to act as gods in a sensible way. He concludes that what the world will be like in the future will not be determined primarily by sensational new scientific advances. Rather, it will depend on how well all of us can organize our efforts and agree on common objectives.

Herbicide resistance in plants has taught us once again in a very dramatic way that plants, especially weeds, have great ecological adaptability and are always evolving or compensating for adverse environmental conditions. We must never become complacent. No matter how marvelous are our inventions or how spectacular are our successes, we are destined to eat bread by the sweat of our brow. The price of success and survival against our pests is eternal vigilance.

The cartoon shown here is not intended to imply a pessimistic attitude on the part of the authors. We do not believe that we need to return to the former days of hand hoeing and labor-intensive farming because of herbicide resistance. Neither is the question asked entirely in jest or with "tongue in cheek." Rather, we need to keep the hoe and all other weed control methods in good repair and available.

Within the area of plant protection with chemicals, much has been done and more can be done in the future to avoid and ameliorate resistance to pesticides. We have learned that we cannot simply develop better and more powerful pesticides to counteract resistance. We must, at the same time, find ways to utilize better the pesticides we have, while not depending entirely or too much on any one, or even all the pesticides we have, to do the total job. We must retain or return to all earlier means of pest control as we develop better, more efficient, and more economical chemicals.

"Is this for those herbicide-resistant weeds?"

With apologies to S. Harris (American Scientist, July-August 1980, p. 419), who drew and captioned this cartoon differently.

The discovery of triazine-resistant weeds was not the first such development. As Chapters 2 and 3 show, there have been several cases of intraspecific resistance to 2,4-D and other herbicides, and many examples of evolution toward herbicide tolerance. Triazine resistance has been covered most extensively in this book because of the importance of this group of herbicides, the numbers and relatively wide distribution of triazine-resistant weeds, the extensive research and knowledge available on the nature of triazine resistance, and the scientific and practical applications available. Much more research is needed on mechanisms of weed resistance to other herbicides (e.g., trifluralin resistance in *Eleusine indica*) and their applications, as well as to the triazine herbicides.

Since the first report of a triazine-resistant weed was published in 1970, the number and areas of distribution have increased markedly. Even since Chapters 2 and 3 were first completed, there have been additional resistant species and areas of infestation reported. At the present time, triazine resistance has been confirmed in at least 18 genera and 30 species of weeds (see Table 18.1), located in at least 23 states (mostly in the northern United States), four provinces of Canada, and seven countries of Europe, and the list is expanding every year. An impressive compendium of all the confirmed reports appears in Appendix Tables A1 and A2.

Recent unconfirmed reports from Hungary[1a] indicate that several additional weed species, including *Echinochloa crus-galli, Amaranthus blitoides, Fallopia convolvulus, Hibiscus trionum, Reseda lutea, Raphanus raphanistrum, Setaria glauca,* and *Equisetum arvense,* may have developed or are developing resistance following continuous use of atrazine alone. Further work is needed to determine if these are the result of shifting populations of more tolerant spe-

## Table 18.1  Confirmed Distribution of Triazine-Resistant Weeds
### (As of November 1981)

| Genus | Species | Located in Number of: States (US) | Provinces (Canada) | Countries (Europe) |
|-------|---------|---------------------|--------------------|--------------------|
| **Broadleaf Species** | | | | |
| *Amaranthus* | *arenicola* | 1 | | |
| *Amaranthus* | *hybridus* | 7 | | 1 |
| *Amaranthus* | *lividus* | | | 1 |
| *Amaranthus* | *powellii* | 1 | 1 | |
| *Amaranthus* | *retroflexus* | 1 | 1 | 4 |
| *Ambrosia* | *artemisiifolia* | | 1 | |
| *Atriplex* | *patula* | | | 2 |
| *Bidens* | *tripartita* | | | 1 |
| *Brassica* | *campestris* | | 1 | |
| *Chenopodium* | *album* | 5 | 2 | 4 |
| *Chenopodium* | *ficifolium* | | | 1 |
| *Chenopodium* | *missouriense* | 1 | | |
| *Chenopodium* | *polyspermum* | | | 2 |
| *Chenopodium* | *strictum* | | 1 | |
| *Erigeron* | *canadensis* | | | 1 |
| *Galinsoga* | *ciliata* | | | 1 |
| *Kochia* | *scoparia* | 11 | | |
| *Polygonum* | *convolvulus* | | | 1 |
| *Polygonum* | *lapathifolium* | | | 1 |
| *Polygonum* | *persicaria* | | | 1 |
| *Senecio* | *vulgaris* | 3 | 1 | 1 |
| *Solanum* | *nigrum* | | | 3 |
| *Stellaria* | *media* | | | 1 |
| **Grass Species** | | | | |
| *Bromus* | *tectorum* | 5 | | |
| *Echinochloa* | *crus-galli* | 1 | 1 | 1 |
| *Panicum* | *capillare* | 1 | 1 | |
| *Poa* | *annua* | 1 | | 1 |
| *Setaria* | *faberi* | 1 | | |
| *Setaria* | *lutescens* | | 1 | |
| *Setaria* | *viridis* | 1 | | |
| **Total** 18 genera | 30 species | 23 states[a] | 4 provinces[a] | 7 countries[a] |

[a]These numbers are not equal to the column totals. More than one triazine resistant biotype has occurred in some states, provinces and countries.

cies or if they are actually resistant biotypes of species previously controlled. In France, also, there are unconfirmed reports of new resistant biotypes of additional weed species developing. Obviously, this concluding chapter of our book is not the final chapter on the subject.

There has been very little evidence of dual resistance: where the same weeds have developed resistance to two or more different types of herbicides. When isolated chloroplasts from triazine-resistant weeds are treated with other herbicides, they show a slightly higher level of tolerance to most PS II herbicides or those with nitrogen in the ring (Chapter 10). There have been some claims that triazine-resistant weeds have also developed resistance to urea-type herbicides.

As many plant scientists had earlier assumed that both of these types of PS II herbicides were affecting the same binding site, it would not be surprising if such a dual resistance occurred rather readily in the same plants. Fortunately, these similar acting but different herbicides are seldom used together or in rotation in selective weed control in crops. Triazine and urea herbicides are often used in combinations or rotation along railroads, roadsides, and other noncropland sites for nonselective control of weeds. The possibility of overlapping domains on the same binding site are discussed in Chapter 10, and a very theoretical consideration has recently been published by Shipman.[2]

Recent but yet unverified reports indicate that some of the annual grasses along railroads in southern Michigan and possibly in other northeastern states are resistant to previously effective rates of both types of herbicides. However, it is in such situations as railroads, roadsides, rangeland, industrial sites, and no-till culture that herbicide resistance will be most serious.

We are fortunate that we have a large number of herbicides with various modes of action. This will probably become more important in future years, especially if weeds develop resistance to more than one type of herbicide.

In terms of the seriousness of herbicide resistance in weeds, we are fortunate for the following reasons:

1 Weeds normally require the full growing season to reproduce and only complete one life cycle per year.
2 Weeds do not move about as freely as do insects and disease pathogens.
3 Apparently, almost all triazine herbicide resistance and possibly some other weed resistances are maternally inherited and thus not transmitted by pollen.
4 There are usually several different types of herbicides which can be used in combination or as alternative means of weed control. Cross resistances between chemically unrelated groups, so prevalent with other pesticides, is obviously a rare event with herbicides.
5 Crop rotations, which usually result in the use of different herbicides, avoid or delay the development of herbicide resistance.
6 Cultivation or other tillage tends to delay or prevent resistant weeds from developing.

**7**  Any resistant weeds that develop will be diluted and competed against by all the soil banks of normal susceptible seed.

**8**  Resistant weeds are likely to be less competitive compared to the susceptible biotypes of the same species.

For these and other reasons, herbicide-resistant weeds should be manageable or retained within a reasonable limit. At least in the United States, the total area of land or crops infested with triazine-resistant weeds is still relatively small and does not seem to be expanding rapidly. On the other hand, we understand that the area of infested corn increased almost exponentially in Hungary and Austria after triazine-resistant weeds first appeared, until about 80 to 100% of the corn cropland is now affected. We are not able to determine all the reasons for this difference, but it can be assumed that the growers were less flexible in their use of available herbicides, tillage, crop rotations, and so on. It may also be due somewhat to the less responsive research and extension programs compared to those in the other countries.

Some additional interesting and probably significant characteristics of triazine resistance include:

**1**  Species that have developed resistance to triazines were normally very sensitive and easy to control with these herbicides prior to the resistance.

**2**  Resistance is generally absolute (e.g., no effect at extremely high rates of herbicide).

**3**  Triazine-resistant biotypes have generally shown cross resistance to all *s*-triazine herbicides.

**4**  There also seems to be related degrees of resistance or tolerance in these biotypes to the asymmetrical triazinones (e.g., metribuzin), ureas, and many other N-containing photosynthetic inhibitors.

**5**  The resistant biotypes seem to develop an even greater degree of sensitivity than the triazine-susceptible biotypes to herbicides having other modes of action (e.g., dinoseb and ioxynil).

**6**  The $I_{50}$ values for each triazine herbicide is quite constant between all sensitive plants. Between resistant weeds, the $I_{50}$ values vary with the species.

**7**  Some data suggest that there could be more than one chloroplast binding site for PS II herbicides.

**8**  A very low frequency of the triazine-resistant biotype was present among the species population when the triazine herbicide was first applied.

**9**  There is no evidence that the herbicides were the direct causes of resistance or that they have had any mutation effect on the natural susceptible population.

**10**  Resistance has developed mostly in areas where triazine herbicides have been used alone and repeatedly for several years, especially where little or no tillage has been used.

**11** Once resistant plants have developed and infested an area with resistant seeds, we will virtually never be able to eradicate the source of resistance.

**12** Other agronomic and ecological factors will interact with herbicides to influence the occurrence and rate of resistance.

In spite of all the research conducted on this subject in recent years, there are many important questions unanswered and much opportunity for further good work to be done to more completely understand the occurrence and nature of resistance. Many of the factors mentioned above, as well as others commented on in this book, should be researched.

We hope that this book has engendered an awareness of the need for early detection of resistance, for rapid verification of the resistance, for understanding the resistance or mechanisms of tolerance and their genetics. In Chapters 2 and 3 on distribution, there are too many "folk tales" on the appearance of resistance, and the quick action by farmers to hand, mechanically, or chemically rogue the putative culprits. We hope that farmers can instead be induced to quickly contact the extension or research specialists of both the public and chemical sectors. They, in turn, must be set up with quick assays and be able to plant seed and verify the accuracy and extent of the resistance. This way we will know ever so much more about when and how resistance begins, and have better advance warning about the herbicides next in line for tolerance and resistance problems. The war against weeds is no different than any other. Early detection of the enemy, his plans and strategies, are paramount for constructing reliable defense systems.

Consideration should be given to establishing a voluntary compliance to follow certain recommended procedures (e.g., herbicide combinations, alternative herbicides, cultivation, crop rotations), when herbicide-resistant weeds do develop in order to control the spread and severity of the infestations. Even limited legal enforcement could be justified in cases where the resistant biotype is recognized as a serious and growing problem. A temporary quarantine of an area may be desirable if the problem is new and relatively localized.

## 2 WEED ECOLOGY AND HERBICIDES

Much of the research reported in this book points to the greater need for studying population biology in agriculture. As was described in Chapter 17, we do not even know how much we reduce weed seed production by various herbicides and, thus, have no idea of the true percent kill imposed. Additional information is needed as a "data base." We should then have continuing follow-up studies to catch changing patterns of resistance as an early warning system for research, extension, farmers, and industry. The kinds of studies by Putwain (see Fig. 17.5), where increasing tolerance was found by correlating the spray history with all concerned, are of great importance, but apparently have not been done before. How much longer will we have untreated control areas? How many other countries or areas of the world have the excellent long-term studies as have been reported in Denmark (Chapter 4)? Without such studies it would have been hard not to come to the conclusion that herbicides are to blame for all or most of the changing distributions of weeds. So many other agricultural practices have changed over this same time period. On the other hand, it is obvious that the continuing application of herbicides to populations with very low frequencies of resistant biotypes would exert great selection pressure favoring these plants, leading rapidly to a relatively resistant population.

None of our authors seemed to have the data base necessary to answer the question about whether biotypes of *Amaranthus powellii* and *A. hybridus* were always resistant to triazines and took over from triazine-sensitive *A. retroflexus,* or whether they too were triazine sensitive. It is unfortunate that weed scientists or taxonomists failed to confirm adequately the identity of some of the early resistant biotypes (e.g., *Amaranthus* spp.) with voucher specimens. Why is it that the two presumably minor *Amaranthus* species within a population of *A. retroflexus* would become resistant more quickly (if they had been sensitive)? Will many or most herbicide susceptible species evolve resistance? Bradshaw (Chapter 15) found that about one-third of the plant species developed resistance to heavy metal mine tailings.

There are other questions we can ask the population geneticists. Is it more than a chance correlation that weeds, no matter how pestiferous, introduced from the Old World to the New World, or vice versa, have developed less resistance in their newer habitat (Chapters 2 and 3)? Could this be due to a narrower genetic base of the few introduced plants, as we have hypothesized in this book (see Chapter 3)?

Triazine-resistant weed material has become quite fashionable as plants to study, both for academic and applied research. Many additional publications on it are appearing as this book goes to press. Some will undoubtedly cast doubts on information in this book; hopefully, most will confirm and extend it. It is obvious that the resistant plant material is and will be useful for a long time to come, as will other resistant and tolerant weed material that is bound to appear.

The research and events leading up to this book are not merely a story of natural and scientific developments, but also involve a human drama with plenty of morals in it worth considering and learning from. It is of a young researcher who bucked the "establishment" or company philosophy to study the ecology and physiology and, with much effort, managed to coopt the biochemists into studying the biochemical mode of action of triazine resistance. Legend has it that pressures on journals by scientific peers even delayed the first publications of the triazine story and there was much loud discussion at scientific meetings. Openness and objectivity are essential from all in such phenomena. Although it is not surprising that the "chemical" community wished away the problem, had they instead supported the research efforts, we might now have both fewer resistant weed problems and more herbicide-resistant crop varieties. Both could benefit the herbicide and seed producers.

## 3 POSITIVE USES OF RESISTANCE

The great biochemist Ephraim Racker has often noted that "it isn't so bad if you fall flat on your face in science; as long as you pick something useful off the floor when you get up." Much of this book has described the agronomic practices that have "tripped" us into helping bring about resistance, especially to one of the best and most essential groups of herbicides. What can "we pick up off the floor?" Chapter 13 has described one way we might transfer the triazine-resistant genes from resistant weeds to crops. The paraquat-tolerant Kent perennial ryegrass was originally considered a "weed" problem, as it was not possible to control the plant with paraquat before direct-drilling cereals. How this was "picked up" and used as genetic material for breeding and selecting even more tolerant perennial ryegrass varieties which could then be treated with the herbicide for weed control has been detailed in Chapter 12. There are many other possibilities that might be worth considering. The triazine-resistant *Stellaria media* described in Chapter 3 does not seem to compete with corn. Could this *Stellaria,* after being further weakened by other genes, be bred to make a good erosion-preventing cover crop in triazine-treated orchards and vineyards? This might be especially useful in hillside plantations.

The sections on breeding for resistance (Chapters 12 to 14) have well pointed out the cost and registration advantages of selection for resistance and tolerance in crops. The chemical industry may well look askance at such an approach. By the time a herbicide has proven itself in the field and the breeders become interested, at least half its patent life has elapsed. When the breeders finally have a variety ready to release years later, the herbicide is off patent, and is often by this time available from completely different manufacturers, usually at lower cost. This is one reason why industry continues looking for more and better herbicides to solve such problems instead of more uses for those presently available. Such a documented case worth describing, as it illustrates the situation, is the widely used soybean herbicide metribuzin. Only after release was it found that there was considerable varietal variation in metribuzin tolerance. The initial discovery was made by a farmer treating a field of the sensitive variety Semmes.[4] In spite of the fact that this imposes a limitation on the breadth of metribuzin use, the manufacturer was not at that time interested in participating in a program to select for metribuzin-tolerant varieties. Times have changed, and they now are interested. The most logical source of selections for herbicide resistance would be the large seed companies with their plant breeders and the economic incentive to provide varieties having features that the farmer will want. Metribuzin tolerance is inherited by a single recessive gene,[5] so transfer to other varieties should be simple. In such research the seed companies might sometimes be at cross purposes with the primary chemical manufacturers. The farmers would like their crops tolerant to the least expensive herbicides. There seems to be no evidence in the published literature that the seed companies have yet had any direct involvement in selection for resistance. The screening for metribuzin tolerance was left to the public agricultural research establishment.[5,6] Their results merely state which varieties could be used.

Soybeans are of special interest in screening for herbicide tolerance, as they have differential tolerances to many herbicides (see Appendix Table A1). The only case of a soybean breeder being presently concerned with metribuzin tolerance is E. E. Hartwig at Stoneville, Mississippi.[4] 'Tracy' was released as being more tolerant to 2,4-DB than other soybean cultivars, but it is sensitive to metribuzin. He has made a selection from 'Tracy' which is supposed to be tolerant to 10 times the rate of metribuzin required to kill Tracy. This cultivar is being released as 'Tracy M.' Preliminary research indicated that the mechanism of selectivity is based on differential rates of metabolism.[7]

The onus of screening varieties for resistance and for breeding has been left to the public sector, which has its limitation. The extreme case can be seen with the selection for resistance using the sophisticated cell culture techniques. In Chapter 17 there is a description of why selection for resistance, with its extremely low frequency, would be nigh impossible in the field but would be possible with cell cultures. There are now three clear-cut cases where we have a crop bred for resistance using cell culture techniques (Chapter 14). These were performed in university laboratories and are of great importance because they showed that "it can be done." In each case, tobacco was chosen as the crop primarily because it was the easiest tissue with which to work. No market survey was first performed to see if tobacco needed another herbicide. Thus we now have tobacco lines that are fully resistant to picloram, bentazon, and phenmedipham. To our knowledge, no one has yet made yield trials, and no one is attempting registration or commercial development of any of these tobacco plants. No one seems to be interested even in studying the physiological and biochemical modes of resistance to these three herbicides in the well-characterized genetic material. The research establishment of the company producing one of the herbicides did not know of the tobacco resistance to their herbicide a year after publication in the scientific literature.

The appearance of varietal differences has also made researchers cognizant of the need to check cultivars for tolerance to herbicides which are widely used on the crop before release. An example in cotton was a cultivar developed at Texas A&M University that "grew up" on trifluralin. When released and grown in areas of the state that used prometryn for weed control, it was found to be sensitive to this herbicide.[4]

All of the above examples point to the need for a more active interfacing among the chemical industries, the seed companies, and the breeders. Especially for the more promising herbicides, this should start at least by the time the company is in the late stages of development, prior to the release of the herbicide for commercial use. Preliminary varietal studies should be made for tolerances useful for standard varietal selection. If the selectivity of the herbicide is insufficient or too variable in a major crop to allow wide use under field conditions, it would be well to also make cell culture selections for resistance in amenable systems. The chemical industry has been provided a level of patent protection that the seed industry has learned to live without while remaining profitable. If research gets to the point of transferring herbicide resistance from noninterbreeding species by the techniques of genetic engineering, even the seed breeder will probably be protected by full patent coverage under a recent U.S. Supreme Court ruling. Crop varieties selected for herbicide resistance, and developed jointly between the chemical and seed industries with the cooperation of the academic community, might go far to help agriculture at minimal cost with maximal gain to all involved. For this to occur, interactions among the three communities will have to become much stronger, with greater lines of open communication, than have been evident so far.

A first gleam of light in such interaction and cooperation between a chemical company and a seed company comes from a closely related field, herbicide protectants. In 1980, a seed company began marketing protectant-coated sorghum seed of varieties best suited for use with the protectant/herbicide combination produced by the chemical company. This may be an ideal "tie in" for both the farmer and the corporation, as the chemical concern happens to own the seed company. At the present time, the chemical company is providing the protectant to other sorghum seed companies for treating their seed in order for their customers also to be able to use the herbicide. The same chemical company and others are actively researching and developing other chemical protectants to expand the range and effectiveness of herbicides available for various crops. In the practical sense, such protectants, safeners, or antidotes provide a form of chemical resistance, allowing otherwise phytotoxic herbicides to be used safely, but only on treated plants. This method of chemically emparting resistance to desirable crop plants has distinct advantages. As the characteristic is not carried genetically, interbreeding between closely related plants will not result in the spread of resistance. Rather, all other undesirable plants, even of the same species growing as volunteers which would be unwanted in breeding stock, and so on, could be eliminated.

The few important and rather striking successes in this area following relatively modest efforts and investments should, and undoubtedly will, lead to much future research into the mechanisms of action, structure-activity relationships, and crop selectivity or specificity to chemical protectants. Somewhat related research to discover chemical or other means to counteract the resistant mechanisms in herbicide-resistant weed biotypes could also lead to promising new and patentable products and significantly extend the useful life of our more valuable herbicides.

It has often been considered that mode-of-action studies (Chapters 8 to 11) were of academic interest only, and would have little bearing on or application to agronomic problems. There are abundant data to negate that view, at least vis-a-vis mode-of-triazine-resistance studies. Now that we have a better understanding of a few of the systems, we see that there are many more questions that we can consider in planning treatments and the types of herbicides worth developing, as well as what we must watch for. From research on the mode of triazine resistance and inheritance, we learned about the low transmission rates (by seed only and not by pollen) and the low rate of developing resistance. We have also been able to make suggestions as to how to delay resistance.

Now that we know that paraquat tolerance is related to enhanced activity of a few rather nonspecific yet ubiquitous enzyme systems that remove toxic free radicals, we can ask the worrisome question: Will those strains developing paraquat tolerance have a cross tolerance to other herbicides which similarly "short circuit" photosynthetic electron transport into releasing toxic free radicals? We may have a similar problem on our hands with other herbicides that are degraded through some of the more general pathways, such as glutathione or cytochrome $P_{450}$ systems. Such a cross tolerance could be very problematic. It is already a problem with insecticides and fungicides. Cross-resistance predictions on possible rates of evolution toward resistance and tolerance have led to industrial decisions not to release otherwise promising newly developed insecticides and fungicides. We do not yet have sufficient information to make knowledgeable decisions in these areas with herbicides.

Chapters 9 and 10 describe less yield per plant under a variety of conditions —the unfit triazine-resistant biotypes have less $CO_2$ incorporation, $O_2$ evolution, electron transport and competitive growth rates. They describe the molecular basis for the decrease in photosynthetic capacity. Skilled researchers continue trying to transfer the triazine-resistant genome from species to species. What yields should they expect? This is not clear, as yield is measured per unit area and various seeding densities must be planted to attain the optimum for maximal yield. The yield per plant may be less, but not the yield per unit area. In many areas of the world, light is not the limiting reaction in photosynthesis; thus a light reaction with lower efficiency may not have deleterious consequences. Unfortunately, the registration of the triazine-resistant rapeseed varieties contain no yield data.[8,9] Even if yield is diminished, the effect may be offset by better weed control, or by savings in fuel and far more expensive herbicides. Thus the basic photosynthesis work should not discourage breeders from evaluating the quality of the resistant plant. In the past, scientists have tried to correlate yield potential with plastid efficiency. On the whole, such correlations have been poor or nonexistent. Ultimately, the essential questions will likely be: (1) how closely tied is the herbicide-resistance trait to photosynthetic efficiency, plant growth rate, yield, and/or crop quality, and (2) how many back-crosses or what other methods are required to obtain all the desirable characteristics.

The modern technology of genetic engineering, including cell and tissue cultures and transfer of recombinant DNA, has stimulated greatly the interest and the probability of systematic and rapid transfer of triazine resistance or other plastid gene information from a weed to even unrelated crop plants without requiring years of laborious breeding. The product should end up with less undesirable traits. One of the exciting features about the triazine resistance characteristic is the easy and convenient genetic marker, which can be also used to develop transfer mechanisms and technology needed for other purposes, as well as to study the process of photosynthesis and its inhibition.

Recent research data, not included in Chapter 10, have led to the suggestion that the triazine binding site of the thylakoids is a 32-kilodalton polypeptide[10] which is encoded on chloroplast DNA although the interpretations of the elegent experiments are questioned.[12]

Another significant advance toward the successful transfer of triazine resistance was recently reported by Jain et al.[13] They fused X-ray-irradiated (functionally enucleated) protoplasts from triazine-resistant *Solanum nigrum* leaves with *Nicotiana sylvestris* (tobacco) protoplasts, and regenerated them to callus. Regenerated plants (morphologically *N. sylvestris*) were self-pollinated, and seeds were bioassayed on atrazine medium. Although some of the seeds survived doses toxic to normal *N. sylvestris,* the differential selectivity was not the same as between resistant and susceptible *S. nigrum.*

While most research on comparative fitness and competitive ability supports the general conclusion that the herbicide-resistant weed biotypes are less vigorous, adaptable, and fit compared to their respective normal or susceptible taxa, there may be some exceptions. Some unpublished reports[1a] indicate that both triazine-resistant *Chenopodium album* and *Amaranthus retroflexus* have greater germination dynamics compared to the susceptible biotypes. The entire question of physiological, biochemical, and genetic differences between the various biotypes from the many locations where herbicide resistance has developed need to be thoroughly investigated. Special effort should be made to identify resistant biotypes which have superior germination, growth rate, photosynthetic efficiency, or other advantages. Subsequent research effort to transfer herbicide resistance from weeds to crops should then concentrate on using germ plasm also having these desirable characteristics.

We have, indeed, come a long way in the past five years, approximately the period of time most of the information in this book has been generated. It is expected that there will be equally exciting and eventually more useful breakthroughs over the next five years.

## REFERENCES

1   T. P. Carney, *Instant Evolution: We'd Better Get Good at It,* University of Notre Dame Press, Notre Dame, Ind., 1980.

1a  P. Solymosi, Institute of Plant Protection, Budapest, Hungary, personal communication, 1980.

2   L. L. Shipman, Theoretical study of the binding site and mode of action for photosystem II herbicides, *J. Theor. Biol.,* **90,** 123 (1981).

3   B. Lovelidge, "Rogue" grass defies paraquat, *Arable Farming,* **1**(9), 9 (1974).

4   E. F. Eastin, Texas Agricultural Experimental Station, Beaumont, Tex., personal communications, 1980.

**5**  C. J. Edwards, Jr., W. L. Barrentine, and T. C. Kilen, Inheritance of sensitivity to metribuzin in soybeans, *Crop Sci.,* **16,** 119 (1976).

**6**  E. F. Eastin, J. E. Sij, and J. P. Craigmiles, Tolerance of soybean genotypes to metribuzin, *Agron. J.,* **72,** 167 (1980).

**7**  E. F. Eastin, Movement and fate of metribuzin in Tracy and Tracy M soybeans, *Proc. South. Weed Sci. Soc.,* **34,** 263 (1981).

**8**  W. D. Beversdorf, J. Weiss-Lerman, and L. R. Erickson, Registration of triazine resistant *Brassica napus* germplasm, *Crop Sci.,* **20,** 289 (1980).

**9**  W. D. Beversdorf, J. Weiss-Lerman, and L. R. Erickson, Registration of triazine resistant *Brassica campestris* germplasm, *Crop Sci.,* **20,** 289 (1980).

**10**  R. Pfister, K. E. Steinback, G. Gardner, and C. J. Arntzen, Photoaffinity labeling of an herbicide receptor protein in chloroplast membranes, *Proc. Natl. Acad. Sci. U.S.A.,* **78,** 981 (1981).

**11**  L. McIntosh, K. Steinback, L. Bogorad, and C. J. Arntzen, Identification of the 32 k dalton triazine binding polypeptides of thylakoid membranes as a chloroplast gene product, *Plant Physiol. (Supplement)* **67,** Abstr. 356 (1981).

**12**  J. Gressel, Triazine herbicide interaction with a 32,000 $M_r$ thylakoid protein—Alternative possibilities. Plant Sci. Lett., 1982 (in press).

**13**  M. Jain, D. Aviv, D. G. Davis, E. Galun, and J. Gressel, Conferring herbicide tolerance on tobacco by cybridization with atrazine resistant *Solanum nigrum, Plant Physiol. (Supplement)* **67,** Abstr. 866 (1981).

# Appendix

**Table A1  Intraspectific Differences in Tolerance and Resistance to Herbicides–by Herbicide**[a]

| Herbicide | Species | Type of Difference | Notes | Chapter or Reference[b] |
|-----------|---------|--------------------|-------|-------------------------|
| *Amides* | | | | |
| Alachlor | *Zea mays* | Tolerance | In inbred lines and hybrids | 1 |
| Diphenamid | *Lycopersicon esculentum* | Tolerance | Selected in mutagen-treated seedlings | 2 |
| Propachlor | *Sorghum bicolor* | Tolerance | Differential response among 40 cultivars | 3 |
| *Benzoic Acids* | | | | |
| Chloramben | *Cucumis sativus* | Tolerance | Differential response among cultivars | Chap. 8 |
| Dicamba | *Cirsium arvense* | Tolerance | Differential response among varieties | 4 |
| | *Kochia scoparia* | Tolerance | Variation among self-pollinated selections | 5 |
| | *Sonchus* hybrids | Tolerance | Variation among crosses and backcrosses | Chaps. 2, 8 |
| TBA | *Cardaria chalepensis* | Tolerance | Natural variation among biotypes | Chap. 2 |
| *Benzonitriles* | | | | |
| Bromoxynil | *Glycine max* | Tolerance | Japanese cultivars most sensitive | 6 |
| Ioxynil | *Matricaria perforata* | Tolerance | Natural variation among populations | Chap. 3 |

| Herbicide | Species | Type of Difference | Notes | Chapter or Reference[b] |
|---|---|---|---|---|
| *Bipyridyls* | | | | |
| Difenzoquat | *Triticum aestivum* | Tolerance | Differential response among 56 cultivars | Chap. 8 |
| Diquat | *Lolium perenne* | Tolerance | Differences among genotypes | Chap. 11 |
| | *Zea mays* | *Resistance* | In an artificially selected albino | Chap. 11 |
| Paraquat | *Conyza linifolia* | *Resistance* | Appeared after repeated applications | Chaps. 3, 11 |
| | *Erigeron philadelphicus* | *Resistance* | More sensitive to diquat | Chap. 3 |
| | *Glycine max* | *Resistance* | Selected in cell cultures | Chap. 14 |
| | *Lolium perenne* | Tolerance | Selected and bred from strain differences | Chaps. 11, 12 |
| | *Nicotiana tabacum* | *Resistance* | Selected in muta-genized cell cultures | Chap. 14 |
| | *Poa annua* | *Resistance* | Selected by repeated applications | Chaps. 3, 11 |
| | *Triticum aestivum* | Tolerance | Differences among cultivars | Chap. 11 |
| *Carbamates and Thiocarbamates* | | | | |
| Asulam | *Brassica napus* | Tolerance | Developed in callus culture | Chap. 14 |
| Barban | *Avena fatua* | Tolerance | Variation among 214 populations | Chaps. 2, 8 |
| | *Hordeum vulgare* | Tolerance | Differential response among cultivars | Chaps. 2, 8 |
| Butylate | *Setaria viridis* | Tolerance | Differential response among taxa | 7 |
| | *Zea mays* | Tolerance | Differential metabolism among hybrids | Chap. 8 |
| Diallate | *Avena fatua* | Tolerance | Variation among 214 populations | Chaps. 2, 8 |
| Molinate | *Oryza sativa* | Tolerance | Differences among lines and cultivars | 9 |
| Phenmedipham | *Nicotiana tabacum* | *Resistance* | In mutagenized haploid plants | Chap. 14 |

| Herbicide | Species | Type of Difference | Notes | Chapter or Reference[b] |
|---|---|---|---|---|
| Propham | *Avena fatua* | Tolerance | Differences among populations | Chap. 8 |
|  | *Nicotiana tabacum* | Tolerance | Selected in isolated protoplasts | Chap. 14 |
| Triallate | *Avena fatua* | Tolerance | Variation among 214 populations | Chaps. 2, 8 |
| **Diphenyl Ethers** | | | | |
| Nitrofen | *Brassica oleracea* | Tolerance | Based on cuticle thickness | Chap. 8 |
| **Dinitroanilines** | | | | |
| Trifluralin | *Eleusine indica* | Tolerance | Found in treated fields | Chap. 2 |
|  | *Zea mays* | Tolerance | Differences among 52 lines | 8 |
| **Halogenated Aliphatics** | | | | |
| Dalapon | *Agropyron repens* | Tolerance | Differential response among clones | Chap. 2 |
|  | *Cynodon dactylon* | Tolerance | Differential response among clones | Chap. 2 |
|  | *Digitaria* sp. | Tolerance | Repeated use in sugarcane | Chap. 2 |
|  | *Echinochloa crus-galli* | Tolerance | Differences among biotypes | Chap. 2 |
|  | *Lolium perenne* | Tolerance | Differences among cultivars | Chap. 12 |
|  | *Paspalum dilatatum* | Tolerance | Variation among biotypes | Chap. 2 |
|  | *Saccharum* sp. | Tolerance | Variation among clones | 10 |
|  | *Setaria faberii* | Tolerance | Variation among biotypes | Chap. 2 |
|  | *Setaria glauca = lutescens* | Tolerance | Variation among biotypes | Chap. 2 |
|  | *Sorghum halepense* | Tolerance | Variation among ecotypes | Chap. 2 |
| TCA | *Cynodon dactylon* | Tolerance | Biotype differences | Chap. 2 |

| Herbicide | Species | Type of Difference | Notes | Chapter or Reference[b] |
|---|---|---|---|---|
| *Phenoxy Acids* | | | | |
| Chlorfenprop | *Avena sativa* | *Resistance* | Differential response among cultivars | Chap. 8 |
| 2,4-D | *Arabidopsis thaliana* | *Resistance* | Selected mutants | 11 |
| | *Cardaria chalapensis* | Tolerance | Biotype differences in the field | Chap. 2 |
| | *Centaurea repens* | Tolerance | Differential response among clones | 12 |
| | *Chenopodium album* | Tolerance | Found in treated cornfields | Chap. 3 |
| | *Citrus sinensis* | Tolerance | Selected in callus culture | 13 |
| | *Cirsium arvense* | Tolerance | Differential response among ecotypes | Chaps. 2, 3 |
| | *Commelina diffusa* | Tolerance | Biotypes selected by repeated applications | Chap. 2 |
| | *Convolvulus arvensis* | Tolerance | Clonal differences | Chap. 2 |
| | *Cyperus esculentus* | Tolerance | Differential response among varieties | 14 |
| | *Daucus carota* (wild) | Tolerance | Biotype variations selected along roadsides | Chaps. 1, 8 |
| | *Daucus carota* (cultivated) | *Resistance* | Selected in cell culture | Chap. 14 |
| | *Erechtites hieracifolia* | Tolerance | Biotypes selected by repeated applications | 15 |
| | *Fragaria X ananassa* | Tolerance | Differential metabolism among cultivars | Chap. 8 |
| | *Kochia scoparia* | Tolerance | Selected by repeated field applications | Chap. 2 |
| | *Linum usitatissimum* | Tolerance | Differences among cultivars | 16 |
| | *Lotus corniculatus* | Tolerance | Selected by repeated field applications | Chap. 12 |
| | *Malus sylvestris* | Tolerance | Differential degradation among cultivars | Chap. 8 |
| | *Nicotiana sylvestris* | *Resistance* | Selected in haploid cell cultures | Chap. 14 |

| Herbicide | Species | Type of Difference | Notes | Chapter or Reference[b] |
|---|---|---|---|---|
| | *Ranunculus* spp. | Tolerance | After 9 repeated treatments | Chap. 3 |
| | *Saccharum* sp. | Tolerance | Differences among clones | 17 |
| | *Sonchus* hybrids | Tolerance | Hybrids with increased tolerance | Chaps. 2, 8 |
| | *Sorghum bicolor* | Tolerance | Variation among hybrids | 18 |
| | *Taraxacum officinale* | Tolerance | After 9 repeated treatments | Chap. 3 |
| | *Trifolium repens* | Tolerance | After 9 repeated treatments | Chap. 3 |
| | *Zea mays* | Tolerance | Differential translocation among cultivars | Chap. 8 |
| 2,4-DB | *Glycine max* | Tolerance | Differences among cultivars | 6 |
| | *Trifolium repens* | Tolerance | Selected in cell culture | Chap. 14 |
| Dichlorprop | *Polygonum lapathifolium* | Tolerance | Biotype variations | Chap. 3 |
| MCPA | *Cirsium arvense* | Tolerance | Clonal response related to past treatment | Chap. 3 |
| | *Matricaria perforata* | Tolerance | Population response related to past treatment | Chap. 3 |
| *Phenyl ureas* | | | | |
| Chloroxuron | *Glycine max* | Tolerance | Differences among cultivars | 6 |
| Diuron | *Chlamydomonas reinhardii* | Resistance | Selected in mutagenized cells | 19 |
| | *Euglena gracilis* | Resistance | Selected strain | 20 |
| | *Gossypium hirsutum* | Tolerance | Differential metabolism among cultivars | Chap. 8 |
| | *Saccharum* hybrids | Tolerance | Differential metabolism among hybrids | Chap. 8 |
| | *Sorghum bicolor* | Tolerance | Differences among 40 cultivars | 3 |
| Fluometuron | *Gossypium hirsutum* | Tolerance | Differential uptake among cultivars | Chap. 8 |

| Herbicide | Species | Type of Difference | Notes | Chapter or Reference[b] |
|---|---|---|---|---|
| Linuron | *Ambrosia artemisiifolia* | Tolerance | | Chap. 2 |
| | *Sorghum bicolor* | Tolerance | Differences among 40 cultivars | 3 |
| Metoxuron | *Poa annua* | Tolerance | Artificially selected in field | Chap. 3 |
| Norea | *Sorghum bicolor* | Tolerance | Differences among 40 cultivars | 3 |
| Siduron | *Agrostis palustris* | Tolerance | Differences among cultivars | 21 |
| | *Hordeum jubatum* | Tolerance | Controlled by three dominant genes | Chap. 2 |

*Triazines and Triazinones*

| Herbicide | Species | Type of Difference | Notes | Chapter or Reference[b] |
|---|---|---|---|---|
| Ametryn | *Cynodon dactylon* | Tolerance | Differences among genotypes | 22 |
| | *Saccharum* sp. | Tolerance | Differences among seedling crosses | 23 |
| Atrazine[c] | *Amaranthus arenicola* | Resistance | Selected in corn | Chap. 2 |
| | *Amaranthus hybridus* | Resistance | As *A. retroflexus* below | Chaps. 3, 5, 10 |
| | *Amaranthus lividus* | Resistance | Selected in continuous corn | Chaps. 2, 3 |
| | *Amaranthus powellii* | Resistance | Selected in continuous corn | Chaps. 2, 5 |
| | *Amaranthus retroflexus* | Resistance | Biotypes with plastid differences selected after repeated applications | Chaps. 2, 3, 5, 10 |
| | *Ambrosia artemisiifolia* | Resistance | Selected by repeated field applications | Chap. 2 |
| | *Atriplex patula* | Resistance | Selected by repeated field applications | Chap. 3 |
| | *Bidens tripartita* | Resistance | Resistance at plastid level | Chap. 3 |
| | *Bromus tectorum* | Resistance | Resistance at plastid level | Chap. 3 |
| | *Brassica campestris = rapa* | Resistance | Maternally inherited plastid resistance | Chaps. 2, 13 |

| Herbicide | Species | Type of Difference | Notes | Chapter or Reference[b] |
|---|---|---|---|---|
| | *Chenopodium album* | *Resistance* | Worldwide distribution after repeated field applications; plastid-level resistance | Chaps. 2, 3, 5, 10 |
| | *Chenopodium ficifolium* | *Resistance* | Selected in continuous corn | Chap. 3 |
| | *Chenopodium missouriense* | *Resistance* | After repeated applications in corn | Chaps. 2, 5 |
| | *Chenopodium polyspermum* | *Resistance* | Plastid-level resistance selected in vineyards | Chap. 3 |
| | *Chenopodium strictum* var. *glaucophyllum* | *Resistance* | Selected in continuous corn | Chap. 2 |
| | *Cucumis sativus* | Tolerance | Differences within world collection | Chap. 8 |
| | *Cynodon dactylon* | Tolerance | Differences among selections | 24 |
| | *Cyperus esculentus* | Tolerance | Differences among varieties | 14 |
| | *Digitaria sanguinalis* | *Resistance* | Related to herbicide degradation | Chaps. 3, 8 |
| | *Echinochloa crus-galli* | Tolerance | Selected by repeated applications in corn | Chap. 3 |
| | | *Resistance* | Not always at plastid level | Chap. 3 |
| | *Erigeron canadensis* | *Resistance* | Not controlled with 9 kg/ha atrazine | Chap. 3 |
| | *Galinsoga ciliata* | *Resistance* | Not controlled by 5 kg/ha | Chap. 3 |
| | *Glycine max* | *Resistance* | Selected in cell culture | Chap. 14 |
| | *Kochia scoparia* | *Resistance* | Selected along railroad beds and cornfields | Chap. 2 |
| | *Linum usitatissimum* | Tolerance | Quantitatively inherited | Chap. 12 |
| | *Polygonum convolvulus* | *Resistance* | After 3 yr atrazine | Chap. 3 |
| | *Polygonum lapathifolium* | *Resistance* | Plastid level resistance | Chap. 3 |
| | *Polygonum persicaria* | *Resistance* | Plastid level resistance | Chap. 3 |

| Herbicide | Species | Type of Difference | Notes | Chapter or Reference[b] |
|---|---|---|---|---|
| | *Saccharum officinarum* | *Resistance* | Cultivar differences based on degradation | Chap. 8 |
| | *Senecio vulgaris* | *Resistance* | Selected by repeated applications in corn and horticultural crops | Chaps. 2, 3 |
| | *Setaria faberii* | *Resistance* | In continuously treated corn | Chap. 2 |
| | Setaria lutescens | *Resistance* | Selected in corn | Chap. 2 |
| | Setaria viridis | *Resistance* | In continuously treated corn | Chap. 2 |
| | | Tolerance | Following repeated treatments; variety differences | Chap. 3; 8 |
| | *Solanum nigrum* | *Resistance* | Plastid level resistance | Chap. 3 |
| | *Sorghum bicolor* | Tolerance | Differential response among cultivars | 18 |
| | *Stellaria media* | *Resistance* | Selected by repeated applications in corn | Chap. 3 |
| | *Veronica persica* | Tolerance | Found after 10 yearly applications in grapes | Chap. 3 |
| | **Zea mays** | *Resistance* | Differential metabolism among lines | Chaps. 8, 12 |
| | *Zoysia japonica* | Tolerance | Differential metabolism among cultivars | Chap. 8 |
| Metribuzin | *Glycine max* | Tolerance | Differential metabolism among cultivars | Chap. 8 |
| | *Lycopersicon esculentum* | Tolerance | Differential metabolism among cultivars | Chap. 8 |
| | *Solanum tuberosum* | Tolerance | Differential metabolism among cultivars | Chap. 8 |
| | *Triticum aestivum* | Tolerance | Cultivar TAM W.101 tolerant | 25 |
| Prometryn | *Gossypium hirsutum* | Tolerance | Differences among cultivars | 26 |
| Propazine | *Setaria viridis* | Tolerance | Differences among taxa | 7 |
| | *Sorghum bicolor* | Tolerance | Differences among 40 cultivars | 3 |
| Simazine[c] | *Amaranthus bouchonii* | Tolerance | Related to spray history | Chap. 3 |

| Herbicide | Species | Type of Difference | Notes | Chapter or Reference[b] |
|---|---|---|---|---|
| | *Amaranthus graecizans* | Tolerance | Related to spray history | Chap. 3 |
| | *Brassica napus* | Tolerance | Differences among cultivars | Chap. 12 |
| | *Capsella bursa-pastoris* | Tolerance | Related to spray history | Chap. 3 |
| | *Chenopodium album* | Tolerance | Related to spray history | Chap. 3 |
| | *Matricaria perforata* | Tolerance | Variation among populations | Chap. 3 |
| | *Panicum capillare* | Resistance | Selected along treated railroads | Chap. 2 |
| | *Poa annua* | Resistance | Selected after yearly applications of 10 kg/ha | Chap. 3 |
| | *Senecio vulgaris* | Resistance | Biotypes with plastid differences selected after repeated applications | Chaps. 2, 3, 6 |
| | *Sinapis alba* | Tolerance | Differences among cultivars | 27 |
| | *Triticum aestivum* | Tolerance | Differences among cultivars | 27 |
| Terbutryn | *Triticum aestivum* | Tolerance | Selected in mutagen-treated seedlings | 2 |
| *Uracils* | | | | |
| Bromacil | *Cynodon dactylon* | Tolerance | Variation among clones | 24 |
| Terbacil | *Lolium multiflorum* | Tolerance | Followed 12 years of repeated use | Chap. 2 |
| | *Vaccinium angustifolium* | Tolerance | Variation among 26 selected clones | 28 |
| *Miscellaneous* | | | | |
| Amitrole | *Agropyron repens* | Tolerance | Variation among ecotypes | 29 |
| | *Cirsium arvense* | Tolerance | Variation among ecotypes | Chaps. 2, 8 |

**Table A1** *(Continued)*

| Herbicide | Species | Type of Difference | Notes | Chapter or Reference[b] |
|---|---|---|---|---|
| | *Nicotiana tabacum* | Tolerance | Selected in green cell cultures | Chap. 14 |
| Bentazon | *Cirsium arvense* | Tolerance | Differential metabolism among ecotypes | Chap. 8 |
| | *Glycine max* | Tolerance | Differential metabolism among cultivars | Chap. 8 |
| | *Nicotiana tabacum* | Resistance | In mutagenized haploid plants | Chap. 14 |
| CP 49814 | *Cucumis sativus* | Tolerance | Differential response among cultivars | 30 |
| Fluridone | *Gossypium hirsutum* | Tolerance | Glanded cultivars most tolerant | 31 |
| Glyphosate | *Cirsium arvense* | Tolerance | Variation among ecotypes | 32 |
| | *Prunus persica* | Tolerance | Variation among cultivars | 33 |
| Picloram | *Nicotiana tabacum* | Resistance | Selected in cell cultures | Chap. 14 |
| | *Sorghum bicolor* | Tolerance | Differential response among cultivars | 35 |
| Pyrazon | *Beta vulgaris* | Tolerance | Differential metabolism among inbred lines | Chap. 8 |

[a] Complied by K. I. N. Jensen and J. Gressel

[b] Additional information can be found in either the chapter cited or in the following list of references.

[c] Biotypes resistant to one *s*-triazine are often resistant to others as well as triazinones.

## REFERENCES FOR TABLE A1

1. D. B. Narsaiah and R. G. Harvey, Differential responses of corn inbreds and hybrids to alachlor, *Crop Sci.*, **17**, 657 (1977).
2. M. J. Pinthus, Y. Eshel, and Y. Shchori, Field and vegetable crop mutants with increased resistance to herbicides, *Science,* **177**, 715 (1972).
3. F. R. Miller and R. W. Bovey, Tolerance of *Sorghum bicolor* (L.) Moench to several herbicides, *Agron. J.,* **61**, 282 (1969).
4. J. H. Hunter and L. W. Smith, Environment and herbicide effects on Canada thistle ecotypes, *Weed Sci.,* **20**, 163 (1972).
5. A. R. Bell, J. D. Nalewaja, and A. B. Schooler, Response of kochia selections to 2,4-D, dicamba and picloram, *Weed Sci.,* **20**, 458 (1972).

6. L. M. Wax, R. L. Bernard, and R. M. Hayes, Response of soybean cultivars to bentazon, bromoxynil, chloroxuron and 2,4-DB, *Weed Sci.,* **22,** 35 (1974).

7. L. R. Oliver and M. M. Schreiber, Differential selectivity of herbicides on six *Setaria* taxa, *Weed Sci.,* **19,** 428 (1971).

8. J. L. Davis, J. R. Abernathy, and A. F. Wiese, Tolerance of 52 corn lines to trifluralin, *Proc. South. Weed Sci. Soc.,* **31,** 123 (1978).

9. E. P. Richard and J. B. Baker, Response of selected rice (*Oryza sativa*) lines to molinate, *Weed Sci.,* **27,** 219 (1979).

10. R. W. Millhollon and R. J. Mathern, Tolerance of sugarcane varieties to herbicides, *Weed Sci.,* **16,** 300 (1968).

11. E. P. Maher and S. J. B. Martindale, 2,4-D resistant mutants of *Arabidopsis, Arabidopsis Inf. Serv.,* **15,** 15 (1978).

12. J. D. Stallings and H. P. Cords, Herbicidal susceptibility of some clonal lines of Russian knapweed (*Centaurea repens*), *Abstr. Weed Sci. Soc. Am.,* No. 184 (1969).

13. J. Kochba, P. Spiegel-Roy, and S. Saad, Selections of *Citrus sinensis* ovular callus lines with increased tolerance to sodium chloride and 2,4-D, *IAEA Res. Coord. Meet.,* Skiernewiccze, Poland (1978).

14. J. Costa and A. P. Appleby, Response of two yellow nutsedge varieties to three herbicides, *Weed Sci.,* **24,** 54 (1976).

15. S. N. Hanson, Weed control practices and research for sugarcane in Hawaii, *Weeds,* **10,** 192 (1962).

16. R. K. Randon, The response of flax to rates and formulations of 2,4-dichlorophenoxyacetic acid, *Agron. J.,* **41,** 213 (1949).

17. J. A. B. Nolla, Injury to sugarcane from 2,4-D, *Proc. Int. Soc. Sugar Cane Technol.,* **7,** 178 (1950).

18. O. C. Burnside and G. A. Wicks, Competitiveness and herbicide tolerance of sorghum hybrids, *Weed Sci.,* **20,** 314 (1972).

19. J. C. McBride, A. C. McBride, and R. T. Togasaki, Isolation of *Chlamydomonas reinhardii* mutants resistant to the herbicide DCMU, Plant Cell Physiol. (Spec. issue) Photosynthetic Organelles, 239 (1977).

20. D. Laval-Martin, G. Dubertret, and R. Calvayrac, Photosynthetic properties of a resistant strain of *Euglena gracilis* Z., *Plant Sci. Lett.,* **10,** 185 (1977).

21. H. J. Hopen, W. E. Splittstoesser, and J. D. Butler, Intraspecies bentgrass selectivity to siduron, *Proc. Am. Soc. Hort. Sci.,* **89,** 631 (1966).

22. G. Ramirez-Oliveras, Differential tolerance to ametryn among six genotypes of giant bermuda grass (*Cynodon dactylon* var. *aridus*) and their diallel progeny, *J. Agric. Univ. Puerto Rico,* **61,** 49 (1977).

23. R. V. Osgood and D. J. Heinz, Selecting sugarcane seedlings for ametryn tolerance, *Annu. Rep. Hawaiian Sugar Plant. Assoc.,* 44 (1976).

24. L. G. Anderson and W. R. Kneebone, Differential responses of *Cynodon dactylon* L. selections to three herbicides, *Crop Sci.,* **9,** 599 (1969).

25. Anonymous, Herbicide resistant wheat, *Weed Sci. Soc. Am. Newslett.,* **7(4),** 12 (1979).

26. J. L. Davis, J. R. Abernathy, and J. R. Gibson, Cotton varietal response to selected s-triazine compounds, *Proc. South. Weed Sci. Soc.,* **33,** 294 (1980).

27. A. Karim and A. D. Bradshaw, Genetic variation in simazine resistance in wheat, rape and mustard, *Weed Res.,* **8,** 283 (1968).

28. K. I. N. Jensen, I. V. Hall, and E. Kimball, Differential response of lowbush blueberry selections to terbacil, *Res. Rep. Expert Comm. Weeds (Cana. East. Sect.)* **451** (1980).

29.  S. Y. Haddad and G. R. Sagar, A study of the response of four clones of *Agropyron repens* to root and shoot applications of aminotriazole and dalapon, *Proc. 9th Br. Weed Control Conf.,* 142 (1968).

30   A. E. Maitland, R. D. Sweet, and P. A. Minges, Differential response of cucumber varieties to experimental herbicide CP 49814, *HortScience,* **5**, 363 (1970).

31   J. R. Abernathy, J. W. Keeling, and L. L. Ray, Response of 130 cotton varieties to fluridone, *Proc. South. Weed Sci. Soc.,* **31**, 77 (1978).

32   W. J. Saidak and P. B. Marriage, Response of Canada thistle varieties to amitrole and glyphosate, *Can. J. Plant Sci.,* **56**, 211 (1976).

33   P. B. Marriage and S. U. Khan, Differential varietal tolerance of peach (*Prunus persica*) seedlings to glyphosate, *Weed Sci.,* **26**, 374 (1978).

34   C. J. Scifres and R. W. Bovey, Differential responses of sorghum varieties to picloram, *Agron. J.,* **62**, 775 (1970).

## Table A2  Intraspecific Differences in Tolerance and Resistance to Herbicides– by Family and Species

| Family | Species | Common Name | Herbicide | Type of Difference |
|---|---|---|---|---|
| Amaranthaceae | *Amaranthus arenicola* Johnst. | | Atrazine | *Resistance* |
| | *Amaranthus bouchonii* Thell. | | Simazine | Tolerance |
| | *Amaranthus graecizans* L. | Prostrate pigweed | Simazine | Tolerance |
| | *Amaranthus hybridus* L. | Smooth pigweed | Atrazine | *Resistance* |
| | *Amaranthus lividus* L. | Livid amaranth | Atrazine | *Resistance* |
| | *Amaranthus powellii* S. Wats. | Powell amaranth | Atrazine | *Resistance* |
| | *Amaranthus retroflexus* L. | Redroot pigweed | Atrazine | *Resistance* |
| Caryophyllaceae | *Stellaria media* (L.) Cyrillo | Chickweed | Atrazine | *Resistance* |
| Chenopodiaceae | *Atriplex patula* L. | Spreading orach | Atrazine | *Resistance* |
| | *Beta vulgaris* L. | Red beet | Pyrazon | Tolerance |
| | *Chenopodium album* L. | Common lambsquarters | Atrazine 2,4-D Simazine | *Resistance* Tolerance Tolerance |
| | *Chenopodium ficifolium* Sm. | Figleaved goosefoot | Atrazine | *Resistance* |
| | *Chenopodium missouriense* Aellen | | Atrazine | *Resistance* |
| | *Chenopodium polyspermum* L. | Manyseeded goosefoot | Atrazine | *Resistance* |
| | *Chenopodium strictum* Roth var. *glaucophyllum* Aellen | Late-flowering goosefoot | Atrazine | *Resistance* |
| | *Kochia scoparia* (L.) Roth | Kochia | Atrazine 2,4-D Dicamba | *Resistance* Tolerance Tolerance |
| Chlorophyceae | *Chlamydomonas reinhardii* Dangeard | Chlamydomonas | Diuron | *Resistance* |

375

| Family | Species | Common Name | Herbicide | Type of Difference |
|---|---|---|---|---|
| Commelinaceae | *Commelina diffusa* Burm. f. | Spreading dayflower | 2,4-D | Tolerance |
| Compositae | *Ambrosia artemisiifolia* L. | Common ragweed | Atrazine<br>Linuron | *Resistance*<br>Tolerance |
| | *Bidens tripartita* L. | Three-parted beggartick | Atrazine | *Resistance* |
| | *Centaurea repens* L. | Russian knapweed | 2,4-D | Tolerance |
| | *Cirsium arvense* (L.) Scop. | Canada thistle | Amitrole<br>Bentazon<br>2,4-D<br>Dicamba<br>Glyphosate<br>MCPA | Tolerance<br>Tolerance<br>Tolerance<br>Tolerance<br>Tolerance<br>Tolerance |
| | *Conyza linifolia* | | Paraquat | *Resistance* |
| | *Erechtites hieracifolia* (L.) Raf. | American burnweed | 2,4-D | Tolerance |
| | *Erigeron canadensis* L. | Horseweed | Atrazine | *Resistance* |
| | *Erigeron philadelphicus* L. | Philadelphia fleabane | Paraquat | *Resistance* |
| | *Galinsoga ciliata* (Raf.) Blake | Hairy galinsoga | Atrazine | *Resistance* |
| | *Matricaria perforata* Merat | Scentless mayweed | Ioxynil<br>MCPA<br>Simazine | Tolerance<br>Tolerance<br>Tolerance |
| | *Senecio vulgaris* L. | Common groundsel | Simazine | *Resistance* |
| | *Sonchus* hybrids | Sowthistles | 2,4-D | Tolerance |
| | *Taraxacum officinale* Weber | Dandelion | 2,4-D | Tolerance |
| Convolvulaceae | *Convolvulus arvensis* L. | Field bindweed | 2,4-D | Tolerance |
| Cruciferae | *Arabidopsis thaliana* (L.) Heynh. | Mouseearcress | 2,4-D | *Resistance* |
| | *Brassica oleracea* L. | Cabbage | Nitrofen | Tolerance |
| | *Brassica campestris* L. = *rapa* | Birdsrape mustard | Atrazine | *Resistance* |
| | *Brassica napus* L. | Rape | Asulam<br>Simazine | Tolerance<br>Tolerance |

**Table A2** (Continued)

| Family | Species | Common Name | Herbicide | Type of Difference |
|--------|---------|-------------|-----------|--------------------|
| | *Capsella bursa-pastoris* (L.) Medic. | Shepardspurse | Simazine | Tolerance |
| | Cardaria *chalapensis* (L.) Hand-Maz. | Lenspodded hoary cress | 2,4-D, TBA | Tolerance |
| | *Sinapis alba* L. | White mustard | Simazine | Tolerance |
| Cucurbitaceae | *Cucumis sativus* L. | Cucumber | Atrazine | Tolerance |
| | | | Chloramben | Tolerance |
| Cyperaceae | *Cyperus esculentus* L. | Yellow nutsedge | Atrazine | Tolerance |
| | | | 2,4-D | Tolerance |
| Ericaceae | *Vaccinium angustifolium* Ait. | Lowbush blueberry | Terbacil | Tolerance |
| Euglenaceae | *Euglena gracilis* Z. | Euglena | Diuron | *Resistance* |
| Gramineae | *Agropyron repens* (L.) Beauv. | Quackgrass | Amitrole | Tolerance |
| | | | Dalapon | Tolerance |
| | *Agrostis palustris* Huds. | Creeping bentgrass | Siduron | Tolerance |
| | *Avena fatua* L. | Wild oats | Barban | Tolerance |
| | | | Diallate | Tolerance |
| | | | Propham | Tolerance |
| | | | Triallate | Tolerance |
| | *Avena sativa* L. | Oats | Chlorfen-prop | *Resistance* |
| | *Bromus tectorum* L. | Downy brome | Atrazine | *Resistance* |
| | *Cynodon dactylon* (L.) Pers. | Bermudagrass | Ametryn | Tolerance |
| | | | Atrazine | Tolerance |
| | | | Bromacil | Tolerance |
| | | | Dalapon | Tolerance |
| | | | TCA | Tolerance |
| | *Digitaria sanguinalis* (L.) Scop. | Large crabgrass | Atrazine | Tolerance |
| | *Digitaria* sp. | | Dalapon | Tolerance |
| | *Echinochloa crus-galli* (L.) Beauv. | Barnyardgrass | Atrazine | Tolerance |
| | | | Atrazine | *Resistance* |
| | | | Dalapon | Tolerance |

**Table A2** *(Continued)*

| Family | Species | Common Name | Herbicide | Type of Difference |
|---|---|---|---|---|
| | *Eleusine indica* (L.) Gaertn. | Goosegrass | Trifluralin | Tolerance |
| | *Hordeum jubatum* L. | Foxtail barley | Siduron | Tolerance |
| | *Hordeum vulgare* L. | Barley | Barban | Tolerance |
| | *Lolium multiflorum* Lam. | Italian ryegrass | Terbacil | Tolerance |
| | *Lolium perenne* L. | Perennial ryegrass | Dalapon | Tolerance |
| | | | Diquat | Tolerance |
| | | | Paraquat | Tolerance |
| | *Oryza sativa* L. | Rice | Molinate | Tolerance |
| | *Panicum capillare* L. | Witchgrass | Simazine | *Resistance* |
| | *Paspalum dilatatum* Poir | Dallisgrass | Dalapon | Tolerance |
| | *Poa annua* L. | Annual bluegrass | Metoxuron | Tolerance |
| | | | Paraquat | *Resistance* |
| | | | Simazine | *Resistance* |
| | *Saccharum* spp. | Sugarcane | Ametryn | Tolerance |
| | | | Atrazine | *Resistance* |
| | | | 2,4-D | Tolerance |
| | | | Dalapon | Tolerance |
| | *Setaria faberii* Herrm. | Giant foxtail | Dalapon | Tolerance |
| | | | Atrazine | *Resistance* |
| | *Setaria lutescens* (Weigel) Hubb. | Yellow foxtail | Dalapon | Tolerance |
| | | | Atrazine | *Resistance* |
| | *Setaria viridis* (L.) Beauv. | Green foxtail | Atrazine | Tolerance |
| | | | Atrazine | *Resistance* |
| | | | Butylate | Tolerance |
| | | | Propazine | Tolerance |
| | *Sorghum bicolor* (L.) Moench. | Sorghum | Atrazine | Tolerance |
| | | | 2,4-D | Tolerance |
| | | | Linuron | Tolerance |
| | | | Norea | Tolerance |
| | | | Picloram | Tolerance |
| | | | Propachlor | Tolerance |
| | | | Propazine | Tolerance |
| | *Sorghum halepense* (L.) Pers. | Johnsongrass | Dalapon | Tolerance |
| | *Triticum aestivum* L. | Wheat | Difenzoquat | Tolerance |
| | | | Metribuzin | Tolerance |
| | | | Paraquat | Tolerance |
| | | | Terbutryn | Tolerance |

| Family | Species | Common Name | Herbicide | Type of Difference |
|--------|---------|-------------|-----------|---------------------|
| | *Zea mays* L. | Corn | Alachlor | Tolerance |
| | | | Atrazine | *Resistance* |
| | | | Butylate | Tolerance |
| | | | 2,4-D | Tolerance |
| | | | Diquat | *Resistance* |
| | *Zoysia japonica* Steud. | Zoysiagrass | Atrazine | Tolerance |
| Leguminosae | *Glycine max* Merr. | Soybean | Atrazine | *Resistance* |
| | | | Bentazon | *Resistance* |
| | | | Bromoxynil | Tolerance |
| | | | 2,4-D | Tolerance |
| | | | 2,4-DB | Tolerance |
| | | | Chloroxuron | Tolerance |
| | | | Metribuzin | *Resistance* |
| | | | Paraquat | *Resistance* |
| | *Lotus corniculatus* L. | Birdsfoot trefoil | 2,4-D | Tolerance |
| | *Trifolium repens* L. | White clover | 2,4-D, 2,4-DB | Tolerance |
| Linaceae | *Linum usitatissimum* L. | Flax | Atrazine | Tolerance |
| | | | 2,4-D | Tolerance |
| Malvaceae | *Gossypium hirsutum* L. | Cotton | Diuron | Tolerance |
| | | | Fluometuron | Tolerance |
| | | | Fluridone | Tolerance |
| | | | Prometryn | Tolerance |
| Polygonaceae | *Polygonum convolvulus* L. | Wild buckwheat | Atrazine | *Resistance* |
| | *Polygonum lapathifolium* L. | Pale smartweed | Atrazine | *Resistance* |
| | | | Dichlorprop | Tolerance |
| | *Polygonum persicaria* L. | Ladysthumb | Atrazine | *Resistance* |
| Ranunculaceae | *Ranunculus* spp. | Buttercup | 2,4-D | Tolerance |
| Rosaceae | *Fragaria X ananassa* Duchesne | Strawberry | 2,4-D | Tolerance |
| | *Malus sylvestris* Mill. | Apple | 2,4-D | Tolerance |
| | *Prunus persica* Batsch | Peach | Glyphosate | Tolerance |

**Table A2** *(Continued)*

| Family | Species | Common Name | Herbicide | Type of Difference |
|---|---|---|---|---|
| Rutaceae | *Citrus sinensis* Osbeck | Orange | 2,4-D | Tolerance |
| Scrophulariaceae | *Veronica persica* Poir. | Birdseye speedwell | Atrazine | Tolerance |
| Solanaceae | *Lycopersicon esculentum* Mill. | Tomato | Diphenamid | Tolerance |
| | | | Metribuzin | Tolerance |
| | *Nicotiana sylvestris* Spegazzini & Comes | | 2,4-D | *Resistance* |
| | *Nicotiana tabacum* L. | Tobacco | Amitrole | Tolerance |
| | | | Bentazon | *Resistance* |
| | | | 2,4-D | *Resistance* |
| | | | Paraquat | *Resistance* |
| | | | Phenmedi-pham | *Resistance* |
| | | | Picloram | *Resistance* |
| | | | Propham | Tolerance |
| | *Solanum nigrum* L. | Black nightshade | Atrazine | *Resistance* |
| | *Solanum tuberosum* L. | Potato | Metribuzin | Tolerance |
| Umbelliferae | *Daucus carota* L. var. *sativa* D.C. (cultivated) | Carrot | 2,4-D | *Resistance* |
| | *Daucus carota* L. (wild) | Wild carrot | 2,4-D | Tolerance |

[a]Compiled by K. I. N. Jensen and J. Gressel.

[b]See Appendix Table A1 for notes and references.

[c]Biotypes resistant to one *s*-triazine are resistant to others.

**Table B1.** Alphabetical Listing of Herbicides Referred to in This Volume by Common Name and Corresponding Trade Name, Chemical Name, and Manufacturer.

| Common Name | Trade Name | Chemical Name | Manufacturer |
|---|---|---|---|
| Alachlor | Lasso, Alanex | 2-Chloro-2',6'-diethyl-*N*-(methoxy-methyl)acetanilide | Monsanto, Makhteshim-Agan |
| Ametryn | Evik, Gesapax, Ametrex, Crisatrine | 2-(Ethylamino)-4-(isopropylamino)-6-(methylthio)-*s*-triazine | Ciba-Geigy, Makhteshim-Agan, Crystal Chemical Inter-America |
| Amitrole | Amizol, Amino Triazole, Azolan, Weedazol, others | 3-Amino-*s*-triazole | American Cyanamid, Union Carbide, others |
| Asulam | Asulox | Methyl sulfanilylcarbamate | May & Baker, Rhone-Poulenc |
| Atraton | Gesatamin | 2-(Ethylamino)-4-(iso-propylamino)-6-methoxy-*s*-triazine | Ciba-Geigy |
| Atrazine | AAtrex, Vectal, Atrazine, Atranex | 2-Chloro-4-(ethylamino)-6-(isopropylamino)-*s*-triazine | Ciba-Geigy, Fisons, Shell, others |
| Barban | Carbyne, Dualweed, Wheatclene | 4-Chloro-2-butynyl *m*-chlorocarbanilate | Velsicol, Fisons |
| Benzazin | BAS-1770H | 2-Phenyl-3,1-benzoxazinone-(4) | BASF |
| Bentazon | Basagran | 3-Isopropyl-1*H*-2,1,3-benzothiadiazin-4(3*H*)-one 2,2-dioxide | BASF |
| Bromacil | Hyvar, Rout, Uragan | 5-Bromo-3-*sec*-butyl-6-methyluracil | du Pont, Hopkins, Makhteshim-Agan |
| Bromnitro-thymol | — | 2-Bromo-4-nitrothymol-phenol | — |
| Bromofenoxim | Faneron | 3,5-Dibromo-4-hydroxy-benzaldehyde-*O*-(2',4'-dinitrophenyl) oxime | Ciba-Geigy |
| Bromoxynil | Brominal, Buctril, Bronate | 3,5-Dibromo-4-hydro-xybenzonitrile | Union Carbide, May & Baker, Rhone-Poulenc |
| Brompyrazon | Basanor | 5-Amino-4-bromo-2-phenylpyridazin-3-one | BASF |
| Butralin | Amex, Tamex | 4-(1,1-Dimethylethyl)-*N*-(1-methylpropyl-2,6-dinitrobenzenamine | Union Carbide |

| Common Name | Trade Name | Chemical Name | Manufacturer |
|---|---|---|---|
| Butylate | Sutan | S-Ethyl diisobutylthio-carbamate | Stauffer |
| Chloramben | Amiben, Vegiben | 3-Amino-2,5-dichloro-benzoic acid | Union Carbide |
| Chloramben-methyl | Vegiben | Methyl ester of 3-amino-2,5-dichlorobenzoic acid | Union Carbide |
| Chlorbro-muron | Maloran | 3-(4-Bromo-3-chlorophenyl)-1-methoxy-1-methylurea | Ciba-Geigy |
| Chlorfen-propmethyl | Bidisin | Methyl 2-chloro-3-(4-chlorophenyl) propionate | Bayer |
| Chloroxuron | Tenoran | 3-[p-(p-Chlorophenoxy) phenyl]-1,1-dimethylurea | Ciba-Geigy |
| Chlortoluron | Dicuran, | 3-(3-Chloro-p-tolyl)-1,1-dimethylurea | Ciba-Geigy |
| Cyanazine | Bladex | 2-[[4-Chloro-6-(ethyl-amino)-s-triazin-2-yl] amino]-2-methylpropionitrile | Shell |
| 2,4-D | Several | (2,4-Dichlorophenoxy) acetic acid | Several |
| 2,4-F | — | 2-Chloro-4-fluoro-phenoxyacetic acid | — |
| 2,4-DB | Butoxone, Butyrac, Embutox | 4-(2,4-Dichlorophenoxy) butyric acid | Union Carbide, Rhone-Poulenc, May & Baker |
| Dalapon | Dowpon, Radapon, Basfopan | 2,2-Dichloropropionic acid | Dow, BASF, Diamond Shamrock |
| Decazolin | BAS-3490H | 1-(α,α-dimethyl-β-acetoxypropionyl)-3-isopropyl-2,4-dioxode-cahydroquinazoline | BASF |
| Diallate | Avadex | S-(2,3-Dichloroallyl) diisopropylthiocarbamate | Monsanto |
| Dicamba | Banvel | 3,6-Dichloro-o-anisic acid | Velsicol |
| Dichlobenil | Casoron | 2,6-Dichlorobenzonitrile | Thompson-Hayward |
| Dichlorprop | Several | 2-(2,4-Dichlorophenoxy) propionic acid | Several |

| Common Name | Trade Name | Chemical Name | Manufacturer |
|---|---|---|---|
| Diclofop methyl | Hoelon, Illoxan | 2-[4-(2,4-Dichloro-phenoxy)phenoxy]-methyl-propanoic acid | Hoechst |
| Dicryl | Dicryl | N-(3,4-Dichlorophenyl) methacrylamide | FMC |
| Difenzoquat | Avenge, Finaven | 1,2-Dimethyl-3,5-diphenyl-1H-pyrazolium | American Cyanamid |
| Dinoseb | Several | 2-sec-Butyl-4,6-dinitrophenol | Several |
| Dinoterb | DNTBP, Herbogil | 2-tert-Butyl-4,6-dinitrophenol | Rhone-Poulenc |
| Diphenamid | Enide, Dymid | N,N-Dimethyl-2,2-diphenylacetamide | Upjohn, Elanco |
| Dipropetryn | Sancap, Cotofor | 2-(Ethylthio)-4,6-bis (isopropylamino)-s-triazine | Ciba-Geigy |
| Diquat | Diquat, Reglone, Aquacide | 6,7-Dihydrodipyrido[1, 2-a:2',1'-c]pyrazinediium | ICI, Chevron |
| Diuron | Karmex, many others | 3-(3,4-Dichlorophenyl)-1,1-dimethylurea | du Pont, others |
| DNOC | Several | 4,6-Dinitro-o-cresol | Several |
| DSMA | Several | Disodium methanearsonate | Several |
| Endothall | Several | 7-Oxabicyclo[2.2.1] heptane-2,3-dicarboxylic acid | Pennwalt |
| EPTC | Eptam, Eradicane | S-Ethyl dipropylthio-carbamate | Stauffer |
| Ethofumesate | Norton, Tramat | (±)-2-Ethoxy-2,3-dihydro-3,3-dimethyl-5-benzofuranyl methanesulfonate | Fisons, Schering |
| Fluometuron | Cotoran, Lanex, Cottonex | 1,1-Dimethyl-3-($\alpha,\alpha,\alpha$-trifluoro-m-tolyl)urea | Ciba-Geigy, Nor-Am Makhteshim-Agan |
| Fluridone | Brake | 1-Methyl-3-phenyl-5-[3-(trifluoromethyl)phenyl]-4(1H)-pyridinone | Elanco |
| Glyphosate | Roundup | N-(Phosphonomethyl) glycine | Monsanto |

| Common Name | Trade Name | Chemical Name | Manufacturer |
|---|---|---|---|
| Hexazinone | Velpar | 3-Cyclohexyl-6-(dimethylamino)-1-methyl-1,3,5-triazine-2,4($1H,3H$)-dione | du Pont |
| Ioxynil | Actril, Totril, Bantrol, others | 4-Hydroxy-3,5-diiodobenzonitrile | May & Baker |
| Isoproturon | Several | $N$-(4-Isopropylphenyl)-$N',N'$-dimethylurea | Several |
| Linuron | Lorox, Afalon, Linurex, others | 3-(3,4-Dichlorophenyl)-1-methoxy-1-methylurea | du Pont, Hoechst, Makhteshim-Agan, Nippon, others |
| MCPA | Several | [(4-Chloro-*o*-tolyl)oxy]acetic acid | Several |
| Mecoprop | Several | 2-[(4-Chloro-*o*-tolyl)oxy]propionic acid | Several |
| Metamitron | Goltix | 4-Amino-3-methyl-6-phenyl-1,2,4-triazin-5($4H$)-one | Bayer |
| Methabenz-thiazuron | Tribunil | 1,3-Dimethyl-3-(2-benzothiazolyl) urea | Bayer |
| Meto-bromuron | Patoran, Pattonex | 3-(4-Bromophenyl)-1-methoxy-1-methylurea | Ciba-Geiby, BASF, Makhteshim- Agan |
| Metolachlor | Dual, Primagram, Bicep, others | 2'-Chloro-$N$-(2-ethyl-6-methylphenyl)-$N$-(2-methoxy-1-methylethyl)acetamide | Ciba-Geigy |
| Metoxuron | Dosanex, Purival, Deftor, others | $N'$-(3-Chloro-4-methoxyphenyl)-$N,N$-dimethylurea | Sandoz |
| Metribuzin | Sencor, Lexon, Sencoral, Sencorex | 4-Amino-6-*tert*-butyl-3-(methylthio)-*as*-triazin-5($4H$-one) | Mobay, du Pont, Bayer |
| Monuron | Monurex | 3-(*p*-Chlorophenyl)-1,1-dimethylurea | Makhteshim-Agan, Hopkins |
| Morfamquat | Morfoxone | 1,1'-di-(3,5-dimethyl-morpholinocarbonylmethyl)-4,4'-bipyridyldiylium | ICI |
| MSMA | Several | Monosodium methanearsonate | Several |
| Napropamide | Devrinol | 2-($\alpha$-Naphthoxy)-$N,N$-diethylpropionamide | Stauffer |

384

| Common Name | Trade Name | Chemical Name | Manufacturer |
|---|---|---|---|
| Nitrofen | TOK, Nip, Trizilin | 2,4-Dichlorophenyl p-nitrophenyl ether | Rohm and Haas, VEB Chem. Bitt., Shen Hong |
| Norea | Herban | 3-(Hexahydro-4,7-methanoindan-5-yl)-1,1-dimethylurea | Hercules |
| Oxadiazon | Ronstar | 2-tert-Butyl-4-(2,4-dichloro-5-isopropoxy-phenyl)-$\Delta^2$-1,3,4-oxadiazolin-5-one | Rhone-Poulenc |
| Oxyfluorfen | Goal, Koltar | 2-Chloro-1-(3-ethoxy-4-nitrophenoxy)-4-(tri-fluoromethyl) benzene | Rohm and Haas |
| Paraquat | Gramoxone, Paraquat, many others | 1,1'-Dimethyl-4,4'-bipyridinium ion | ICI, Chevron, many others |
| Pendimethalin | Prowl, Herbadox, Stamp | N-(1-Ethylpropyl)-3,4-dimethyl-2,6-dinitro-benzenamine | American Cyanamid |
| Phenmedi-pham | Bentanal | Methyl m-hydroxy-carbanilate m-methylcarbanilate | Nor-Am, Schering |
| Picloram | Tordon, Amdon | 4-Amino-3,5,6-trich-loropicolinic acid | Dow |
| Potassium azide | KZ | $KN_3$ | PPG |
| Prometon | Conquer, Gesafram, Pramitol | 2-4-bis(Isopropylamino)-6-methoxy-s-triazine | Ciba-Geigy |
| Prometryn | Caparol, Gesagard, Prometrex | 2,4-bis(Isopropylamino)-6-(methylthio)-s-triazine | Ciba-Geigy, Makhteshim-Agan |
| Propazine | Milogard, Gesamil, others | 2-Chloro-4,6-bis(isopro-pylamino)-s-triazine | Ciba-Geigy, Makhteshim-Agan, |
| Propham | Chem-Hoe, Triherbide, Beet-Kleen, Premalox | Isopropyl carbanilate | PPG, Shell, May and Baker, others |
| Pyrazon | Pyramin | 5-Amino-4-chloro-2-phenyl-3(2H)-pyridazinone | BASF |
| Pyridate | Fenpyrate | 6-chloro-3-phenylpyridazin-4-yl S-octyl thiocarbonate | — |

| Common Name | Trade Name | Chemical Name | Manufacturer |
|---|---|---|---|
| Secbumeton | Sumitol, Etazine | $N$-Ethyl-6-methoxy-$N'$ (1-methylpropyl)-1,3,5-triazine-2,4-diamine | Ciba-Geigy |
| Siduron | Tupersan | 1-(2-Methylcyclohexyl)-3-phenylurea | du Pont |
| Simazine | Princep, Gesatop, Aquazine, others | 2-Chloro-4,6-$bis$ (ethylamino)-$s$-triazine | Ciba-Geigy, others |
| Sodium chlorate | Several | $NaClO_3$ | Several |
| 2,4,5-T | Several | (2,4,5-Trichlorophenoxy) acetic acid | Several |
| 2,3,6 TBA | Benzac, others | 2,3,6-Trichlorobenzoic acid | Union Carbide, Fisons, others |
| TCA | Several | Trichloroacetic acid (sodium salt) | Several |
| Tebuthiuron | Spike, Graslan | $N$-[5-(1,1-Dimethylethyl)-1,3,4-thiadiazol-2-yl]-$N,N'$-dimethylurea | Elanco |
| Terbacil | Sinbar | 3-$tert$-Butyl-5-chloro-6-methyluracil | du Pont |
| Terbutryn | Igran, Prebane, Terbutrex | 2-($tert$-Butylamino)-4-(ethylamino)-6-(methylthio)-$s$-triazine | Ciba-Geigy, Makhteshim-Agan |
| Triallate | Avadex BW, Far-go | $S$-(2,3,3-Trichloroallyl) diisopropylthiocarbamate | Monsanto |
| Trifluralin | Treflan, many others | $\alpha,\alpha,\alpha$-Trifluoro-2,6-dinitro-$N,N$-dipropyl-$p$-toluidine | Elanco, others |
| Vernolate | Vernam | $S$-Propyl dipropylthio-carbamate | Stauffer |

# Index